39.1
54.94
64.00
158.04

QUANTITATIVE
ANALYSIS

WILLIS CONWAY PIERCE

PROFESSOR OF CHEMISTRY

UNIVERSITY OF CALIFORNIA AT RIVERSIDE

DONALD TURNER SAWYER

ASSISTANT PROFESSOR OF CHEMISTRY

UNIVERSITY OF CALIFORNIA AT RIVERSIDE

QUANTITATIVE

NEW YORK . JOHN WILEY & SONS, INC.

LONDON · CHAPMAN & HALL, LIMITED

EDWARD LAUTH HAENISCH

PROFESSOR OF CHEMISTRY

WABASH COLLEGE

ANALYSIS

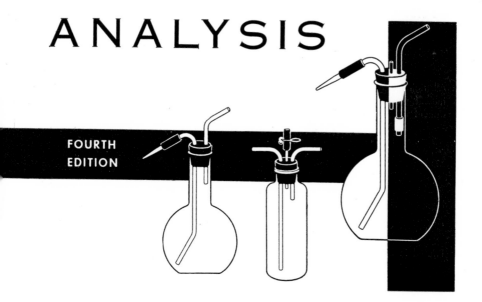

**FOURTH
EDITION**

PREFACE

In the ten-year period since the last revision of this text, the authors, along with other teachers, have made many changes in the analytical chemistry program. The rapid rise of instrumentation for analyses of research, industry, and clinical chemistry has made it important to provide training in instrumental methods of analysis. Since the amount of time available for analytical chemistry is limited, the addition of instrumental analyses to the program has required deletion of some classical chemical methods formerly studied. Consequently the entire program in analytical chemistry has been reorganized to some extent. The first course is still largely devoted to chemical methods, since the student must have a thorough training in the use of basic instruments and methods before he can properly use instrumental methods. The second course is today largely devoted to instrumental methods. In both, emphasis is placed on the physicochemical principles involved in the analyses.

This revision of the text is designed to meet these new conditions. First of all, the organization of the material has been changed, to permit greater flexibility in teaching. The book is in five major sections. Part I, with which the course starts, describes basic analytical tools and the unit operations of analytical processes. Part II deals with the calculations of analyses, that is, stoichiometry, treatment of data, and the equilibria involved in acid-base, precipitation, and oxidation-reduction reactions. In each of the sections dealing with equilibria, there is an elementary chapter for use in the first course and a more advanced treatment for use in the second. The advanced treatment is in separate chapters which can be omitted without loss of continuity. Parts III and IV deal respectively with laboratory methods of volumetric and gravimetric analyses. The material is so arranged that laboratory work may begin with either type of determination and assignments may be alternated between the two, as is done by the authors. Part V is an elementary treatment of three important instrumental methods—colorimetry, potentiometric titrations, and electrodepositions.

In previous editions this text contained all the material used by the authors for a year course. This is no longer possible, for the specialized nature of many instruments and the frequent introduction of new methods combine to make it undesirable to include more details on instrumentation in an elementary text. Most of the theoretical treatment given in the second course is retained, but experimental directions and descriptions of various instruments are to be taken from supplementary books.

Much of the text has been rewritten, to incorporate methods of presentation now used by the authors. Liberal use is made of the Lowry-Brönsted treatment of acid-base equilibria, although a conventional treatment is retained for the hydrolysis of salt solutions. The section on evaluation of data and precision of measurements is new. It is based on statistical methods for use with small populations. Precision is evaluated by the range of the data, and methods are shown for using the range to justify discarding a result. A chapter on competing equilibria in aqueous solutions gives an advanced treatment for use in the second course, based upon combining the equilibrium constants for the various species present and intelligent use of approximations in solving the complicated equations frequently obtained. The treatment of electromotive force of cells and their applications to oxidation-reduction equilibria is simplified. The conventions for the emf of cells follow recommendations[1] of the International Union of Pure and Applied Chemistry.

The selection of experimental material has been re-examined, and the procedures retained are those that have been found to give good results. Several new determinations have been added, including determinations of calcium by titration with a complexing agent, of nickel by precipitation with dimethylglyoxime, and of zinc by electro-

---

[1] For a detailed explanation of these conventions see T. F. Young, *American Institute of Physics Handbook*, McGraw-Hill Book Co., New York, 1957.

deposition; colorimetric determinations by colorimeter or filter photometer; and electrometric titrations with a commercial pH meter. Use of modern equipment is stressed throughout the laboratory procedures. Each procedure is accompanied by explanatory notes. It has been found in preceding editions that these notes enable the student to work with a minimum of supervision.

A departure from practice in preceding editions is the addition of answers to approximately half the problems. Experiments with numerous classes have convinced the authors that it is advantageous to provide some of the answers, to give the student a check on his work. Answers to the remaining problems are available to teachers, upon request to the publisher. As in past editions the problems are designed to provide drill, to illustrate applications of methods described in the text, and to point out methods of analysis not given in the text.

The writers acknowledge with thanks the valuable aid of Professors Donald DeFord and J. N. Pitts, who reviewed the preceding edition and suggested many of the organizational changes now made, of T. F. Young for helpful suggestions regarding the electrochemical sections, and of T. R. Lee and Donald Rosenthal for numerous valuable criticisms. Also we wish to repeat our thanks to those who gave assistance in previous editions. These include W. A. Noyes, Jr., W. E. Vaughan, P. C. Gaines, L. O. Hill, G. W. Schaeffer, W. R. Line, G. A. Perley, G. Frederick Smith, A. A. Danish, N. H. Koenig, A. H. Jaffe, D. P. MacMillan, W. W. Marshall, and the many teachers and students who have submitted criticisms and comments.

CONWAY PIERCE
EDWARD L. HAENISCH
DONALD T. SAWYER

March 1958

CONTENTS

ix

LABORATORY EXERCISES

AND METHODS OF ANALYSIS

CHAPTER 2

General Directions

Equipment

The equipment used in a quantitative analysis course is of three kinds: (1) the desk outfit, which is assigned to the student at the beginning of the course and may be returned at its close; (2) special equipment, which is constructed by each student; (3) general equipment, such as the balance, trip scale, steam bath, drying oven, and hot plate, which must be jointly used by several students. Before the work of the course can be started, it is necessary to prepare the equipment needed. Instructions for the use of the various items are given in later sections.

Desk Outfit

Returnable Apparatus. The assigned desk is equipped with apparatus which must be returned in good condition at the end of the course; it should be carefully inspected before it is accepted, and each item should be checked against the list supplied by the instructor. Note in particular that the glassware is not cracked, that buret and pipet tips are not chipped, that the buret stopcock fits well, and that the crucibles are in good condition. Make a list of any apparatus missing or defective, and present this to the instructor for his approval.

Special Equipment

The apparatus provided includes non-returnable items that will be needed in preparation of special equipment. The instructor will specify that certain of the following items be constructed for later use.

Wash Bottles. Construct from flat-bottomed boiling flasks (Florence flasks), preferably with ring necks, two wash bottles as shown in Fig. 2–1. Either 1000- or 500-ml flasks may be used, as provided in the equipment. One bottle is for ordinary washing with cold water

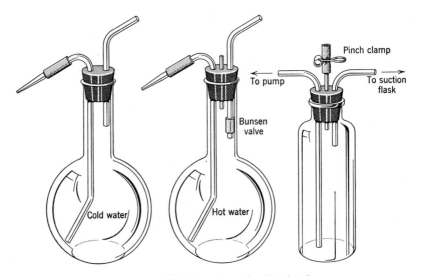

FIGURE 2–1. Wash bottles and safety bottle.

and the other is for hot water and special solutions. The fittings are made from glass tubing of proper size to fit the holes in the rubber stoppers. Bends can be properly made only by using a wing tip on the burner so that the tubing is heated over a distance of 4–6 cm. While heating, constantly rotate the tubing, but take care not to twist it as it becomes soft. After it is soft, remove the tubing from the flame and quickly bend it to the desired angle (if a tube is bent to a more acute angle than is desired, it is not easy to rebend and the piece should be discarded). Place the bent tube on a wire screen, and allow it to cool. Cut the two ends to the desired length, and fire-polish the ends by heating with constant rotation in the flame from a burner (without wing tip). This will remove sharp edges that might cut stoppers or rubber tubing.

The delivery tip is constructed as follows: Hold a short length of tubing in the blue flame from an ordinary burner (without wing tip) so that about 2 cm at the center is heated. Rotate constantly, and, when the tubing is soft, push the ends gently to form a slight bulge as shown in Fig. 2–2a. Now with constant rotation heat the bulge until it collapses as shown in Fig. 2–2b. Remove from the flame; while the collapsed bulge is still red hot, draw out the tubing, slowly at first and then more rapidly as it cools. When it is cold, cut the tip to the proper length, fire-polish until the orifice is about 1 mm in diameter, and set aside to cool. Fire-polish the other end after cutting to the proper length. This procedure produces a thick-walled tip that is

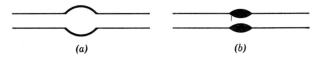

(a)   (b)

FIGURE 2–2.   Construction of wash bottle tip.

less fragile than it would have been had the tubing not been first thickened at the heated portion. This tip is attached by a 5-cm length of flexible rubber tubing. The ends of the glass pieces should be pushed together. The flexibility of the rubber will permit free motion of the tip. The entire fitting should be so compactly made that the tip may be manipulated by the forefinger as the bottle is held in the hand. The rubber stopper should fit the neck of the flask tightly when it is inserted not more than halfway into the flask.

The hot-water wash bottle is provided with a pressure-regulating tube in order to prevent a rush of steam into the mouth at the cessation of blowing. This is a short glass tube, open at both ends, which projects through the stopper. When the bottle is in use this tube is closed by placing the thumb over the upper end; removal of the thumb instantly reduces the pressure within the bottle. If three-hole rubber stoppers are not provided, it will be necessary to drill a hole for the pressure tube. This is done by means of a sharp cork borer which is lubricated by dipping it into dilute sodium hydroxide. The hot-water wash bottle may (optionally) be provided with a Bunsen valve, which serves to keep steam from escaping through the mouthpiece as the contents of the bottle are heated. This valve is made from a 5-cm length of flexible rubber tubing attached to the inner end of the mouthpiece. The lower end of the tubing is closed with a length of glass rod. A longitudinal cut in the rubber tubing acts as the valve and permits blowing into the bottle, but, when pressure is applied in the opposite direction, the edges of the cut are forced together and seal the opening. The neck of the hot-water bottle should be wrapped with asbestos paper or cord. The paper is best applied wet and allowed to dry overnight; there is sufficient adhesive material in the paper to make it cling tightly.

The glass wash bottle is replaced in many laboratories by plastic bottles. Liquid is delivered from the tip by squeezing the sides of the bottle.

Safety Bottle.   In all suction filtrations a safety bottle should be inserted between the water aspirator pump and the suction flask, since a change in water pressure may cause water to back up through the pump and contaminate the filtrate. The safety bottle is also used, as shown on page 50, as protection when cleaning solution is drawn by suction into burets or pipets. Construct the safety bottle as shown in Fig. 2–1. The three-hole stopper carries the two glass tubes that serve as inlet and outlet and a short length of glass tubing which connects to a rubber tube that is closed with a pinch clamp. This serves as a convenient method for breaking the vacuum when apparatus is attached to the suction pump.

Stirring Rods.   Cut six or eight stirring rods from solid glass rod and fire-polish the ends. Lengths should be made for use with the 250-, 400-, and 600-ml beakers provided in the outfit. The rod should, when resting in the beaker, project 5–7 cm beyond the lip. Two of the rods should be fitted with rubber tips, closed at one end (policemen). These are used in filtrations to remove precipitates that adhere to the walls of beakers.

Desiccator.   Clean and dry the desiccator (Fig. 2–3), and fill the lower portion to a depth of 1–2 cm with the desiccant provided. If the plate does not fit securely, fasten it in position with three pieces of cork cut as shown in Fig. 2–3 and wedged in at the sides of the plate. Grease the ground-glass rim lightly with petrolatum (not stopcock grease), set the cover on, and move it around until the greased joint is transparent; a minimum of grease should be used.

The desiccator is a container used to preserve samples, ignited crucibles, and the like, in an atmosphere of low constant humidity. It may be of any form and size; the only requirement is that it be tight. The most widely used style is a two-piece glass vessel as shown in Fig. 2–3. Recently several forms of metal desiccators have been marketed. In some of these the cover fits by a ground joint, as in the glass desiccators; in others the cover is a cap similar to that of an ordinary metal can. Metal desiccators have the advantage of low cost, low breakage, and rapid heat interchange with the surrounding air. The only disadvantage is that most desiccants cannot be placed directly in the chamber, because of reaction with the metal; usually it is advisable to place the desiccant in an evaporating dish which rests on the bottom of the desiccator.

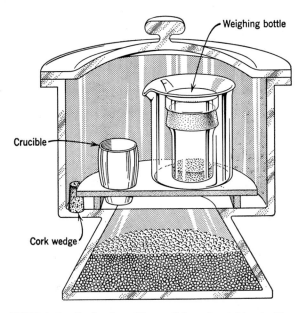

FIGURE 2–3. Desiccator with crucible and weighing bottle.

A desiccator is most efficient when the volume of air contained is kept small. When the cover is removed, for insertion of a sample, much of the air in the chamber is replaced by air of higher humidity from the room. Air circulation in the closed vessel is so slow that considerable time is required to establish equilibrium again, and a sample may absorb an appreciable amount of water from the moist air. Once a sample has been dried in a desiccator, the cover should be lifted only when absolutely necessary.

Care must be exercised in the handling of a desiccator. Ignited crucibles should be placed in the chamber only when they have partially cooled after the removal of the flame. After objects have cooled in the desiccator, the cover should be removed gradually by sliding to one side, to prevent a sudden inrush of air. A partial vacuum exists in the chamber because some air, expanded by the heat from the hot objects when they were first placed in the chamber, has escaped.

Desiccants. The desiccants most commonly used in the analytical laboratory are anhydrous calcium chloride and concentrated sulfuric acid, not because they are best but because of their ready availability. In fact, these are two of the poorer desiccants. Calcium chloride does not remove water at all completely but produces a low enough humidity for student needs. Concentrated sulfuric acid is a fairly good desiccant, but it has the disadvantage that the liquid may be splashed onto the bottom of objects in the desiccator. If it is used, the bottom of the chamber should be filled with glass beads and the acid level kept below the level of the beads. Barium oxide is superior to calcium chloride, but it is not readily available in suitable form. Phosphorus pentoxide is a very good drying agent but is expensive and difficult to handle. When the surface of the solid becomes moist, the powdered particles coalesce, and a film of phosphoric acid forms on the surface to give a hard glassy layer that prevents air from coming into contact with the unused oxide below the surface. The more expensive desiccants such as barium perchlorate (dessichlora), magnesium perchlorate (anhydrone), and calcium sulfate (drierite) are superior to those previously listed and are today available through all supply houses. All these have high drying efficiency and are easily handled.

Cleanliness

The importance of cleanliness in the laboratory cannot be overemphasized. Good analyses cannot be performed with dirty equipment. Furthermore, a dirty or a disarranged desk is usually an indication of a poor or careless technician. The desk and the reagent bottles must be kept clean at all times. All equipment not in use must be kept within the locker, clean and neatly arranged. At the conclusion of each working period, the desk top must be sponged off and dried with a towel. Reagent bottles must be washed frequently. Spilled chemicals must be washed up immediately.

Volumetric glassware is best cleaned by sulfuric acid-dichromate cleaning solution, as described on page 49. All other glassware is cleaned with soap and water (preferably hot) or a soapless detergent, with the aid of a brush. After the apparatus is washed completely free of soap or detergent and rinsed with distilled water from a wash bottle (*never at the distilled-water tap*), it is wiped dry on the outside and is then allowed to drain by being inverted on a towel or placed on a drain board. Use of cleaning solution for general cleaning is not only dangerous, because of the powerful oxidizing properties of the mitxure, but also inefficient.

Reagents

Chemicals are manufactured and sold in degrees of purity ranging from commercial grade to the best reagent grade. The well-trained analyst should know when it is advisable to use chemicals of low purity and when it is necessary to employ the purest grade obtainable. Moreover, he should know, in each analysis, which of the impurities in the chemicals used may affect his results and by what tests he can assure himself of the absence of these impurities. In those operations in which the presence of impurities cannot cause harm, the cheapest grades of chemicals can be employed. For example, cleaning solution is prepared not from pure sodium dichromate and sulfuric acid but from technical grades of these substances.

Reagents are classed as follows:

Technical grade (Tech.). This grade is not purified by the special processes necessary for the elimination of traces of other substances. It is seldom employed in the laboratory for the preparation of analytical reagents.

USP grade. This designation indicates that the standards set up in the *United States Pharmacopœia* are followed. For some uses, USP chemicals are sufficiently pure reagents.

C.P. and analyzed grades. The term C.P. (chemically pure) is somewhat elastic in meaning; there is no general list of specifications to define the term. Because of this, various manufacturers prepare "analyzed" chemicals for use as reagents. Each package of analyzed chemicals has a label giving the manufacturer's limits of certain impurities. Such an analysis does not imply that the composition of the chemical is definitely that of the formula. For example, reagent-grade sodium carbonate may contain a certain amount of sodium bicarbonate but nevertheless be of high purity. Certain chemicals are packaged with a label which gives the "assay" or percentage of the major constituent. Such a label is valuable to the analyst, since he is often interested in the percentage purity rather than the limits of impurities.

The American Chemical Society Committee on Analytical Reagents[1] has established standards for certain reagents. Many chemicals are now prepared to meet these specifications; they are labeled "Conforms to ACS Specifications."

It is never wise to rely on the label of a bottle as a guarantee of the purity of a chemical. Mistakes may occur, not only in testing but also in packaging and in handling after the bottle has been opened. When-

---

[1] *Reagent Chemicals, 1955,* American Chemical Society Committee on Analytical Reagents, American Chemical Society, Washington, D.C.

ever the presence of an impurity in the reagent may lead to trouble, a test should be made. The usual method is to run a blank determination.

In the laboratory, care must be taken to prevent contamination of reagents. No chemical should be returned to a stock bottle. All bottles should be closed except when in use. Liquid reagents should be *poured* from the bottle; under *no circumstances* should a pipet be inserted into a reagent bottle. Particular care should be taken to avoid contamination of reagent bottle stoppers. When liquid is being poured from a bottle, the stopper should never be placed on the desk or shelf; most chemists hold the stopper between two fingers of the right hand so that the plug projects from the back of the hand.

The most common impurity in solid reagents is moisture, which is more or less firmly held by all solids. Very often the adsorption of water causes no difficulty, but, if the weight of a chemical must be accurately known, the chemical must be dried. Directions are given whenever drying is necessary.

Certain reagents are provided as concentrated solutions. Since these solutions are frequently used for the preparation of dilute solutions, the student should know their approximate strength. This is given, for the more common reagents, in Table 2–1.

TABLE 2–I. Strength of Laboratory Reagents

Reagent	Density, g/ml	Per Cent by Weight	Approximate Molarity
HCl	1.18	36	12
HNO_3	1.42	72	16
H_2SO_4	1.83	95	18
$HC_2H_3O_2$ (glacial)	1.057	99.5	17
NH_4OH	0.90	28 (NH_3)	15

Dilute solutions of these reagents are prepared, when needed, by adding the concentrated solutions to water. The concentrations of dilute solutions are usually from 3 to 6 molar; usually the value of the concentration is specified as the molarity (or normality) or as the degree of dilution, which is indicated by the ratio of the volume of concentrated solution to that of water. For example, 1:4 hydrochloric acid indicates a solution prepared by adding 1 volume of concentrated acid to 4 volumes of water.

Distilled water is used for the preparation of all solutions employed in quantitative analysis. Even this water, as it comes from the tap, is not pure but is contaminated by dissolved gases and by material dissolved from the container in which it has been stored. The dissolved gases may be removed by boiling the water for a short time. Occasionally distilled water is found to be contaminated by non-volatile impurities which have been carried over by the steam, in the form of a spray. Also, a still may froth while in operation and badly contaminate the distillate. When a fresh supply of distilled water is obtained, a test should be made for chloride or sulfate ion. If these are absent, the water may be assumed to be reasonably pure.

When water of the highest purity is required, distilled water is redistilled from an alkaline permanganate solution, in silica or block tin apparatus. (The permanganate solution oxidizes nitrogenous matter present.) Redistilled water prepared in this manner is known as *conductivity water.*

The storage vessel employed may have a marked effect on the purity of water or of reagents. Soda-glass or soft-glass vessels are much more readily attacked by reagents than the borosilicate glasses, such as Pyrex. When reagents are to be stored for long periods of time it is desirable to use Pyrex bottles.

Records

One of the important details of all chemical work is the keeping of complete and legible records of experiments. Quantitative analysis provides especially valuable training in keeping records, for the work of the course is of such a nature that all data can be concisely and completely summarized in a notebook. The following suggestions for keeping the notebook should be carefully studied. They are illustrated later by sample notebook pages.

Suggestions for Keeping Notebook

1. Use only bound notebooks, of the size specified by the instructor. Reserve the first page for an index and number all subsequent pages.

2. Record all data in ink as taken. Never use loose paper. Loss of data taken on loose paper has cost many students valuable time.

3. Make no erasures. If a value is to be invalidated, draw a line through it but in such a way that it is still legible. State in the notebook why data are discarded.

4. Head each page with the date and title of the analysis.

5. Strive for clarity so that any other chemist would be able to

interpret your figures. Before starting an experiment, it is advisable to think over the data needed and construct a table in which data and computations can be neatly arranged.

Work of the Course

A list of the determinations to be made will be furnished. Each is to be made in duplicate or triplicate; that is, two or three portions of sample are analyzed with concordant results. The result of the analysis is reported on a card as specified. Sample cards are shown later. The report is graded according to the accuracy and precision of the results.

Suggestions for Work

At best, many hours must be spent in the work of a course in quantitative analysis. Many students, through inefficient planning of their work, require far more time than necessary and have difficulty in completing the laboratory work. Adherence to a few simple rules may prevent much of this loss of time:

1. Study the procedure and notes before beginning an experiment. Know the reason for each operation. Such study should be done outside of laboratory hours. The notes contain many hints and explanations; they are an integral part of the procedure.

2. Prepare a written outline of the procedure, and work from this rather than from the book.

3. *Do not attempt short cuts.* The procedures are based on proved methods.

4. Obtain the necessary chemicals before starting an analysis.

5. Utilize the time spent waiting for the cooling of crucibles, evaporations, ignitions, digestions, and the like by working on other experiments. Lengthy delays caused by slow evaporations can often be eliminated by planning to carry them out on the hot plate or steam bath while other work is in progress.

6. Keep all containers for each individual sample carefully marked, to avoid mixing or interchanging solutions at some stage of the analysis. A great deal of time can be lost by carelessness. Beakers and flasks may be identified by gummed labels or by writing with a pencil on the etched circle provided for that purpose. Weighing bottles are placed in small marked beakers before they are put in the drying oven,

which is usually filled with student samples. Porcelain crucibles are given identification letters or numbers by a special marking ink which fuses into the glaze when the crucible is heated. If special ink is not available, dissolve a few crystals of lead nitrate in about a milliliter of 6 M NaOH. Apply this solution onto the side of the crucible to form the desired letter or symbol (using a pointed stick), allow to dry, and ignite the crucible to dull redness. After cooling, wash off residual salts.

CHAPTER 3

The Analytical Balance and Its Use

Accurate weighing of small samples is prerequisite to every analysis. Consequently the first assignment in this course is to learn how to weigh accurately and rapidly and to gain an appreciation of the principles involved in the use of a balance.

Mass and Weight

Mass and weight, though often used interchangeably, differ in meaning. The *mass* of an object is the amount of matter in the object, as referred to a standard mass of platinum-iridium alloy preserved at the International Bureau of Weights and Measures. Many duplicates of this standard are in existence. The unit of mass is the gram (g), which is $\frac{1}{1000}$ of the mass of the standard; the mass of the standard, 1000 g, is known as a kilogram (kg). It was originally intended that the kilogram be the mass of 1000 cc of water, at the temperature of greatest density, 3.98°C, but because of an error in measurement this objective was not realized; it is now known that the volume of a kilogram of water at 3.98°C is 1000.028 cc.

The *weight* of an object is the force exerted on that object because of gravitational attraction between the body and the earth. The weight is expressed in force units (dynes). Since the force of gravity varies with geographical location and with the elevation above sea level, the weight of an object is a variable, whereas the mass is invariant. The term *weight* is, however, often used instead of the term *mass;* the expression "weight of 1 g" is commonly used and is correct if properly interpreted to mean "a mass whose weight is that of a 1-g mass." This usage has developed because of the method followed in determining the mass of an object, which is commonly done by comparing the weight of the object with the weight of a known mass.

Construction of the Balance

The essential feature of the balance is a simple lever operating on a knife-edged fulcrum at the center of the lever. The object to be weighed is suspended from one end of the lever, and weights of known mass are suspended from the other end. The points of suspension of the object and weights are equidistant from the fulcrum. When the mass of the object is equal to the mass of the weights, the lever, which is called the beam of the balance, remains in a horizontal position. A long pointer, attached to the center of the beam, serves to indicate a displacement of the beam from a horizontal position. This method of weighing is based on the assumptions (1) that the two arms of the lever are of equal length and (2) that the masses of the weights are accurately known.

Types of Balances

Prior to the early part of this century all analytical balances were of one type, employing individual separate weights and a rider for the smaller values. The need for more speed in weighing has led to the development of numerous improvements of mechanical features, to speed up the operation of weighing an object. These newer-type balances are like the old in principle but differ in the method for adding weights. We shall describe the old type in some detail, then point out features of some of the more widely used newer types.

1. Rider Balance. The working parts of a balance are illustrated in Fig. 3–1. The beam B is so made that it gives a maximum of rigidity with a minimum of weight. It is balanced at the center on a knife edge K, which is made of agate. The knife edge rests on a plane agate plate at the top of the main supporting pillar M. Pans PP are suspended from stirrups which hang from agate plates suspended on knife edges $K'K'$. It is essential that the three knife edges $K'KK'$ lie in a plane. When the balance is not in use, the beam and pans are lifted by a beam arrest AA, so that the knife edges are not in contact with their plates. The beam arrest is operated by a knurled screw BC,

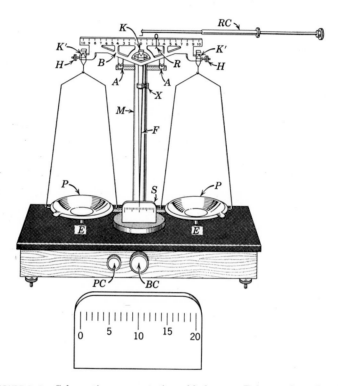

FIGURE 3–1.   Schematic representation of balance.   Below: enlarged scale.

which projects on the front of the balance case.   Beneath each pan there is a pan arrest EE, operated by a pan control PC from the front of the case.   It is used to stop lateral swinging of the pans.   In some models the beam arrest and pan arrest are operated simultaneously by the same control.   Motion of the beam is indicated by the pointer F, which moves before a scale S.   An enlarged diagram of the scale is also shown in Fig. 3–1.

The top side of the beam is graduated in equal divisions, which serve to indicate the effective weight of a rider R, when the latter is placed on one of the division marks.   The rider is a small aluminum or platinum stirrup which may be set at any position upon the beam.   It is used for weights up to 5 or 10 mg, depending on the particular balance.   The rider is manipulated from outside the case by means of a carrier *RC*, which terminates in a hook for engaging the rider.   The rider in Fig. 3–1 is placed at 2.7 mg.

Different manufacturers graduate the beam differently.   One method is shown in Fig. 3–1, with the zero mark at the center and graduation marks to each side of the center.   When the rider is used on the right-hand side, its weight is added to the weight on the right-hand pan.   When the rider is on the left-hand side, its weight is subtracted from the weight on the pan.

In another type of graduation for the rider, the zero position is above the left-hand pan and the graduation marks extend to a position above the right-hand pan.   This method is common for microbalances.   It has the advantage that the length of the rider arm is double the usual length, and the precision in placing the rider is increased. When this method is used, the actual weight of the rider is only half its nominal value.   In using a balance so graduated, care must be taken to have the rider on the beam at all times.

The entire working mechanism is enclosed within a case, in order to protect it from air currents while in operation and from dust.   The front glass section of the case is suspended

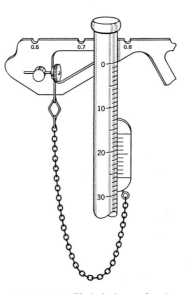

FIGURE 3–2.   Chain balance showing scale and vernier.   The vernier is set at 16.5 mg.

by a sash arrangement which permits it to be raised, thereby giving access to the pans for inserting objects or weights.   The bottom of the case rests upon leveling screws.

2. Chain Balances.   The use of rider balances today is largely confined to instruction.   Labor is too costly to permit its waste in routine weighings, and most commercial laboratories use balances that permit greater speed in weighings than can be obtained with a rider balance. The first stage in the development of high-speed balances was the chain style balance, one type of which is sketched in Fig. 3–2.   The smaller weights and the rider are replaced by a chain, one end of which is fastened to the beam and the other end to a moveable support which is raised or lowered to increase the weight of the chain hanging from the beam. The most common range is 0–100 mg for the chain. The moveable carrier for the chain is mounted on a graduated scale, from which the effective chain weight is read directly. A vernier attachment permits readings to 0.1 mg.

In more recent models of chain balances fractional weights are eliminated by a rider that rests in notches on the top of the beam which is graduated in tenths of a gram.

3. Automatic Balances. Several models of semi-automatic balances are now available, to give even higher speed of weighing than can be obtained from a chain balance. These balances are so constructed that weights are changed by simply turning dials on the front of the case. When the rest point is brought to the proper position, the weight is read directly from the dials. Such balances are of single-pan construction.

Sensitivity of the Balance

Balances vary in the effect produced by a small weight inequality on one side of the beam. They are rated in terms of this effect. In general, the term *sensitivity* is used to describe balance response, but the term is rather loosely used. In this text we shall define sensitivity as the weight required to give a pointer deflection of one scale division. In another usage the sensitivity (sometimes called sensibility) denotes the smallest weight difference that can be detected with accuracy. This usage is generally more convenient for rating balances. Thus analytical balances are rated as having sensitivities of $\frac{1}{10}$ or $\frac{1}{20}$ mg, and a microbalance may have a sensitivity of $\frac{1}{1000}$ mg.

The Operations of Weighing[1]

Weighing an object consists in determining the mass of known weights necessary to counterpoise or balance the object. The condition of equality is recognized by the motion of the pointer; if the equilibrium position of the loaded balance coincides with that of the

---

[1] In modern analytical work the multiple-swing method of determining rest points is not usually used because of its slowness. But in the most exacting determinations this is the preferred method. Microbalances are always used with multiple swings. It is recommended that students first learn this method, regardless of what method is used later.

empty balance, the masses on the two sides are equal, provided the two arms of the beam are of equal length.

When the beam and pans are released, the beam starts to swing with a period, depending on the characteristics of the particular balance. The midpoint of this swing is the point at which the beam would be at rest; however, instead of waiting for the beam to come to rest, we *calculate* the equilibrium point from the amplitudes of successive swings of the pointer. This method is not only more rapid than waiting for the beam to come to rest but also more accurate because, if the beam were allowed to swing until it came to rest, the position of rest would be influenced by vibration, friction, and other external forces.

For the purpose of accurately observing the swings of the balance, the scale S is marked with equally spaced divisions. These are conveniently read[2] by denoting the center mark as 10 and calling the left-hand mark 0. In reading the position of the pointer, the nearest 0.1 division is estimated.

Because of frictional and air resistance, the swings of a balance pointer are damped; that is, they are constantly decreasing amplitude. Suppose, for example, that in a given weighing the equilibrium position is at 10.0 on the scale (exactly at the center). When the beam is put into motion, the following consecutive swings are noted (trace these swings on the scale of Fig. 3–1; the damping is exaggerated, for purpose of illustration):

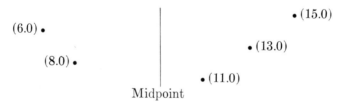

$$(6.0) \bullet \qquad\qquad\qquad \bullet\,(15.0)$$
$$\qquad\qquad\qquad\qquad \bullet\,(13.0)$$
$$(8.0) \bullet$$
$$\qquad\qquad \bullet\,(11.0)$$
Midpoint

It is obvious, from a consideration of these swings, that the equilibrium position is not halfway between any two consecutive swings, such as 15.0 and 6.0 or 6.0 and 13.0. The rest point is, however, halfway between the midpoint of two swings to one side, such as 15.0 and 13.0,

[2] Many chemists prefer to number the center of the scale zero and to call readings to the right and left respectively plus and minus. Thus a reading which according to the convention of this book is 8.0 would become in the other system −2.0 and a reading of 12.0 would become +2.0. The plus-and-minus system has an advantage in that the operator need observe only the number of divisions from the center mark. This system is preferred in weighing by the single-swing method, since usually all rest points are to one side of the center mark.

and a corresponding single swing to the other side, such as 6.0. That is,

$$\text{Rest point} = \frac{\dfrac{15.0 + 13.0}{2} + 6.0}{2} = 10.0$$

or

$$\text{Rest point} = \frac{\dfrac{15.0 + 13.0 + 11.0}{3} + \dfrac{6.0 + 8.0}{2}}{2} = 10.0$$

From the foregoing, the following rule may be formulated: *To find a rest point, observe an odd number of consecutive swings to one side and an even number to the other side. Take the average of the swings to each side. (Either three or five swings may be used.) Add the two averages and divide by two.*

In weighing, the following steps are necessary:

1. Determine the rest point of the unloaded balance.
2. Place an object on one pan, and determine the weight needed to make the rest point of the loaded balance coincide with that of the unloaded balance.

Since weights must be determined to the nearest 0.0001 g, it would prove a tedious procedure to continue the addition of weights, by trial and error, until the two rest points are brought into exact coincidence. It is a much simpler procedure to add weights until the rest point is near that of the unloaded balance, and, by then determining the effect of 1 mg on the rest point position, to calculate by interpolation the exact weight that would be needed to bring the rest point to the true zero. This method is known as *weighing by sensitivity;* the reciprocal of the displacement of the rest point caused by 1 mg is known as the *sensitivity* of the balance. Step 2 consists of three parts:

(*a*) Determine, by trial, the approximate weight of the object.
(*b*) Determine the sensitivity of the balance.
(*c*) Calculate the exact weight of the object.

These steps will be followed in the directions for weighing.

Determination of Rest Point of Empty Balance (Zero Point). Identify the controls and carefully operate each. Determine, by the attached spirit level, whether the balance is level. Carefully lower the beam by turning the beam control knob counterclockwise. Push the pan arrest control in, and start the balance swinging (note 1) by gently fanning an air current onto one pan (note 2). Close the balance case and observe the swings, estimating each point to the nearest 0.1 division (note 3) ; record each swing immediately. When three or five consecutive swings have been recorded, raise the pan arrest, raise the beam arrest, and calculate the zero point as follows:

$$
\begin{array}{lll}
\text{Scale readings} & 5.1 & \begin{array}{l} 14.8 \\ 14.6 \end{array} \\[1em]
\hline
\text{Average} & 5.1 & 14.7 \\[1em]
\text{Zero point} & \dfrac{5.1 + 14.7}{2} & = 9.9
\end{array}
$$

Release the beam and pan arrests, and repeat the observations. The two rest points should agree within 0.1–0.2 unit. Should the rest point not fall between 8 and 12, consult the instructor (note 4).

Notes. 1. The amplitude of the swing should be 3–5 divisions to each side of the center mark. Take care that the rider is not so placed that it can touch the swinging beam.

2. An alternative method of inducing swinging is to touch one pan lightly with a stiff hair, waxed to a match stick or to the handle of the brush that is kept in the set of weights.

3. Care must be taken to avoid parallax in observing the swings. The head should be held steady, in a position directly before the scale, for the entire set of readings.

4. The position of the rest point may be adjusted by screws *HH*, on the ends of the beam. Such adjustment should be made only by the instructor.

Determination of Approximate Weight of Object. Be sure that the beam and pan arrests are raised. Carefully place the object on the center of the left-hand (note 1) pan. Place a weight (note 2) of the estimated amount on the right-hand pan, and partially lower the beam arrest. A large inequality in weight will be shown by a displacement of the pointer to one side. If no inequality is noted, momentarily release the pan arrests and note the direction of swing. Raise the beam arrest, and add or subtract weights as indicated by the direction of the swing. Continue in this manner until it is necessary to employ fractional weights. Add fractional weights until the addition of a weight having the magnitude of the rider is sufficient to *overbalance* the object, then remove this weight, and use the rider to obtain a weight within 1 to 2 mg of the true weight. This point is recognized when the pointer no longer passes either end of the scale when swinging a distance of 5–6 scale divisions.

Notes. 1. The object is always placed on the left-hand pan and should be located near the center of the pan to prevent lateral motion.

2. Place the larger weights near the center of the pan, and arrange smaller weights to both sides, so that the pan will hang in a level position.

Determination of Sensitivity.   When the object is approximately balanced by weights, determine the rest point, as in the determination of the zero point.   Move the rider to a position 1 mg *heavier* if the weights are too light (note 1), or to 1 mg *lighter* if the weights are too heavy, and determine the new rest point (note 2).   The sensitivity is the reciprocal of the rest point displacement effected by a weight change of 1 mg.

Notes.   1. In determining the sensitivity it is desirable but not essential to choose two rest points such that one lies on either side of the zero point.

2. The rider may be moved any known amount and the sensitivity calculated by dividing the weight difference by the scale divisions.

Calculation of Exact Weight.   Select the weight at which the rest point is nearest the true zero, and, by subtraction, determine the number of scale divisions by which the rest point for this weight deviates from the true zero.   Convert the number of scale divisions into weight units (multiply the number of scale divisions by the sensitivity).   To obtain the exact weight add this value to or subtract it from the approximate weight, depending on whether the approximate weight is less or greater than the exact weight.

The use of the method of sensitivity may be illustrated by a numerical example:

Weight on pan	17.360 g	17.360
Weight on rider	0.003 g	0.004
Rest point	11.0	8.5
True zero	9.9	

Calculation.   The weight 17.363 g is too light, since the rest point lies to the right of the true zero.   The difference is 1.1 scale division, 11.0–9.9.   When a weight of 1 mg is added (rider moved from 3 to 4), the rest point changes 2.5 scale divisions, from 11.0 to 8.5.   Thus, since 1 mg causes a deflection of 2.5 scale divisions, the sensitivity is

$$S = \frac{1 \text{ mg}}{2.5 \text{ division}} = 0.4 \text{ mg/division}$$

This sensitivity is now used to compute the exact weight.   The weight 17.363 g is too light by 1.1 scale division.   This is 1.1 division $\times$ 0.4 mg/division = 0.4 mg or 0.0004 g.   The exact weight is 17.363 g + 0.0004 g = 17.3634 g.   Note that the computation is not carried beyond the fourth decimal place.   The balance is not sensitive to weights of this magnitude, and it is therefore misleading to carry the computation further.

Completion of Weighing.   Record (note) the weights and swings, as shown in Table 3–1.   Raise the pan arrest and lock the beam.   Re-

TABLE 3–1. Data for a Typical Weighing

Notebook Page

Determination of true zero:

Scale readings	5.1	13.7
	5.3	13.5
	5.5	
Average	5.3	13.6

$$\text{True zero} \qquad \frac{5.3 + 13.6}{2} = 9.4$$

Weighing:

Weight on pan	
	20
	10
	10
	2
	1
	0.500
	0.200
	0.050
	0.020
	0.010
	43.780

Swings with rider at 3 mg			Swings with rider at 4 mg		
	7.6	13.5		3.6	12.5
	7.8	13.3		3.8	12.3
		13.1			12.1
Average	7.7	13.3		3.7	12.3

$$\text{Rest point} = \frac{7.7 + 13.3}{2} = 10.5 \qquad \frac{3.7 + 12.3}{2} = 8.0$$

Sensitivity = 0.4 mg/division
Scale difference with weight 43.783 = 1.1 division = 1.1 × 0.4 = 0.4 mg
True weight = 43.783 + 0.0004 = 43.7834 g

move the object and the weights, checking each weight against the notebook record as it is removed from the balance case. Remove the rider from the beam, by means of the rider carrier, and push the rider carrier into the balance case until the rider hangs above the center of the beam. Close the balance case.

Note. After more experience is gained, the weights should not be individually recorded as shown in Table 3–1. The experienced analyst does not record either the individual weights or the balance swings but only the final result.

When the same balance is used regularly, it is not necessary to determine the sensitivity as part of each weighing. A table is constructed giving the sensitivity for various weights. In a weighing a single rest point is determined, and the sensitivity table is used to compute the weight required to bring the rest point from the observed position to the true zero.

Weighing by Single Deflection (First Swing). If a balance is so adjusted that the weight on one arm is slightly greater than that on the other, it will swing to a reproducible position when the pan arrest is released. The maximum excursion of the pointer in the first swing is taken as the rest point. The writers recommend the method of first swing for student use. Experience with both general chemistry and quantitative analysis classes has shown that this method gives good results. The average student can make weighings from two to four times faster by single swings than with the method of multiple swings. A further advantage is that no impetus is needed to start the swing.

In using this method the balance is adjusted to give an initial swing of 4–6 divisions to the left (direction chosen arbitrarily). In reading the scale the center is taken as zero. The operations and computations are exactly the same as those used in the method of multiple swings. The data of Table 3–2 illustrate the method.

TABLE 3–2. Data for a Typical Weighing by Single Deflection Method

Notebook Page

Swing of empty balance = 4.2

Weights on Pan	Rider	First Swing (left)
16.37 g	3 mg	2.0
16.37	4	7.4

Sensitivity = 1 mg/5.4 divisions = 0.2 mg/division
Scale difference with weight of 16.373 g = 4.2 − 2.0 = 2.2 divisions
Weight to be added = 2.2 divisions × 0.2 mg/division = 0.4 mg = 0.0004 g
Exact weight = 16.373 g + 0.0004 g = 16.3734 g

Weighing with Rider-Type Chain Balance. Raise or lower the chain to give an initial deflection of about 5 divisions to the left. Adjust the scale vernier to read zero by turning the exterior knob provided for this adjustment. Place the object on the left-hand pan and add weights by trial until they are within 1 g of the balance point.

Close the balance case and adjust the rider to the nearest 0.1 g. Manipulate the chain to give a first swing within 3–4 scale divisions of the swing point for the empty balance. Complete the weighing by sensitivity, just as was done with the rider balance, except that settings of the chain adjustment replace use of the rider.

Many analysts prefer to complete the weighing by actually adjusting the chain position until the first swing agrees with the swing point for the empty balance. A knowledge of the sensitivity of the balance makes it possible to set the chain at the correct position quickly.

Care of the Balance

The balance is a delicate precision instrument which will not function properly if abused. Each student is individually responsible for knowledge of and adherence to the following rules on the use and care of the balance.

1. Whenever the inequality of mass on the two pans exceeds 1 g, be sure that the beam is locked before changing a weight. The pan arrest will usually support a load of 1 g or less.

2. Always release the beam gently. Avoid a sudden jar which might damage the knife edge.

3. Always close the balance case before observing swings. Air currents must be avoided.

4. Do not overload the balance. Most analytical balances are designed for loads of 200 g on each pan.

5. Never weigh any chemical or moist object directly on the balance pan.

6. Never place a hot object on the balance pan. If an object is warm, its weight will be too light because of convection currents set up by the rise of heated air.

7. Handle the weights with forceps. Never use the fingers. Do not use the ivory-tipped weight forceps for any other purpose than to handle weights.

8. Objects to be weighed should be handled with tongs or forceps whenever the shape of the object permits. Round objects, such as weighing bottles, may be handled by the fingers, but care must be taken to prevent weight changes because of moisture from the hand. Dry the hand just before lifting an object to the balance pan. Do not hold any object longer than necessary.

9. Remove all objects, and close the balance case on completion of a weighing.

10. Clean up any material spilled on the pan or within the balance case.

11. Always support the rider on its carrier hook at the completion of a weighing. Suspend the rider directly above the center of the beam so that it cannot interfere with free swinging of the beam. If a chain balance is used, set the chain and rider at zero.

12. Do not monopolize the balance. Respect the rights of others.

13. If the balance is out of order, report the fact immediately.

Laboratory Exercises

The following should be completed and the notebook submitted to the instructor for approval before new assignments are started.

1. Practice obtaining a reproducible rest point by the method of multiple or single swings, as specified by the instructor.

2. Determine the sensitivity of the balance for loads of 0, 10, 20, and 50 g. Place weights of proper denomination on the pans so that they are approximately balanced. Determine the rest point by the proper method. Add 1 mg by the rider or chain, and again determine the rest point. Construct a graph in the notebook showing sensitivity on the y-axis as a function of load on the x-axis. This graph will be useful in making mental computations of the adjustment needed at any weight to bring the rest point into coincidence with the zero point.

3. If you are using a rider balance, test the rider. Place the rider on the 5- or 10-mg division and a corresponding fractional weight on the left-hand pan; then see whether the rest point coincides with the zero point. If the rider is grossly in error, notify the instructor. A difference of 0.1–0.2 mg in the weight of the rider and the comparison weight can be neglected.

4. Weigh separately a crucible and cover, and then reweigh the two together. The results should agree within about 0.0002 g. Record in notebook.

Errors in Weighing

The most common weighing errors, outside personal errors in misreading turning points of the pointer or miscounting the weights, are due to inaccuracies in the values of the weights; these errors may be eliminated by calibration, as described in a later section. Other errors are:

1. The two arms of the beam may not be of the same length.

2. Moisture may condense on an object and change its weight.

3. If the temperature of the object is not that of the surroundings, the weight will be affected.

4. Electrification of glass vessels will cause induced charges in adjacent metal parts, and the balance will be caused to swing in an erratic manner.

5. The balance may not give reproducible swings.

6. Buoyancy due to air may affect the apparent weight.

The effects of these errors and methods for their avoidance are discussed in the following paragraphs.

Inequality of Balance Arms. Most analytical balances are so accurately made that error due to inequality of arms does not exceed one part in several thousand, and this error will not affect ordinary analytical results. Furthermore, since most analyses involve a ratio of weights, a small inequality of the balance arms will not affect the results at all; all weighings made on a particular balance will be in error by a constant factor.

Any error due to balance arm inequality may be eliminated by (1) weighing by substitution or (2) double weighing.

Weighing by Substitution. Place the object on the left pan, and counterpoise it by a suitable *tare* on the other pan. Copper shot placed in a beaker or crude weights may be used as a tare. Observe the rest point after the tare is added. Remove the object, and add weights to the left-hand pan until the rest point coincides with the one originally obtained. Since the object and weights have been placed on the same pan of the balance, their masses have been directly compared without any assumptions being made regarding the lengths of the lever arms.

Double Weighing. Weigh the object as usual, then interchange the weights and object, and reweigh. The arithmetic mean in the two weights is very near to the true weight.

Let W_1 = weight when object of mass M is on the left pan

W_2 = weight when object is on the right pan

L_1 = length of the right-hand arm of balance

L_2 = length of the left-hand arm of balance

$ML_2 = W_1L_1$ and $ML_1 = W_2L_2$ (from first and second weighings)

Multiplying the equations by one another and solving for M gives

$$M^2 = W_1W_2 \quad \text{or, approximately,} \quad M = \frac{W_1 + W_2}{2}$$

Moisture. Condensation of water vapor (from the atmosphere) on dry samples or precipitates may cause a weight gain of as much as several milligrams if the sample is exposed to air. Consequently, it is standard practice to weigh all samples in a stoppered bottle. Usually ignited precipitates are weighed in a covered crucible; if the weighing is made rapidly, moisture pickup will not cause appreciable error. In the most exact work it is advisable to weigh crucibles with ignited precipitates inside stoppered bottles, similar to the weighing bottles used for samples. This is not required in student work.

Glass containers may adsorb appreciable amounts of water vapor, if their surface area is large. Whenever it is necessary to use containers such as U-tubes, the weighings should be made with a *counterpoise*. This is a glass vessel similar to the one that contains the sample, but it is lighter in weight. The counterpoise is placed on the right-hand pan for all weighings. If the amount of moisture adsorbed by the counterpoise is about the same as that of the sample container, the error due to adsorption of water is made negligibly small. To insure that counterpoise and sample vessel have the same moisture content and to avoid electrification of the glass, both are wiped with a damp cloth and placed within the balance case 10–15 minutes before the weighing is done. We have observed that a glass tube weighed on successive days without a counterpoise showed changes of several milligrams, but, when a counterpoise was used, the variations were reduced to a few tenths of a milligram.

Temperature. It is absolutely essential that the object to be weighed be at room temperature. Frequently a student is tempted to weigh an ignited crucible while it is still warm, to avoid waiting. Invariably such weighings are too light because convection currents set up in the balance case tend to push up on the left-hand pan and retard the swings of the balance.

Electrification. When glass vessels are wiped with a dry cloth, they acquire a charge of static electricity. This may cause the swings of a balance to behave in a very erratic manner, since, as the beam moves, the charge on the object induces charges in other parts of the balance. Some time may be required to dissipate a static charge on a glass surface.[3] When large glass vessels are to be weighed, they should first be wiped with a damp cloth, which will aid in dissipating charges, and then allowed to stand until most of the surface moisture is evaporated and equilibrium is attained with atmospheric humidity. Nor-

---

[3] A few crystals of a uranium salt in the balance case will ionize the air and expedite dissipation of charges.

mally no special precautions are taken for weighing bottles which have a small area, except that they are not rubbed or wiped prior to weighing.

Defective Balances. Occasionally it is found that a balance will no longer give reproducible readings. The only solution is a thorough overhauling by a skilled repair man. A variety of defects may affect the accuracy and sensitivity of balances. Most common is a shifting of the knife edges so that they do not rest in identically the same positions each time the beam is lowered.

Every chemist should know the following simple tests to apply to balances prior to use if faulty performance is suspected.

1. Determine the reproducibility of the zero point. A number of consecutive determinations should be made of the zero point, with the beam raised and lowered before each observation. If the knife edges are loose, this will show up, for the edges will not return to the same position each time the beam is lowered; consequently the effective lengths of the beam arms will not remain constant.

2. Determine the sensitivity at no load and at a 50-g load. The sensitivity at 50 g should not be decreased by more than 20 per cent from the sensitivity at no load, and at all loads the sensitivity should be at least 0.4 mg per scale division.

Occasionally it is found that a balance suddenly tends to show a creeping rest point. Usually this is due to uneven warming and expansion of the beam by exposure to sunlight. It is essential to protect the beam from sunlight at all times. A north exposure room is preferred for a balance room. Radiators should not be located too near the balances.

Buoyancy. When an object is weighed in air, the apparent weight is affected by the amount of air displaced. Although this effect is not of appreciable magnitude in most student work, it should be understood, and a buoyancy correction should be applied in the rare cases where it is needed.

According to the Archimedes principle, a body surrounded by a fluid is buoyed up by an amount equal to the weight of fluid displaced. A pound of brass will displace about 2 oz of water when immersed. Consequently, if 1 lb of brass is weighed under water, the apparent weight will be only 14 oz.

Air is a fluid of low density. Under average conditions of temperature, pressure, and humidity, a liter of air weighs about 1.2 g. Therefore an object whose volume is 1 liter will be buoyed up by air to the extent of 1.2 g, or the apparent weight of such an object is too light by 1.2 g.

In a weighing both the object and the weights are displacing air, and there is a buoyant effect on both sides of the beam. The effect of this on the weighing depends on the relative volumes of the object and the weights. We may consider the three cases illustrated in Fig. 3–3. In (a) the object has a smaller density than the weights, and consequently it occupies a larger volume. If the object and weights are exactly

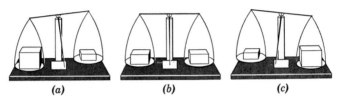

(a)   (b)   (c)

FIGURE 3–3.   Effect of buoyancy in weighing.

balanced in air and the air is then pumped out of the balance case, the object will then appear to be heavier, for the left-hand pan will sink as the buoyant effect of the air is removed. To restore the condition of balance, weights must be added to the right-hand pan. Thus the true weight (weight in vacuo) is greater than the apparent weight (weight in air). In (b) the object and weights are of equal density and occupy the same volume. Removing the air does not affect the equilibrium, or the weight in vacuo is the same as the weight in air. In (c) the density of the object is greater than that of the weights. The weight in vacuo is less than the weight in air.

The relations discussed in the preceding paragraph may be put in the form of an equation.

$$W_v = W_a + \text{buoyancy on object} - \text{buoyancy on weights}$$

or

$$W_v = W_a + \frac{W_a}{D_o} \times \text{density of air} - \frac{W_a}{D_w} \times \text{density of air}$$

where W_a = weight in air (the apparent weight)
W_v = weight in vacuo (the true weight)
D_o = density of object
D_w = density of weights, or approximately 8

Since the density of air at average temperature, pressure, and humidity is about 0.0012 g per milliliter, we have

$$W_v = W_a + \frac{W_a}{D_o} \times 0.0012 - \frac{W_a}{8} \times 0.0012$$

Simplifying, we have

$$W_v = W_a\left[1 + \left(\frac{1}{D_o} - \frac{1}{8}\right)0.0012\right]$$

This equation[4] may be used to compute the weight in vacuo from the weight in air or the reverse, depending on what data are given.

To illustrate the use of the buoyancy correction, we may consider a practical application. It is found that a 100-ml pipet delivers a volume of water weighing 99.60 g in air. The weight in vacuo must be computed, since the density of water is given in tables on the basis of weights in vacuo. Substitution of the weight in air into the preceding equation gives a value of 99.71 g for the weight in vacuo. Handbook tables give a density of 0.9959 g per milliliter for water at the temperature of the experiment, 21°C. Division of the weight of water by its density gives a volume of 99.71/0.9959 = 100.12 ml.

It is not usually necessary to apply a buoyancy correction for the weighings performed in an analysis. Ordinarily the sample or precipitate weighs in the neighborhood of 0.2 to 0.5 g, and the density is of the order of 2 to 4. Therefore the volume of air displaced by the solid is not over 0.1–0.2 ml, and the weight of this air is not over 0.1–0.2 mg, which can be neglected. The buoyancy correction is applied only when large volumes are weighed, as in calibration of glass apparatus by measuring the volume of water delivered or contained.

Weighing of Samples and Precipitates

Weighing Bottles. Samples for analysis are almost universally weighed in weighing bottles. Weighing bottles are small glass bottles having flat ends with straight sides, and ground-glass stoppers. A common form is illustrated in Fig. 3–4.

Samples for analysis are placed in weighing bottles after they have been properly prepared, by grinding, drying, and so on, and kept in a desiccator until needed. In weighing a sample, the bottle, with contained material, is first weighed. Approximately the desired amount of material is removed and quantitatively transferred to a suitable container. The bottle with the remaining material is then reweighed. If no material has been lost, the weight of sample taken is the difference in the two weights. This procedure has the advantage

---

[4] It is obvious that the equation is not strictly rigorous since we have computed the volume by the ratio W_a/D_o rather than the ratio W_v/D_o. The difference between the two volumes so computed is so small that the approximation introduces no error in the buoyancy correction.

that the material is not exposed to the air while the weighings are made, thereby eliminating the possibility of absorbing moisture or carbon dioxide from the air.

In weighing by difference, it is not necessary to determine the zero point of the empty balance; the second weighing is made to the same rest point as the first. It is obvious that this procedure can introduce no error, provided the zero point remains the same in the two weighings. Any error introduced into the single weighing is canceled in taking the difference.

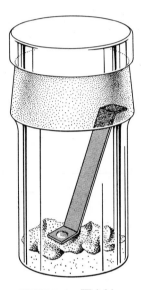

FIGURE 3–4.   Weighing bottle with spoon.

In removing portions of sample from the weighing bottle, be sure that none of the removed sample is lost. The usual analytical technique is to hold the bottle over the container that is to receive the sample, remove the stopper, incline the bottle, and rotate it back and forth to pour out portions of the sample slowly. Continue the process by successive trials until the approximate amount of sample desired is removed. Stopper the tube and reweigh.

Time can be saved by using a small metal spoon for removing portions of the sample. For use with non-corrosive samples the spoon may be a strip of aluminum, bent as shown in Fig. 3–4. It is left in the bottle for all weighings. To remove the sample, hold the bottle over the intended sample receptacle, remove the stopper, and take out one or more spoonfuls of the powder. With a little practice the spoon can be used to measure the desired amount of sample. After removing the sample, replace the spoon, stopper the bottle, and reweigh.

Liquid Samples.   A clean dry weighing bottle is accurately weighed, then the approximate amount of liquid desired is introduced, from a graduated pipet, and the bottle is stoppered and reweighed. If a liquid sample is large, it may be necessary to make a correction for the buoyancy of the air, as explained previously.

Precipitates.   The final precipitates obtained in gravimetric analyses are weighed after drying or ignition. Frequently these precipitates will readily take up carbon dioxide or water vapor from the air. The crucibles containing the precipitates should be weighed with the cover on.   In extreme cases, where the absorption from air is very rapid, the weighing must be very quickly done as follows. After the first heating, an approximate weight is determined.   After subsequent heatings, the approximate weight is placed on the balance pan before the crucible is removed from the desiccator.   After the crucible is in the balance case, it is only necessary to observe the rest point; from this and the predetermined sensitivity of the balance, the exact weight is calculated.   Although all weighings in crucibles are made by difference, a considerable time interval may elapse between the weighing of the empty crucible and the weighing of the sample; it is therefore necessary to determine the zero point of the balance before each weighing.

Theory of Balance Sensitivity

The factors in the construction of a balance which influence its sensitivity can readily be understood by the application of the principles of simple levers.   When a lever is in a position of equilibrium, the force moments that tend to cause clockwise rotation must equal those that tend to produce rotation in a counterclockwise direction.   It may be recalled from elementary physics that "force moment" is defined as the product of the force (mass $\times$ acceleration of gravity g) and the lever arm (the perpendicular distance from the fulcrum to the vertical line passing through the point of application of the force).

In Fig. 3–5 the dotted lines represent the equilibrium position of the unloaded balance.   The solid lines indicate the condition of equilibrium attained after a small weight m has been added to the right-hand pan.   In this position the force producing clockwise rotation is caused by the attraction of gravity for the

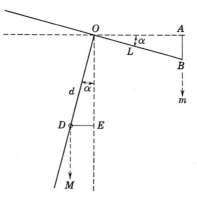

FIGURE 3–5.   Sensitivity of balance.

mass m.   The counterclockwise rotation is produced by the gravitational attraction for the mass of the beam M, which acts through its center of gravity D, located d cm below the knife edge.

Application of the force moment principle gives

$$Mg\overline{DE} = mg\overline{OA}$$

By simple trigonometry,

$$\overline{DE} = \overline{OD} \sin \angle DOE$$

$$\overline{OA} = \overline{OB} \cos \angle AOB$$

Since their respective sides are mutually perpendicular to each other, the angles DOE and AOB are equal and are designated as α in the following. Further, it can be seen from the figure that $\overline{OD}$ is the distance d and $\overline{OB}$ is the length L of the balance arm. The use of these substitutions reduces the first equation to

$$Mgd \sin \alpha = mgL \cos \alpha$$

or

$$\frac{\sin \alpha}{\cos \alpha} = \tan \alpha = \frac{mL}{Md}$$

For small angles the tangent is approximately equal to the angle itself (measured in radians). Thus, the final relation between the displacement or sensitivity of the balance and the weight causing the displacement is

$$\alpha = \frac{mL}{Md}$$

The displacement caused by unit weight has previously been designated as the sensitivity. From the foregoing equation it may be seen that the sensitivity increases with length of the beam but decreases with increase in the mass of the beam-pan system and with increase in the distance from the fulcrum to the center of gravity. This distance may be varied somewhat at will, for there is on the pointer a bob whose position is adjustable. If the bob, shown as X in Fig. 3–1, is raised, the sensitivity is increased, but the *period* of the balance is correspondingly increased. Increase in length of the beam also increases the period.

In the previous discussion it is assumed that the beam is rigid and that the knife edges are in a plane. As the load is increased, the sensitivity may decrease very rapidly because of a bending of the beam, which brings the pan knife edges below the plane of the beam knife edge. High-quality balances show less variation of sensitivity with load than cheaper ones because their better construction minimizes the amount of distortion of the beam. Some balances even show increased sensitivity with load. This is achieved by constructing the balance so that the pan knife edges are slightly above the plane of the beam knife edge. As load is applied, the distortion of the beam brings the three knife edges into the same plane.

Weights

Analytical weights are provided in sets usually going to 50 or 100 g. Each set has a regular progression such that every weight can be replaced by the sum of weights of lower denomination. In older sets the units are 50, 20, 20, 10, etc. This order has the disadvantage that there must be duplicates for certain denominations, which may

TABLE 3–3. Bureau of Standards Tolerances for Analytical Weights*

Denomination	Class M	Class S	Class S-1
100 g	0.50 mg	0.25 mg	1.0 mg
50	.25	.12	.60
30	.15	.074	.45
20	.10	.074	.35
10	.050	.074	.25
5	.034	.054	.18
3	.034	.054	.15
2	.034	.054	.13
1	.034	.054	.10
500 mg	.0054	.025	.080
300	.0054	.025	.070
200	.0054	.025	.060
100	.0054	.025	.050
50	.0054	.014	.042
30	.0054	.014	.038
20	.0054	.014	.035
10	.0054	.014	.030

* These are acceptance tolerances for new weights.   Maintenance tolerances for class M and class S weights which have been in use are approximately twice as large as the values shown here.

lead to confusion and requires that the duplicates be legibly marked. Today's weights are quite generally made on a 5–3–2–1 progression, which eliminates any duplications and makes it possible to replace a weight below 50 g by the next two smaller ones.

Analytical weights are available in three grades, classes M, S, and S-1. The National Bureau of Standards[5] has developed acceptance standards for each of the three grades. The tolerances for the three grades are shown in Table 3–3.

[5] National Bureau of Standards Circular No. 547, August 1954.

Riders. Each balance is equipped with a rider which agrees with the graduation marks on the beam, and it cannot be used with a rider of a different denomination. For a balance with the zero mark on the beam above the central knife edge, the mass of the rider is the mass indicated by the graduation mark directly above the pan knife edge. The most common sizes are 5 and 10 mg, although both 6- and 12-mg riders are frequently used. Since riders of these sizes are so similar in appearance, it is necessary to test the mass of a rider, as described on page 28, before it is installed. In the calibration of weights the rider is not calibrated, since the correction to be used would vary with the position on the beam. It is tested against a calibrated weight. If it is too light, it must be discarded. Heavy riders can be adjusted to the proper weight by successively filing off small portions until the true mass is reached.

For chain balances with notched beams and others for which the graduations extend across the beam from left knife edge to right knife edge, the mass of the rider is one-half the mass marked above the right knife edge.

Calibration of Weights

Since the weight of an object is determined from the sum of the masses of small weights which just balance the object, it is essential in exact work that the masses of the individual weights be accurately known. Therefore every chemist should calibrate his weights before using them and should recheck his calibration at least once a year to detect any change in mass which may occur because of corrosion. In general, calibration is not required of students in the first semester's work because it is very time consuming for an inexperienced worker. After one semester of experience a student should be able to calibrate his weights in a single laboratory period, if the single-swing method of weighing is used.

Weights may be calibrated on either a relative or an absolute basis. In relative calibration some one of the weights is selected as standard, and corrections are determined to make the masses of all other weights agree with that of the selected standard. That is, the correction for the 5-g weight is the value to make its mass exactly half that of the 10-g weight. In absolute calibration a weight of known mass is provided as the standard, and correction values are determined to make all weights agree with the standard one.

Method for Calibration. A modification of the T. W. Richards[6] method is used for the relative calibration. To eliminate the effect of inequality of balance arms, the method of substitution is employed. Weights from another set, which need not be calibrated, are used as tares. All the fractional weights, when taken together, should constitute a gram. It is convenient to add an extra 10-mg weight[7] from another set[8] to supplement the other small weights. The borrowed 10-mg weight is used as a preliminary standard. It is denoted in the tabulation as $\overline{0.01}$. The comparison begins with this weight and proceeds upward.

Procedure. Place the $\overline{0.01}$ weight on the left pan, counterpoise with a suitable weight, and observe the rest point. This does not need to agree with the zero point. Replace the $\overline{0.01}$ weight with a 10-mg weight from the set, and determine the new rest point. Record the data as in Table 3–4. From the sensitivity and the difference is the two rest points, compute the difference in weights between $\overline{0.01}$ and the 10-mg weight. In the data of Table 3–4, 0.01 is 0.6 mg heavier than $\overline{0.01}$. Its relative value is then 0.0106 g, if a value of 0.0100 g is assumed for the preliminary standard. In the same manner obtain a relative value for 0.01′ if there is a duplicate 10-mg weight in the set being calibrated.

Place 0.01 and 0.01′ on the left pan, and counterpoise by a suitable weight on the right. Obtain the rest point. Replace 0.01 and 0.01′ by 0.02 and obtain the new rest point (see comparison 3 of Table 3–4). In the data of Table 3–4, 0.02 is 0.6 mg lighter than 0.01 + 0.01′, whose combined relative values are 0.0211 g. Therefore the relative value for 0.02 is 0.0205 g.

Proceed in this manner with all the comparisons indicated in Table 3–4 (data for comparisons 5–13 are omitted from the table). In comparison 15 a Bureau of Standards calibrated weight is introduced to give the basis for an absolute calibration. This step is omitted if the calibrated weight is not available.

The relative values of column B in Table 3–4 are the values of all weights in terms of the preliminary standard, the borrowed 0.01 weight. It may be noted that these values differ widely from the nominal values for the various weights. The reason is that any error

---

[6] T. W. Richards, *J. Am. Chem. Soc.*, **22**, 144 (1900).

[7] If the balance has a 5- or 6-mg rider, the borrowed weight is 5 mg; it is then used as the preliminary standard.

[8] It is desirable to use a class **M** 10-mg weight as the preliminary standard if one is available.

TABLE 3–4. Data for Calibration of Weights*

(Notebook Page)

Weights Compared	Rest Point	Sensitivity	Difference	Relative Value A	Relative Value B	Aliquot Part of 10-g Weight	Correction in mg
A = standard							
B = weight of unknown value							
1. A. 0.01	−5.0	0.17		0.0100			
B. 0.01	−1.5	0.17	+0.0006		0.0106	0.0102	+0.4
2. A. 0.01	−5.0	0.17		0.0100			
B. 0.01′	−2.0	0.17	+0.0005		0.0105	0.0102	+0.3
3. A. 0.01, 0.01′	−4.0	0.17		0.0211			
B. 0.02	−7.6	0.17	−0.0006		0.0205	0.0204	+0.1
4. A. 0.02, 0.01, 0.01, 0.01′	−3.0	0.17		0.0516			
B. 0.05	−7.2	0.17	−0.0007		0.0509	0.0509	0.0
. . .							
14. A. 5, 2, 2′, 1	−2.0	0.17		10.1834			
B. 10′	+1.0	0.17	+0.0005		10.1839	10.1844	−0.5
15. A. 10′	−3.0	0.17		10.1839			
B. 10 (Bur. Stand.)	0.0	0.17	+0.0005		10.1844	10.1844	Standard
16. A. 10, 10′	−3.0	0.20		20.3683			
B. 20	−10.0	0.20	−0.0013		20.3670	20.3688	−1.8
17. A. 20, 10, 10′, etc. . . .	−5.0	0.20		50.9187			
B. 50	+5.0	0.20	+0.0020		50.9207	50.9220	−1.3

* Rest points were determined by the single-swing method—the + and − signs refer to scale positions measured from the center mark.

in the value of the preliminary standard is greatly magnified in the comparisons; for example, the 10-g weight is 1000 times the mass of a 0.01-g weight. It is better therefore to take one of the larger weights as the final standard. In Table 3–4 the Bureau of Standards 10-g weight is taken as the reference.[9] If a certified weight is not used, one of the 10-g weights of the set is taken as the reference. In Table 3–4 the preliminary value for the reference weight is 10.1844 g. Aliquot values are computed from this for all the other weights of the set, based on ratios of their nominal values. For example, the aliquot value for the 1-g weight is one-tenth of 10.1844 g, or 1.0184 g; the aliquot value of the 20-g weight is twice 10.1844 g, or 20.3688 g.

After the aliquot values are computed, the correction value for each weight is determined from the relation:

$$\text{Correction} = \text{relative value} - \text{aliquot value}$$

For the 20-g weight the correction is 20.3670 − 20.3688, or −0.0018 g (−1.8 mg). When computed in this manner, the correction has the proper sign, and in the use of the weights it is to be added algebraically to their nominal values. That is, the nominal value plus the correction gives the true value. The 20-g weight has a nominal value of twice that of the 10-g weight. Adding −1.8 mg gives its true value of 19.9982 with reference to the 10-g weight as standard. The correction is negative because the weight does not weigh so much as its nominal value.

After the corrections are computed, prepare a card showing the corrections in milligrams for each weight of the set. Keep this card with the weights for ready use. If any weight has an abnormally large correction, report it to the instructor. Such weights should be replaced if possible.

Calibration of Chain.   Determine the zero point of the balance. Compare the chain with calibrated weights, at 10-mg intervals, from 10 mg to the full value of the chain by placing the weight on the left-hand pan and setting the chain to bring the rest point into coincidence with the zero point. The correction for each weight is given by the relation:

$$\text{Chain correction} = \text{correct weight} - \text{scale reading}$$

For each setting of the chain the correction is added to the scale reading to give the true weight.

---

[9] If there is a correction to the face value of the reference weight, it may be taken into account by the method of Blade. See *Ind. Eng. Chem., Anal. Ed.*, **11,** 499 (1939).

Construct a graph showing the correction value of the chain at the experimental points, and connect the points by straight lines. Correction values for any chain reading can then be determined by reference to this graph.

Calibration of Notch-Beam Rider. Determine the zero point of the balance. Place a calibrated 0.1-g weight on the left-hand pan, and set the rider at the 0.1-g mark. Determine the rest point. From the deviation between zero and rest point and the sensitivity of the balance, compute the weight difference between rider and weight. From the corrected value of the weight and the weight difference, compute the effective weight of the rider. The correction is given by the relation,

$$\text{Rider correction} = \text{true weight} - \text{nominal value}$$

Determine in a similar manner the corrections for other settings of the rider. The correction values are to be added to nominal values of the rider at the various settings to give the true weight.

Questions

1. Explain the functions of each of the following parts of the balance illustrated in Fig. 3–1: (a) K, (b) BC, (c) PC, (d) F, (e) RC.

2. Why are computations of rest points from swing data not carried beyond the nearest $\frac{1}{10}$ division?

3. How much should a rider weigh to be used on each of the following balances:

   (a) The zero mark is over the central knife edge, and the 6 mark is over the knife edge which supports the pan.

   (b) The zero mark is over the central knife edge, and the beam is graduated to 12, but the 10 mark is over the knife edge that supports the pan.

   (c) The zero mark is over the knife edge supporting the left pan, and the 10 mark is over the knife edge for the right pan.

4. In which of the following cases is it satisfactory to assume that 10 is the zero point of the balance rather than to determine the rest point from swing data:

   (a) An object is to be weighed to the closest 2 mg.

   (b) A series of samples is to be weighed to the closest 0.1 mg by successive removals from a weighing bottle.

   (c) The weight of a precipitate is to be determined to the closest 0.1 mg by the difference in weights of a crucible empty and filled with the precipitate; at least 6 hours elapse between the weighings.

(*d*) The weight of a piece of metal for a density determination is to be found to the closest 0.1 mg.

5. What is the *sensitivity* of a balance? What is the advantage of weighing by sensitivity rather than by adjustment of the rider until a point of balance is obtained?

6. Describe a procedure for determining the sensitivity curve of a balance. What should be the general shape of this curve?

7. If the method of swings is used, why is it desirable that the sensitivity of the balance be at least 0.5 mg per division?

8. If the method of first deflection is used, why is it desirable that the sensitivity of the balance be around 0.2 mg per division?

9. What effect does raising the bob on a balance pointer have on the sensitivity? On the period?

10. How does the mass of the balance system affect the sensitivity of the balance? How is this reflected in the construction of the balance?

11. What is the effect on the sensitivity of a balance of:

(*a*) Lowering the bob on the pointer?
(*b*) Increasing the total load?
(*c*) Changing the adjustment of one of the screws on the end of the beam?

12. What maximum load should be placed on each pan of the analytical balance? What harm will be caused if the balance is greatly overloaded?

13. What injury might be done to a balance if an object is placed on the pan while the beam is unlocked?

14. Why is it permissible to leave the beam arrest lowered while changing fractional weights?

15. Describe the method of weighing by substitution. When should this method be used? What other method of weighing can also be used in such circumstances?

16. Criticize and correct the statement, "The weight in vacuo of an object is always greater than the weight in air."

17. If the density of a metal in vacuo is the same as in air, what can be concluded about the density of the metal and the density of the weights used in the experiment?

18. What effects will be noticed in weighing a glass vessel if the surface is electrified? How may these effects be avoided?

19. Explain the following swing readings obtained when a sample of an anhydrous material was weighed on a watch glass.

8.0	13.4
8.3	13.5
	13.6

20. A large bulb used to weigh gases had a weight of 35.5063 g at one time and 10 hours later weighed 35.5092 g. Offer a possible explanation, and suggest a method of avoiding the discrepancy.

21. Why is it necessary to apply the buoyancy correction in the calibration of volumetric apparatus for volumes of 10–250 ml? Why is it usually not necessary to apply the buoyancy correction in the weighing of sample for analysis and in the weighing of precipitates obtained in analysis?

22. Describe a proper technique for weighing each of the following:
   (a) A piece of platinum foil.
   (b) Solid $BaSO_4$ which is non-hygroscopic.
   (c) Solid $Na_2CO_3 \cdot 10H_2O$ which is highly efflorescent.
   (d) A sample of vinegar.
   (e) An absorption tube for CO_2 which is filled with about 100 ml strong KOH solution.

23. Why is relative calibration of weights satisfactory for most analytical purposes?

24. In the calibration of weights why are the values redistributed in terms of one of the larger weights?

25. Why are not buoyancy corrections made during calibration of weights?

26. The correction for a 2-g weight is minus. Is this weight heavier or lighter than its nominal (face) value in relation to the 10-g weight which was selected as a standard for the calibration?

27. In which of the following cases would an absolute rather than a relative calibration of weights be necessary?
   (a) In a large industrial laboratory where the samples and precipitates are weighed by different operators using various sets of weights.
   (b) In a gravimetric analysis in which the sample and precipitate are weighed with the same set of weights but on different balances.
   (c) In a volumetric analysis in which a solid primary standard is used and the sample is a weighed portion of a liquid.
   (d) As in (c), but the sample is a measured volume of liquid.
   (e) In a calibration of volumetric glassware in which weighings are made to the closest 0.5 mg.

Problems

In the following problems which involve the computation of weight from swing or deflection data, it is assumed that the object is on the left pan and the rider on the right half of the beam unless otherwise specified.

1. The rest point with the rider at 4 is 10.5 and at 3 is 7.0. Is the rider on the right or left half of the beam?

2. Find the weight of an object to the closest 0.1 mg from the following data:
   Swings of the unloaded balance: 8.3, 11.4, 8.5
   Weights on the pan: 10 g, 5 g, 2 g, 1 g, 500 mg, 50 mg, 20 mg, 5 mg

Swings with the rider at 1 mg: 14.0, 10.3, 13.8
Swings with the rider at 2 mg: 8.9, 11.8, 9.1   *Ans.* 18.5773 g.

3. A weighing is done on a notched-beam chain balance by the swing sensitivity method. From the following data determine the weight of the object to the closest 0.1 mg.

Swings of the empty balance:	8.3	11.4
	8.5	

Weights on right pan: 10 g, 5 g, 3 g
Beam rider at 0.6 g

Swings with the chain at 76 mg:	10.3	14.0
		13.8
Swings with the chain at 77 mg:	8.9	11.8
	9.1	

Ans. 18.6773 g.

4. Determine the mass of an object to the closest 0.1 mg from the following:

Zero point of the balance	10.5
Rest point for 22.5130 g	10.0
Rest point for 22.5135 g	8.5

List the weights on the pan assuming that the balance has a 10-mg rider.

5. The swings of an empty balance are: 8.3, 13.1, 8.6. The rest point with a mass of 16.56 g on the pan and the rider at 6 is 13.0; with the rider at 7 the rest point is 11.2. Where would the rider have to be placed to make the rest point and zero point coincide?   *Ans.* 7.2.

6. The following data were obtained with the object on the left pan and the rider on the left half of the beam:

Zero point	9.7
Mass on right pan	24.83 g
Rest point with rider at 4	9.1
Rest point with rider at 5	11.4

What is the mass of the object to the closest 0.1 mg?

7. The zero point of a balance is 14.1 when the single-deflection method is used. With a load of 12.48 g and the rider at 4 the deflection is 8.5. If the sensitivity is 0.15 mg per division, where should the rider be set? *Ans.* 3.2.

8. The first deflection of a balance with a 100-mg chain is 5.5; from the following deflections calculate where the chain should be set to complete the weighing.

With 27.343 g the deflection is 15.0.
With 27.344 g the deflection is 7.5.

9. In a series of weighings by difference, the zero point of the balance is assumed to be 10.0. Actually the zero point as determined by swings is 8.8; the sensitivity at this load is 0.53 mg per division. What error is made in each individual weighing? What error is made in the weight of the sample?

10. A balance has graduations on the beam from zero at the center to 10 over the right knife edge. By mistake a 5-mg rider is used on this balance and the weight is recorded as 17.7134 g. What is the correct weight of the object?
          Ans. 17.7117 g.

11. Could weighings accurate to 0.1 mg be made on a balance whose sensitivity is 1.4 mg per division if rest point data are reproducible to 0.1 division?

12. If the left-hand arm of a balance is 15.02 cm in length and the right-hand arm is 15.00 cm, what will be the apparent weight of a 20.0000-g object?
Ans. 20.0267 g.

13. A sample of platinum (density = 21.45 g/ml) weighs 63.5781 g in air with brass weights (density = 8 g/ml). Compute the weight in vacuo.
Ans. 63.5721 g.

14. A sample of aluminum (density = 2.7 g/ml) has a weight in vacuo of 21.5068 g. What would the sample weigh in air if gold weights (density = 19.3 g/ml) are used? Repeat the calculation using brass weights.

15. What is the value of a buoyancy factor for converting weights of mercury (density = 13.5 g/ml) determined in air with brass weights to weights in vacuo?
Ans. 0.99994.

16. 500.0 mg of NaCl is weighed in air with brass weights. What absolute error is made in the total weight of the sample if buoyancy is neglected? Is this error detectable on the ordinary analytical balance? (Density of NaCl = 2.2 g/ml.)

17. The density of water in vacuo at 21°C is 0.99802 g per milliliter. What is the weight in air with brass weights of 50.00 ml of water? *Ans.* 49.8487 g.

18. A 0.5000-g sample of density 4.5 g per milliliter yields a precipitate of silver chloride (density = 5.6 g/ml) which weighs 1.3950 g. Compute the weight in vacuo of the sample and of the precipitate. Compute the ratio of the uncorrected and corrected weights of sample and precipitate. According to your results, is it necessary to apply buoyancy corrections to these data? Brass weights are used.

19. An object weighed in air with brass weights had a mass of 30.0075 g; when weighed under water at 23°C, the mass was 25.8753 g. What is the density in vacuo of the substance? What would the substance weigh if it were immersed in alcohol (density = 0.90 g/ml)?
Ans. 7.244 g/ml; 26.2797 g in alcohol.

20. Two weights, one of aluminum (density = 2.7 g/ml) and one of copper (density = 8.92 g/ml), are each adjusted to 10.0000 g in vacuo. The aluminum weight is placed on the left-hand pan of a balance and the copper one on the right. If the zero point of the balance is 10.3 and the sensitivity is 0.22 mg per division, where must the rider be set to make the rest point and zero point coincide? The balance beam is graduated to 10 mg on both sides of the central knife edge.

21. On the basis of the 10-mg weight the preliminary values of the 2-g and 10-g weights are 1.9962 and 10.0152 g, respectively. What is the final correction for the 2-g weight in terms of the 10-g weight? Is the 2-g weight lighter or heavier than it should be? *Ans.* −6.8 mg; lighter.

CHAPTER 4

The Laboratory Tools and Operations of Quantitative Analysis

Quantitative analyses require precision tools and careful techniques in the use of these tools. Since these tools and techniques are used repeatedly throughout the course, we have assembled in this chapter detailed instructions to which the student will refer as needs arise for applications of the various methods. These instructions are in three parts: (1) operations of volumetric analysis, (2) operations of gravimetric analysis, and (3) preparation of samples for analysis. The first part should be studied before undertaking any volumetric analyses and the second before undertaking any gravimetric analyses. The material of the third part applies to all types of analyses.

I. Operations of Volumetric Analysis

Volumetric analyses require accurate measurements of liquid volumes, of the volumes of solutions that react with one another, and of the concentrations of solutions. We shall consider first the tools used for these measurements, then the operations required in using these tools.

Measuring Instruments

The instruments used for accurate measurements of liquid volumes are the buret, the transfer pipet, and the volumetric flask. In addition, when precision measurements are not needed, the graduated cylinder and the graduated pipet are used.

Buret. The buret is a long glass cylinder of uniform bore. It is used to measure precisely the volume of one solution required to react with a fixed volume of another solution, in the operation of titration. The volume withdrawn from a buret is measured by graduation marks etched into the glass tube. Burets are available in various capacities, ranging from a fraction of a milliliter for use in microdeterminations to 50 or 100 ml. The most widely used size is 50 ml. The graduation marks for this size are in 0.1-ml units. In high-quality instruments the fractional marks extend halfway around the barrel and the integral marks completely encircle it. The etched marks should be narrow and sharply defined.

Delivery from a buret is controlled by a ground-glass stopcock, connected to a capillary tip. The tip should be of such bore that 50–100 seconds is required for delivery of 50 ml. If withdrawal is too rapid, drainage from the walls is not complete.

Weight Buret. Increased accuracy in titrations can be obtained by weighing, rather than measuring, the volume of reagent required. For this purpose, a weight buret (Fig. 4–1) is employed. The body of a weight buret is made short so that the entire buret can be suspended from a hook just below the pan support of the balance.

FIGURE 4–1. Weight buret.

The buret has a glass stopper at the top, to prevent evaporation of liquid, and a guard cap which fits over the delivery tip.

Transfer Pipet. The fixed-volume pipet is used for rapid and accurate measurement of fixed volumes of liquids. Essential features are a long delivery tip that should reach to the bottom of the volumetric flasks used, an enlarged bulb at the center to give the desired capacity, and a narrow stem. The graduation mark, which gives the level to which the pipet should be filled to deliver a given volume, is on the stem, about halfway between the bulb and the end. The tip should be tapered and ground smooth. The size of the orifice should be such that delivery is slow, to permit drainage from the walls. Minimum delivery time for a 5-ml pipet is 15 seconds; for a 50-ml pipet, 30 seconds.

Volumetric Flask. The volumetric flask is a thin-walled flat-bottomed flask with a long narrow neck. It is used to dissolve solids in a known volume or to dilute solutions to an accurately known volume from which carefully measured fractions (aliquots) can be taken for analysis. The graduation mark, for the volume *contained* in the flask, is placed on the neck. The flask has a ground-glass stopper so that it may be inverted without loss of liquid.

Cleaning

All volumetric apparatus must be completely free of grease when used, since the least trace of grease causes aqueous solutions to adhere to the walls in drops and thereby prevents complete drainage. Quite frequently adequate cleanliness can be obtained by washing the apparatus with a detergent solution. If this is not sufficient, then "cleaning solution" is required. The instructor will specify the method to use. The most widely used cleaning agent is a solution of sodium dichromate in concentrated sulfuric acid. Because of its oxidizing power, the solution, particularly when hot, removes grease quickly and completely. Cleaning solution, as the mixture is called, is not a general solvent for cleaning all apparatus, only for cleaning volumetric apparatus. Flasks, beakers, and similar apparatus are best cleansed with soap and water (better still are detergents) unless they contain organic tarry residues, in which case the oxidizing properties of cleaning solution may be useful. Other mixtures for cleaning are described in handbooks of chemistry.

Preparation of Cleaning Solution. Stir about 20 g powdered technical sodium dichromate with just enough water to make a thick paste. Add 300 ml of technical-grade concentrated sulfuric acid. Store in a glass-stoppered bottle. It is unnecessary to remove the residue of undissolved salt by filtration, but clear solution should be decanted from the bottle each time it is used. The solution may be repeatedly used until the reddish color of dichromate has been replaced by the green of chromic ion.

Caution: *Do not allow the solution to come into contact with you or your clothing, for it will cause very bad burns. If any is spilled on the floor or desk, neutralize immediately with commercial-grade*

*sodium bicarbonate; then wash completely with water. If it is spilled
on the skin, wash in running water as soon as possible.*

Cleaning Buret.   Inspect the stopcock plug.   If it is well greased,
the plug turns easily, and the surface between the plug and barrel
appears to be *transparent*.   If the plug needs greasing, remove it and
wipe clean; also wipe out the inside of the barrel.   Both parts must
be dry.   Apply a thin layer of grease at each end of the plug as demon-

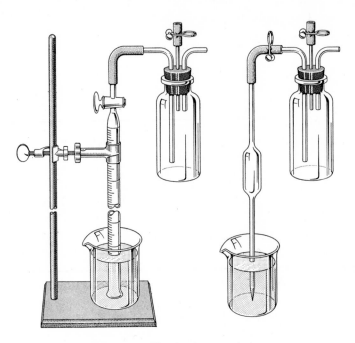

FIGURE 4–2.   Cleaning buret and pipet.

strated by the instructor.   A good-grade stopcock grease, *not vaseline,*
must be used.   Replace the plug in the barrel and turn back and
forth.   If too much grease is applied, the tip will become clogged.
If this happens, remove the plug, clean out the lump of grease by the
method specified by the instructor, and then regrease.

Heat about 100 ml of cleaning solution in a 250-ml beaker to a temperature of 60–70°C (note 1).   Use a thermometer; take care not to
exceed 70°C.   Clamp the buret in an inverted position with the opening
reaching nearly to the bottom of the beaker, which contains cleaning
solution, as shown in Fig. 4–2.   Attach a safety bottle to the buret tip
by rubber tubing, and draw cleaning solution into the buret by suction

until the level is slightly past the final graduation mark. Do not allow the solution to reach the stopcock, where it would remove the grease. Close the stopcock, and allow the filled buret to stand 3–5 minutes. Open the stopcock, raise the buret above the liquid level, and allow it to drain thoroughly. Remove the beaker of cleaning solution, replace by a beaker of tap water, and flush out the buret by drawing water to the 50-ml mark and allowing it to drain out. Remove from the clamp; wash several times with tap water and finally with distilled water. When all trace of cleaning solution is removed, clamp the buret in an upright position, fill with water, wipe off the outside, open the stopcock, and allow to drain in order to test for cleanliness. If any drops of water adhere to the inner wall, the buret must be recleaned. Fill the cleaned buret with distilled water, and leave it in this condition until needed (note 2). If the buret is left empty, it will quickly become contaminated with a film of grease.

Notes. 1. Cleaning solution may be used without warming if desired, but then the solution should stand in the buret for at least one-half hour, and preferably overnight.

2. Burets will remain clean for long periods if always stored in this way when not in use. Optionally the buret may be left empty and inverted, with the open end dipping into a beaker of water. If it is stored in this manner, less rinsing is required before the buret is ready for use, since the barrel is left in a dry condition.

Cleaning Pipet. Clamp the pipet in a vertical position with the tip dipping into a beaker of cleaning solution, as shown in Fig. 4–2. Draw warm cleaning solution into the pipet until the level is within an inch of the top. *Caution. Be sure to use a safety bottle.* Close the rubber connecting tubing with a pinch clamp, and allow the solution to stand 3–5 minutes. Raise the pipet, allow to drain, wash thoroughly with tap water, and rinse with distilled water. Test for cleanliness by filling with water, allowing the pipet to empty, and observing whether drops form on the side within the graduated portion. Lack of cleanliness above the volume mark does not affect the accuracy, since solution delivered from a pipet does not come in contact with this portion. Pipets must be recleaned frequently since they quickly become contaminated with organic matter deposited from the breath.

Cleaning Flask. Pour 50–100 ml warm cleaning solution into the flask, stopper tightly, and manipulate so that all portions of the wall are repeatedly brought into contact with the solution. Continue for 3–5 minutes, then pour the cleaning solution back into its container, rinse the flask repeatedly with tap water, and finally rinse with distilled water. It is important that the neck of the flask above the graduation mark be clean because, when solutions are diluted in the flask, drops of water might adhere to an unclean wall, thus invalidating the measurement of volume.

Calibration

Necessity for Calibration of Volumetric Apparatus. The results of volumetric analyses are all based on accurate measurements of volumes. In precise work it is never safe to assume that the volume contained or delivered by any instrument is exactly that amount indicated by the graduation mark. All volumetric apparatus should be either purchased with a calibration certificate or calibrated by the analyst. In addition to proving the accuracy of the apparatus, the operation of calibration serves to provide the beginning student with drill in the use of volumetric apparatus before he undertakes analyses.

Method for Calibration. Calibration is usually performed by measuring the amount of water delivered or contained by the apparatus. This measurement may be made either by weighing the water and from its density calculating the volume, or by measuring the water in an apparatus which has previously been calibrated. Water is used as calibration liquid because of its ready availability, and because it is similar in viscosity and speed of drainage to the dilute solutions ordinarily employed in volumetric analysis.

Standard Volume. The unit of volume employed in all scientific work is the *liter*. The liter is defined as the volume occupied by 1 kg of water at the temperature of greatest density, 3.98°C. At the time of introduction of the metric system, it was intended that the liter be 1 cubic decimeter, or 1000 cc, thereby interrelating the units of length and volume; but, because of experimental errors, this ideal was not attained, and in reality the liter is 1000.028 cc.

The liter is an inconveniently large unit for laboratory use, since most of the volumes encountered in titration are not greater than 0.05 liter. Because of this, the practical unit is chosen as 0.001 liter, which is known as the milliliter (ml). For all practical purposes the milliliter is identical with the cubic centimeter (cc), for the difference is only 28 parts per million.

Change of Volume with Temperature. The unit of volume is independent of temperature; if a liter container were constructed of some material which did not expand on application of heat, it would hold a liter at any temperature chosen, but the weight of material contained would change because of the variation in density with temperature. Practically, however, no such containers are available. Glass, which is used for all volumetric apparatus, does expand slightly when heated. A vessel which holds 1.0000 liter at 15°C holds a volume of 1.00025 liters at 25°C. Although small in magnitude, the change is measurable, and in precise work it is necessary to define the temperature at which apparatus is calibrated. The Bureau of Standards has chosen 20°C as the normal temperature, and apparatus calibrated by the Bureau will contain or deliver the stated volume at this temperature. In this course all apparatus will be calibrated at room temperature. If desired, the values obtained may be corrected to 20°C by use of the equation,

$$V_{20} = V[1 + 0.000025(20 - t)]$$

where V is the observed volume, V_{20} is the volume at 20°C, and t is the observed temperature. This correction is not necessary in student work.

Another effect of temperature is of importance in all volumetric work, namely, the expansion of liquids when heated. As previously noted, the change in volume of a glass vessel is only 0.25 part per 1000 for a change in temperature from 15 to 25°C. Over the same temperature interval the volume of water changes 2 parts per 1000; that is, a liter of water at 15°C will occupy a volume of 1.002 liters at 25°C. This effect is important in connection with the use of standard solutions.

Technique of Reading Volumetric Apparatus. When water is held in a glass vessel, a *meniscus,* whose radius of curvature is a function of the diameter of the column, is formed at the upper surface. With clear solutions the lowest portion of this meniscus is chosen as the point of observation, since it is the most easily reproducible point for visual measurement. Obviously it is immaterial, in a column of given constant radius, what portion of the meniscus is chosen for observation, provided the same point is used for all readings; when translucent or opaque solutions are employed, it is necessary to make the reading at the upper edge rather than at the bottom of the meniscus, since the bottom cannot be seen.

The bottom of the meniscus lies some distance behind the graduation mark on the outer surface of the container; it is necessary (to avoid errors of parallax) to hold the eye on a level with the plane of the graduation mark and the meniscus. The effect of the eye position on the reading is shown in Fig. 4–3. If the eye is below the meniscus, the reading is too high; that is, a graduation mark is read which lies

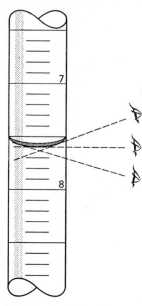

below the true value. A similar but opposite error occurs when the eye is above the meniscus. In all readings some device must be employed to insure absence of parallax. In reading apparatus whose graduation marks extend completely around the cylinder, such as the pipet and volumetric flask, hold the eye in such a position that the graduation on the back of the cylinder appears to merge with the line on the front. Similarly, if a buret has marks which extend halfway around the cylinder, it may be turned until the two ends of the line appear to merge as the eye is moved to a level position. Many burets have the smaller divisions marked by short lines only. In reading them some external means must be employed for avoiding parallax. A simple device is a loop of paper held about the barrel with the top edges together at the ends. This loop is moved to a position slightly below the meniscus, and the eye is brought to level by moving it until the top edge of the rear portion of the paper is just visible over the top of the front portion. The reading is made with the eye in this position.

FIGURE 4–3. Effect of parallax in reading buret.

Calibration of Buret. See that the buret is clean, that the stopcock is well greased, and that the capillary tip permits free flow of liquid. Withdraw water until the level is at or just below the zero mark. Wait 10–20 seconds for drainage (note 1), and then read the meniscus, estimating its position to the nearest 0.01 ml. Test for tightness of the stopcock by allowing the buret to stand for 5 minutes and rereading. There should be no noticeable change. During this interval weigh, to the nearest milligram (note 2), a stoppered Erlenmeyer flask of 125-ml capacity.

After tightness is assured, completely fill the buret with distilled water which is at room temperature (note 3), open the stopcock, and allow the liquid to flow rapidly until all air bubbles are expelled from the tip. Continue to withdraw liquid until the level is at or slightly below the zero mark. After allowing time for drainage, read to the nearest 0.01 ml. Touch the tip to the wall of a beaker to remove the pendent drop of water. Place the weighed flask beneath the tip, and

TABLE 4-1. Data for Buret Calibration

Initial Reading, ml	Final Reading, ml	Apparent Volume, ml	Initial Weight, g	Final Weight, g	Weight Water, g	Temperature, °C	Volume Water, ml	Correction, ml
0.05	15.10	15.05	43.993	59.014	15.02	26	15.08	+0.03
0.00	25.30	25.30	43.752	68.913	25.16	26	25.26	+0.04
0.12	35.26	35.14	44.730	79.750	35.02	26	35.16	+0.02
0.15	45.31	45.16	44.892	89.922	45.03	26	45.21	+0.05

withdraw approximately 15 ml of water, taking care to avoid wetting the neck of the flask (note 4). Remove the last drop from the tip by touching it to the wall of the flask. Stopper the flask, allow time for drainage, and read the buret. Weigh the flask with water to the nearest milligram. Record the results as shown in Table 4–1. Empty the weighing flask, wipe the neck dry, stopper, and reweigh. Refill the buret (note 5), read, withdraw approximately 25 ml into the flask, read the buret, and reweigh the flask. Proceed in the same way with the 0–35-ml and 0–45-ml intervals.

Calculate the actual volume of water delivered in each observation by multiplying the weight of water in air by the volume occupied by 1 g of water at the temperature of withdrawal (note 6). These data are given in Table 4–2. Subtract the apparent volume, as given by

TABLE 4–2. Volume Occupied by 1 g of Water Weighed in Air with Brass or Stainless Steel Weights at Various Temperatures

Temperature, °C	Volume, ml	Temperature, °C	Volume, ml	Temperature, °C	Volume, ml
10	1.0016	17	1.0023	24	1.0036
11	1.0017	18	1.0025	25	1.0038
12	1.0018	19	1.0026	26	1.0041
13	1.0019	20	1.0028	27	1.0043
14	1.0020	21	1.0030	28	1.0046
15	1.0021	22	1.0032	29	1.0048
16	1.0022	23	1.0034	30	1.0051

the buret readings, from the actual volume in order to obtain the buret correction for each interval. Repeat the calibration. Duplicate results should agree within 0.04 ml (note 7). If the agreement for any interval is not within this range, repeat that observation. Plot the average buret corrections for each interval as shown in Fig. 4–4.

Notes. 1. After liquid is withdrawn from a buret, some time must be allowed for drainage. In general, 10–20 seconds is sufficient. The magnitude of the drainage effect may be noted by rapidly withdrawing water from a buret and then reading the position of the meniscus at 10-second intervals until the readings become constant.

2. Since the buret readings are accurate to only 0.01 ml, it is unnecessary to know the weight of water more precisely than to 0.01 g. In weighing, the nearest milligram is determined in order to insure accuracy to the nearest centigram. Weighing to the nearest milligram requires no determination of rest point.

3. The temperature must be known since the density of water varies with its temperature. Only water that is at room temperature should be used. A frequent

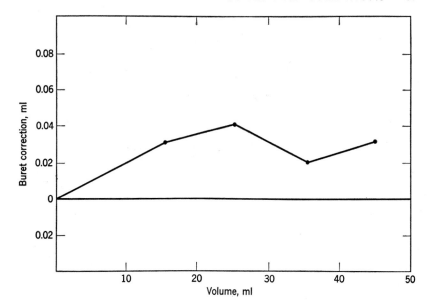

FIGURE 4–4. Calibration curve for buret.

source of error in calibration is inaccuracy of the thermometer used for determining the temperature.

4. The portion of the flask that comes in contact with the cork should not be allowed to become wet, for the cork may absorb water which will be lost by evaporation before the weighing.

5. In practice, a buret is so filled that the initial reading is at or near the zero mark. Calibration is consequently performed in the same manner.

6. The buoyancy correction has been used in compiling the data of Table 4–2. If the true density of water, as given in handbooks, is used for computing buret corrections, it is necessary to correct all observed weights for buoyancy, according to the methods on page 32.

7. Since each reading may be in error by 0.01 ml, the two readings for one calibration may be in error by 0.02 ml. Should the two readings for the duplicate determination also be in error by 0.02 ml, but in the opposite direction, the total deviation between the results may be 0.04 ml. Any deviation of this size or greater between the two results is probably due to an accidental error, such as leakage from the buret or insufficient time for drainage. Notice that nothing has been said about the possibility of an error in weighing. The reason is that the weighing is made to the nearest milligram, whereas the accuracy needed is only to the nearest 0.01 g.

Calibration of Volumetric Flask "to Contain." The Exax-grade volumetric flasks now used in most college chemistry laboratories are manufactured to a tolerance of about 0.1 per cent. In most student work a calibration of the flasks is not necessary. Should such be required, it is essential to have available a large-capacity balance capable of weighing over 500 g with a sensitivity of 0.1 g.

To calibrate a flask, clean, clamp in an inverted position, and allow to drain until thoroughly dry, wiping off the drops that adhere to the rim. Weigh to 0.1 g by substitution. (Put the flask on the right-hand pan and add crude weights or metal shot to the left-hand pan to balance. Remove the flask and add weights to the right-hand pan to bring the pointer to the original balance position.) Fill the flask just to the mark and reweigh. From the weight and temperature of the water compute the true volume of the flask. The difference between the true and nominal volumes is the correction, which should be added algebraically to the nominal value for the flask.

Calibration of Pipet to Deliver. Read the instructions for use of a pipet, page 59. Clean the pipet and rinse thoroughly. Weigh a small stoppered flask to the nearest milligram. Fill the pipet with distilled water and deliver the water into the weighed flask, following instructions carefully. Stopper the flask and weigh. From the weight and temperature of the water compute the true volume delivered by the pipet. The correction is the difference between the true and nominal values. Empty the flask and repeat the calibration. Duplicate results should agree within 1 part per 1000 (0.01 ml in 10).

Relative Calibration of Flask and Pipet. If the flask is not calibrated, it is desirable to make a relative calibration of flask and pipet, so that the amount delivered by the pipet is an exactly known aliquot portion of the liquid held in the flask. When the flask and pipet are used for measuring out aliquot portions of a sample, there is no necessity for having an absolute calibration of either.

Procedure. Before starting the calibration, make certain that the pipet stem will reach entirely to the bottom of the flask; otherwise the two are incompatible. Clean and dry the flask by drainage. Read the instructions of page 59 for use of the pipet. Deliver 5 or 10 portions of distilled water (as required to fill the flask) from the pipet into the flask. Mark the position of the meniscus by pasting a gummed label halfway around the stem of the flask, so that the upper edge of the label is at the bottom of the meniscus. Empty and dry the flask and repeat the determination. For a 250-ml flask the meniscus position should not deviate more than 1 mm in two calibrations.

Unit Operations of Volumetric Analyses

The success of a volumetric analysis will depend on the care and skill with which the tools are employed in the operations in-

volved in the analysis. These operations are described in the
following sections.

1. Use of Transfer Pipet. The pipet is a convenient and very accurate instrument for measuring fixed volumes of solutions. Properly
used, the volume delivered is reproducible to
better than 1 part per 1000, but if care is not
taken there may be variations of several times
this amount. Precautions must be taken to
avoid (1) dilution of the solution delivered by
a pipet, (2) transfer of liquid adhering to the
outside of the tip, and (3) erratic retention of
liquid in the pipet after delivery.

The pipet must be clean so that no drops
adhere to the walls after delivery. Careful
rinsing is necessary after using cleaning solution to insure its complete removal.

The transfer of a measured volume (usually
an aliquot portion of a sample) by a pipet
requires the following steps.

Rinsing the Pipet. Shake out residual water
from the tip of the cleaned pipet and dry the
outside of the tip with a clean cloth. Insert
the tip into the liquid to be used and by
suction draw up enough liquid to bring a small
amount into the bulb. Hold the liquid in the
pipet by placing the forefinger of the right
hand over the end of the stem. See Fig. 4–5.

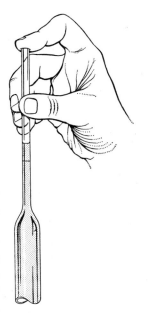

FIGURE 4–5.   Method of
holding a pipet.

Withdraw the pipet from the liquid, and shake gently so as to wet all
portions of the bulb with the contained liquid. This is to remove any
adhering water. Drain out and discard the rinsing liquid. Repeat.

The instructor of the course will specify the method to be used
for sucking liquid into the pipet. This may be done either (1) by
mouth or (2) by an aspirator bulb. Most industrial laboratories
today require the latter method because of the safety hazard if
dangerous chemicals are sucked into a pipet by mouth.

Filling the Pipet. Insert the pipet in the liquid, with the tip at or
very near the bottom of the container. Apply suction to draw liquid
to a point above the graduation mark. Quickly place the forefinger of the right hand on the end of the stem, so as to make an air-
tight seal that prevents liquid from running out of the tip. In doing
this the stem of the pipet is held between the thumb and second
finger, in such a position that the forefinger is free to close the end
of the stem as shown in Fig. 4–5. With a little practice it is easy

to manipulate the pipet and so regulate pressure that flow of liquid is controlled. With the liquid level above the mark, remove the pipet from the liquid and wipe off the outside of the stem with a clean towel. The purpose is to prevent accidental transfer of liquid that is adhering to the outside of the tip. Now place the tip against the wall of the container from which liquid was withdrawn and release pressure just enough to permit excess liquid to flow back, thereby bringing the meniscus exactly to the etched graduation mark. Hold the pipet level with the eye for this adjustment. When the level stands at the mark, quickly move the pipet to the discharge container, release pressure on the forefinger, and allow the liquid to flow out. Avoid splashing or loss of liquid.

Draining the Pipet. When most of the liquid is discharged and only a little remains in the tip, touch the tip to the wall of the receiving vessel. The pendent drop will be drawn from the tip by capillary action. At the same time some of the residual liquid inside the tip will flow out, but some will remain. Do not shake or blow out this residual amount. Drainage with the tip touching the container wall gives more reproducible delivery than if residual liquid is blown out.

If further portions of liquid are desired, pipet out immediately, without further rinsing. If the pipet has been allowed to stand for a time after delivery, it should then be rinsed before taking another portion. The reason is that on standing some of the liquid may have evaporated, thereby changing the concentration of the residual amount adhering to the pipet.

2. Quantitative Transfer of Solutions. Almost all analyses require at some stage the transfer of solution from one container (usually a beaker) to another. This transfer must be accomplished so that none of the solution is lost by spilling or by remaining as a residue in the original container.

The procedure for quantitative transfer of a solution from a beaker to a volumetric flask is illustrated in Fig. 4–6. Rinse a clean funnel with a stream of water from a wash bottle and place in the flask. Hold a stirring rod vertically above the funnel and pour the solution from the beaker down the rod, as illustrated, keeping the lip of the beaker against the rod. Use of the rod will prevent loss of liquid by creeping down the side of the beaker. After all the liquid is out, rinse the beaker while still holding it in place, in the position shown on the right of Fig. 4–6. Direct the stream of water so that it rinses all portions of the beaker walls. After rinsing, remove the beaker and withdraw the funnel, rinsing its stem both inside and out as withdrawn.

3. Aliquot Portions of Samples. It is frequently necessary to take a known fraction of a sample, called an aliquot portion, for a determination. This is done by diluting the solution of the sample to a known volume, then withdrawing from this a smaller known volume. The entire sample is transferred to a volumetric flask, as directed in the

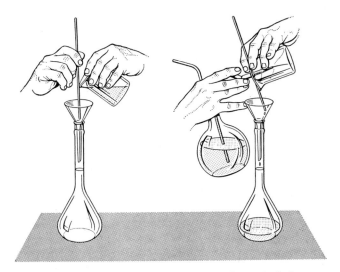

FIGURE 4–6. Transfer of solution to volumetric flask.

preceding section. The flask is usually not more than three-fourths full at this point. Pour in distilled water, from a small beaker, until the liquid level is slightly below the graduation mark on the stem of the flask. If the solution has been heated, cool the flask in tap water until it is at room temperature. Now add water drop by drop from a pipet to bring the level just to the graduation mark. Hold the mark level with the eye for this adjustment. Stopper the flask and mix the contents thoroughly by inverting and shaking the bulb sideways. This operation should be repeated several times, with the flask returned to an upright position after each shaking.

After mixing is completed, withdraw the desired aliquot portions by a pipet that has been calibrated relative to the flask.

4. Use of Buret. The buret is used in titrations and is designed to measure the total volume of solution delivered and permit drop-by-drop delivery of the solution as the end point is approached. In using the buret it is essential that (*a*) all the solution withdrawn from the buret enter the titration vessel, (*b*) there be no dilution or contamination of the reagent solution delivered from the buret, and (*c*) the buret drain cleanly. The detailed directions below must be followed carefully to insure that these conditions are met.

Filling the Buret. The cleaned buret when not in use is filled with distilled water. Place the buret in its stand, open the stopcock, and allow the water to flow into a beaker. When all the water has run out except that standing in the tip, remove the buret from its holder and pour the desired solution directly into the buret from the reagent bottle (note 1). Hold the buret in the left hand, the bottle in the right. Do not lay the stopper down, which might lead to contamination. Hold it between the second and third fingers of the right hand while pouring. Wipe the lip of the bottle with a clean towel before liquid is poured.

Pour about 5 ml of solution into the buret for a rinse, to remove residual water. Hold a finger over the top and manipulate the buret so that the rinse solution wets all portions of the walls. Run the rinse solution out through the tip. Repeat the rinsing, then fill the buret to a point 2–4 cm above the top graduation mark, and replace in the stand. Open the stopcock and allow liquid to discharge rapidly at first so as to displace air bubbles and fill the tip completely. After the tip is filled, withdraw liquid slowly until the meniscus level is at or just below the zero graduation mark. Touch the tip to the side of a flask to remove the pendent drop. The buret is now ready for use. Allow a little time for drainage before taking a reading.

Notes. 1. If the bore of the buret is so narrow that it is difficult to pour into it from a bottle, transfer the reagent solution from the bottle to a small, dry beaker and pour from the beaker into the buret. Alternatively a small funnel may be used in filling the buret from the reagent bottle. Care must be taken that the funnel is absolutely clean and dry. Consult the instructor.

5. Titration of Solutions. The process of determining the volume of reagent solution required to react with a solution of another substance is known as *titration*. The reagent solution is added from a buret until an end point is obtained, which marks completion of the operation. The end point is an abrupt change in some physical or chemical property of the sample solution. In most of the titrations of this course the end point is a color change in a dye that is added to serve as an indicator. This dye reacts with the titrating reagent as soon as it is present in excess of the amount required to react with the sample. It is always desirable in titrations to have the end point occur at the stoichiometric or equivalence point, the point at which just sufficient reagent is added for complete reaction with the sample. As we shall see later, however, the end point and equivalence point (EP) do not always coincide. It is sometimes necessary to determine by another measurement an indicator blank, i.e., the excess volume of reagent that is needed after the equivalence point to give an end point. When this is done, the volume used for the indicator blank is subtracted from the total titration volume to give the net volume of reagent needed to reach the equivalence point. Proper technique is essential for rapid and accurate titrations. The beginning student should initially follow the instructions below; later, as skill is attained, modifications may be made to fit individual situations.

The sample solution for a titration is usually prepared in or transferred to a 250- or 500-ml wide-mouth conical flask. This vessel is easy to manipulate, without danger of loss of solution.

The first step in a titration is to fill the buret, as instructed in the preceding section. Read and record the meniscus position (notebook). Add the proper amount of the designated indicator to the sample solution. Place the conical titration flask on the table beneath the buret and adjust the buret height so that the tip just barely extends into the neck of the flask. The table or bench should be of proper height so that the operator can sit in a position convenient for resting both elbows upon the table as he manipulates the buret stopcock and flask during the titration. Place a towel, a piece of white paper, or a porcelain tile beneath the titration flask to give a good background for observation of the end point.

Turn the buret so that the stopcock handle is on the right-hand side as the operator faces the buret. Manipulate the stopcock with the thumb and first two fingers of the left hand, held so that the butt end of the stopcock fits into the palm and the fingers encircle the barrel of the buret as shown in Fig. 4–7. Hold the neck of the flask between the thumb and first two fingers of the right hand. Lift the flask barely off the table and swirl the contents continuously during the titration by a gentle rotary motion. Avoid any back-and-forth motion that might cause the liquid to splash.

At the start of a titration open the stopcock wide and allow the solution to flow from the buret as rapidly as the tip size permits. From time to time close the stopcock and continue swirling the solution until mixing is complete and the local color change in the indicator has disappeared. With a little experience the course of a titration can be judged by the rate at which color disappears as the solution is swirled. As the end point is approached, decrease the rate of adding solution, and finally add just a drop at a time. At this stage let the drop form slowly on the buret tip until it is almost large enough to fall; then remove the drop by touching it to the side of the flask. The last drop added should give a permanent color change in the solution. Read and record the meniscus position.

If the substance titrated is a solution of a weighed sample or an aliquot portion that has been measured from a pipet, it is important

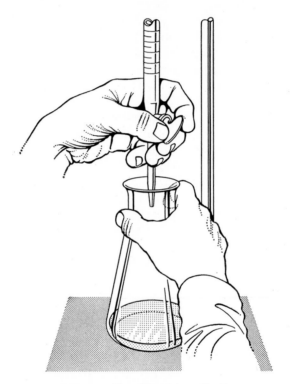

FIGURE 4–7.   Use of buret in a titration.

to avoid overrunning the end point, which invalidates that determination. In many titrations it is convenient to use two burets, one containing the sample solution and the other the reagent. If this method is chosen, start the titration by withdrawing some 35–45 ml of the sample solution into the flask, titrate with the reagent until the end point is passed, then back titrate with the other solution. When the end point is reached, note and record the volumes of each solution. Back titrations can also be made with a reagent which is not itself involved in the analysis but which reacts with the reagent solution. From the amount of such a reagent required for back titration, we can compute how much titrating reagent was added in excess; deduction of this excess amount from the total volume gives the net volume of reagent required for reaction with the sample.

The amount of solution used in a titration depends on two factors, the strength or concentration of the solution and the amount of sample used. By judicious control of these two quantities, the inherent uncertainty of an analysis can be materially reduced. To illustrate, let us consider the uncertainty in the titration of a sample. Each buret reading has an uncertainty of about 0.01 ml, since the position of the meniscus can be estimated to the nearest 0.01 ml. For the two readings (before and after the titration) the total uncertainty in the buret reading is thus 0.02 ml. In addition, there is the uncertainty in locating the end point. If we assume that one drop causes a sharp color change, the uncertainty is on the average about half a drop or 0.02–0.03 ml. Thus the total uncertainty in the titration is some 0.05 ml. If the size of the sample were such that the titration required only 1 ml, the uncertainty of 0.05 ml would be 5 per cent. The same degree of uncertainty in a titration volume of 50 ml amounts to only 0.1 per cent or 1 part per 1000. Ordinarily analytical conditions are chosen to require titration volumes of 35–45 ml, as large as is safe to use without running the danger of having to exceed 50 ml, which would require refilling the buret in the course of a titration.

2. Operations of Gravimetric Analysis

Gravimetric analyses require separation of a specific substance from other materials of a sample, in such a manner that the separated substance can be isolated in a pure form and weighed. The most common method for effecting the separation is by precipitation. Such separations require the unit operations of (1) precipitation, (2) filtration, (3) washing the precipitate free of soluble impurities, (4) drying or igniting the precipitate, and (5) weighing.

Precipitation

The first step in a gravimetric analysis by precipitation is to form an insoluble compound containing the desired ion of the sample. A reagent is usually added that will effect this precipitation. The person who first devises a particular method selects the precipitant that gives (1) a precipitate of low solubility, (2) a precipitate of definite chemical composition which is relatively free of impurities, and (3) a precipitate that can be freed of water by drying or igniting, and converted into a definite compound for weighing. Many of the precipitates used for the separations of qualitative analysis do not have all these qualities and consequently are not used in the separations of quantitative analysis.

Precipitations are usually made in low-form lipped beakers because solids are easily removed from such vessels. Each beaker is equipped with a stirring rod of such length that it projects 4–6 cm from the edge when resting on the bottom of the beaker. The beaker should be of such size that it is not more than two-thirds full at the completion of the operation. The precipitating reagent is usually added from a buret or pipet, seldom from a beaker or graduated cylinder. A pipet is used for precipitations that need not be made exceedingly slowly, but a buret is preferable for those analyses that require very slow addition of reagent. In such precipitations, the buret is mounted above the beaker, in a slanting position, and the stopcock is opened just enough to allow a dropwise flow of solution. If the precipitation is to be made in hot solution, the beaker rests on a stand so that it may be heated.

Mechanical stirring is recommended for precipitations requiring as much as 10 minutes, since this insures better mixing than can be achieved when the solution is stirred by hand. Many types of good mechanical and magnetic stirrers are available for this purpose. Usually an all-glass propeller or agitator is best, to prevent possible reaction with the solution.

Before the addition of reagent the approximate amount required should be calculated, on the basis of previous knowledge of the composition of the sample. About 5–10 per cent in excess of the calculated amount should be added (as a dilute solution), and then the stirring should be interrupted and the precipitate allowed to settle. A drop of reagent is then added to the clear supernatant liquid; if a further precipitate is formed, more reagent should be added. After filtration the liquid should again be tested for complete precipitation.

In many analyses the precipitate first obtained is dissolved in acid or other suitable solvent, and a second precipitation is made. The purpose of the reprecipitation is to obtain a purer precipitate.

Digestion

After precipitation is complete, the solution is, in most analyses, placed on a steam bath and allowed to stand for a time before filtration. The purpose of this operation, which is known as digestion, is to improve the filterability and purity of the precipitate. A discussion of the changes that occur during digestion is given in Chapter **18**.

Digestion is not always necessary or desirable. Some precipitates, such as magnesium ammonium phosphate, are unstable in hot solution and cannot be digested. Some are too soluble in hot solution. In general, digestion is of no benefit for gelatinous precipitates, such as hydrous ferric oxide. With this substance a few minutes' boiling of the solution will effect all the coagulation that could be obtained from prolonged digestion. The analytical procedure will in all cases specify the time and temperature of digestion.

Filtration

Beginning students encounter more difficulty in filtration than in any other operation of gravimetric analysis. Improperly performed, this operation can waste many hours' time; at best, it is tedious and slow. The student should learn proper techniques at the start and should never attempt short cuts, for they only lead to trouble.

Filtering Media.   The filtering media used depend on the nature of the precipitate and on the treatment needed to convert the precipitate to a form suitable for weighing. The most common filters are:

1. Paper.
2. Gooch crucible.
3. Sintered-glass crucible.
4. Porous-porcelain crucible.
5. Munroe crucible.

Filter paper is suitable when the precipitate is not easily reduced by the action of the carbon of the paper and when a high ignition temperature is needed. Paper *must* be used for gelatinous precipitates, since they clog the pores of the various other types of filters. The Gooch crucible, the sintered-glass crucible, and the porous-porcelain crucible are all used for precipitates easily reduced by the carbon of paper. When only drying is desired, sintered-glass or porous-porcelain crucibles are recommended because they require less preliminary preparation than the Gooch crucible. When the precipitate must be ignited to dull redness, the Gooch or porous-porcelain crucible must be used, since sintered-glass crucibles will not withstand high temperatures. Even with the porcelain crucibles, ignition must be carefully performed, as described subsequently. The Munroe crucible,[1] because of its expense, is seldom employed in student work. It is in construction similar to the Gooch crucible (described below), but the filtering mat is of platinum. Consequently, it may be heated to the highest attainable ignition temperatures.

Filter Paper. A special grade of "ash-free" paper must be used for all precipitates that are to be ignited and weighed. Ordinary paper such as that used in qualitative analysis may contain several milligrams of ash per circle, enough to invalidate completely the results of an analysis. Ash-free paper is acid-treated in manufacture to remove most of the inorganic constituents, and the weight of ash per circle of paper is negligible.

Quantitative paper is manufactured in several degrees of porosity. It is important to select the proper grade for a given precipitate. Use of too coarse a paper will permit small crystals to pass through the filter, and use of too fine a paper will make filtration unduly slow. In Table 4–3 there are given the manufacturers' recommendations for

TABLE 4–3. Quantitative Filter Papers

W	S&S	JG	Porosity	Speed	Use for	Approximate Weight of Ash per 9-cm Circle
41	589–1	801	Coarse	Very rapid	Gelatinous precipitates	<0.1 mg
40	589–2	802	Medium	Rapid	Ordinary crystalline precipitates	<0.1
42	589–3	803	Fine	Slow	Finest crystalline precipitates	<0.1

Whatman (W), Schleicher and Schull (S&S), and J. Green (JG) quantitative papers. Other makes are just as reliable as these, which have been listed because of their general availability. In addition to the three degrees of porosity shown in Table 4–3, all manufacturers also provide quantitative paper for special purposes, such as papers to withstand moderate suction or the action of alkalies.

Technique of Filtration with Paper. The size of the paper should be governed by the amount of precipitate to be retained and not by the

---

[1] The Munroe crucible, made of platinum with a perforated base on which a mat of platinum is formed, is used for very high-temperature ignition with reducible precipitates and for other special applications. See W. B. Snelling, *J. Am. Chem. Soc.*, **31**, 456 (1909), and O. O. Swett, *ibid.*, **31**, 928 (1909).

amount of filtered liquid. The paper should be about one-third filled with the precipitate at the end of the filtration. The funnel should match the paper in size. The folded paper should come within 1–2 cm of the top of the funnel, but never closer than 1 cm. The funnel should be of exactly 60° angle. It should have a long stem, with a constriction at the upper end of the stem. This constriction will facilitate maintenance of a solid column of water in the stem, thereby accelerating the speed of filtration. The funnel should be held in a rigidly mounted funnel holder, at proper height for the stem to extend into the beaker which receives the filtrate. The stem should touch the side of the beaker so that a column of liquid runs down the side; this arrangement prevents splashing.

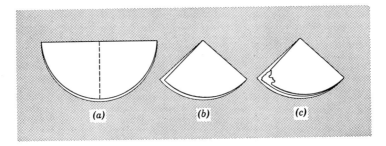

FIGURE 4–8. Method of folding filter paper.

The operation of folding the paper and fitting it into the funnel is most important. The circle of paper is commonly folded exactly in half and folded again in quarters. The folded paper is then opened so that a 60°-angle cone is formed, with three thicknesses of paper on one side and a single thickness on the other. The paper is adjusted to fit the funnel exactly, and the corner of the outside fold is torn off as shown in Fig. 4–8, in order to give a tighter fit. The paper is then placed in the funnel, moistened, pressed down tightly to the sides of the funnel, and filled with water. The stem of the funnel will fill with a continuous column of liquid if the paper fits properly. This liquid column promotes a gentle suction which accelerates filtration. Much time can be saved in the operation of filtration if care is taken to choose good funnels (with 60° angles and with long stems) and to fit the paper to the funnel correctly.

The experienced analyst often folds the paper so that it does not fit the funnel tightly except at the top. To do this, the second fold is made so that the angle is slightly greater than 60°. During filtration, liquid passing through the paper is free to run down the side of the funnel between the paper and glass. Care must be taken that papers so fitted do not pull loose from the funnel, thereby allowing air to enter and the liquid column to break. Funnels with 58° angles are available to use with 60° paper cones, so that the paper fits only at the top. Another type of funnel has grooves extending radially from the bottom, to provide more rapid filtration.

After the paper is fitted and the receiving vessel is in place, the filtration may be started. The beaker of solution is held just above the funnel. A stirring rod is held against the lip of the beaker, to guide the column of liquid into the funnel and to prevent drops of liquid from running down the outside of the beaker (see Fig. 4–6). Solution is poured into the filter until it is filled nearly to the top of the paper. *Care must be taken not to overrun the paper.* If the liquid runs through the paper very rapidly, the beaker is held in the pouring position, and liquid is added to the funnel as rapidly as it can be accommodated. In most filtrations some time is needed for the filled paper to empty, so that several filtrations may be in progress simultaneously. Care must be taken to prevent mixing of filtrates. Each beaker with solution is numbered, and the receiving beaker carries a corresponding number. Care must be taken to prevent loss of solution on the outside of the beaker lip as the beaker is set upright after pouring.

The precipitate is retained in the beaker during filtration. When no more liquid can be poured off the precipitate without transferring the precipitate to the paper, the filtration is interrupted, and the receiving beaker is replaced by an empty one. Thus, if any precipitate passes through the filter, it will not be necessary to refilter the entire bulk of solution. The last of the mother liquor is removed from the precipitate in the operation of washing, described in a later section.

Whenever the solubility of the precipitate permits, it is advantageous to filter solutions while hot. Because of lowered viscosity, hot solutions will run through a paper much faster than cold solutions. Instructions are given in the analytical procedures when the filtration can be made with hot solutions.

Use of Suction.   When suction is applied to the ordinary filter paper, the paper ruptures at the apex of the cone. If a perforated platinum filter cone is put into the funnel beneath the paper to reinforce the apex, or if a small cone of specially hardened paper is used, a moderate degree of suction may be employed to reduce the time required for filtration. The papers are fitted to the funnels (containing the cones) in the customary manner, and the funnel is inserted into a suction flask

(Fig. 4–9), which is then connected to the water aspirator pump through a safety bottle (to prevent water from backing up from the pump and contaminating the filtrate). The stem of the funnel must project below the side arm of the flask to prevent loss of the filtrate. Speed of filtration may be controlled by a screw clamp on the side tube which is inserted in the rubber stopper along with the funnel. If

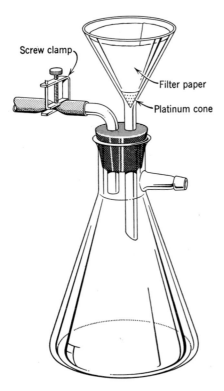

FIGURE 4–9.   Filtration with suction, using platinum filter cone.   A cone of hardened filter paper may be used instead of the platinum cone.

the filtrate is to be subjected to further analysis, care must be exercised to rinse it completely out of the suction flask, or to arrange the apparatus so that the filtrate is caught in a beaker. To do this, the funnel is mounted in the top section of a vacuum desiccator (or bell jar), the beaker is placed in the lower portion of the desiccator beneath the funnel, and the assembly is evacuated through an opening in the lower portion.

Filtration by suction is advantageous when the precipitate is crystalline, but suction cannot be employed for gelatinous precipitates. With such precipitates suction draws particles into the pores of the paper, clogging it until no liquid will pass.

Macerated Filter Paper. Addition of a suspension of macerated filter paper is often advantageous in the filtration of gelatinous precipitates, which tend to clog the pores of filter paper. When a suspension of paper fibers is added to such a precipitate, the fibers serve to keep the gelatinous precipitate in a condition more permeable toward water. Macerated filter paper tablets may be purchased ready for use, or the papers may be prepared by disintegrating ashless filter paper with concentrated hydrochloric acid for a few minutes, washing out the acid, and storing in distilled water. The macerated paper is added immediately after precipitation and the solution is stirred to insure thorough mixing. The amount of paper added should be such that the volume of the paper is approximately the same as that of the precipitate.

Gooch Crucible. As previously stated, the Gooch crucible is advantageous (1) when the precipitate is readily reduced by carbon of the paper on ignition, and (2) when extremely high ignition temperatures are not necessary.

The Gooch crucible is a porcelain thimble with a perforated base. The filtering medium is a mat of asbestos fiber which covers the bottom of the crucible. To prepare a mat for quantitative use, proceed as follows. Place the crucible in its holder, as shown in Fig. 4–10. The receiving test tube is omitted. Prepare a suspension of acid-washed asbestos fiber in a beaker of water. With the suction disconnected, pour enough of the suspension into the crucible to make an asbestos mat at the bottom. After a moment, when some of the fiber has settled, turn on the suction and drain all water from the crucible. With continued suction slowly pour asbestos suspension into the crucible until a mat ½–1 mm in thickness is formed. With a properly prepared mat one can just discern the perforations in the bottom of the crucible if it is held with the bottom toward the light and is observed from the open end.

When a suitable mat is formed, wash with water, with suction applied, to remove loose particles. The water should be poured into the crucible gently, with the aid of a stirring rod. If a porcelain plate is supplied, place this on top of the mat before washing (an advantage of the plate is that it prevents disruption of the mat by liquids poured into the crucible). Wash the mat until the following test for loose fibers indicates their complete removal. Test by collecting a portion of the wash water in a clean test tube, as shown in Fig. 4–10, and observing the presence or absence of fine asbestos fibers as the tube is viewed in a strong light. Remove the crucible from the holder, and heat in a muffle furnace at the temperature to be used for drying or igniting the precipitate. Cool in a desiccator and weigh. Replace in the holder, and repeat the process of washing, drying, and weighing. The weight should remain constant within 0.3 mg. When this constancy is attained, the crucible is ready for use.

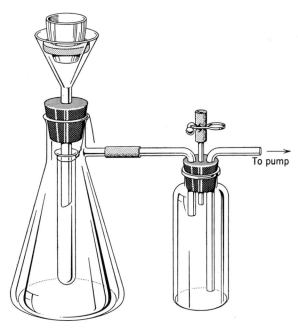

FIGURE 4–10. Filtration with suction, using porous porcelain crucible and test tube for collecting samples of wash liquid.

In filtration with a Gooch crucible, clean the filter flask, place the crucible in its holder, turn on the suction (a safety bottle must be used), and pour the liquid through the filter in a continuous stream. Take care to prevent rupture of the mat by the liquid; pouring should be done with the aid of a stirring rod, and the stream should be directed against the center of the mat. Suction should be applied before the filtration begins, to hold the fibers in place. The bulk of the precipitate is retained in the beaker, as directed for filtration with paper.

Gooch crucibles are cleaned by scraping out the mat and adhering precipitates and re-forming another mat. Since it takes considerable time to prepare a mat and bring it to constant weight, it is often desirable to use a crucible for several filtrations before removing the precipitate and making a new mat. Successive filtrations in the same crucible can be made when the precipitate already present is very insoluble, so that weighable amounts of it do not pass into solution when a new filtration is made.

Crucibles with Permanent Mats. Filtering crucibles with permanent mats are rapidly replacing Gooch crucibles, because of the saving in time. They are of two general types, glass and porcelain. Either may be used when the precipitate need only be dried; but if high temperature ignition is required, the porcelain crucible must be used, since glass will not withstand temperatures near its softening point.

The general technique of filtration with permanent mats is the same as that described for Gooch crucibles, except that no mat is prepared. The crucible is cleaned and brought to constant weight at the ignition temperature that is to be used for the precipitate. Filtration and washing are done as described on page 73. Ignition must be carried out in a muffle furnace. Preferably the crucible should be placed in an ignition thimble, to prevent possible contamination from the floor of the furnace.

Filtering crucibles, both glass and porcelain, are cleaned by dissolving the precipitate in a suitable reagent. The following are commonly used.

Precipitate	Reagent
Barium sulfate	Conc. sulfuric acid or cleaning solution
Silver chloride	Ammonia solution
Fats, organic compounds	Cleaning solution or ignition (for porcelain)
Metal sulfides	Aqua regia
Manganese dioxide	Conc. hydrochloric acid

Cleaning may be done either by immersing the crucible in a beaker of the reagent or by "back washing" with the reagent. In back washing the liquid is poured through the mat in the reverse direction. The instructor will provide directions for this operation, based on the type of crucible in use.[2]

Washing

Immediately after filtration all precipitates must be washed free of soluble impurities. If the precipitate is allowed to dry in the paper, it will cake in such a manner that complete washing will be impossible.

[2] Apparatus for cleaning ignition crucibles is described in *Ind. Eng. Chem., Anal. Ed.,* **16,** 277, 539 (1944).

In order to reduce solubility losses, the quantities of wash liquid should be as small as possible. Most efficient washing is obtained, with a given volume of water, if the wash liquid is added in small portions and each portion is allowed to drain completely before the next one is added. Since a certain amount of liquid is always retained by the precipitate, the fraction of the impurity removed in a single washing is the ratio of the volume retained to the total volume used. If, for example, 1 ml is retained in the precipitate, the fraction of the impurity retained in a washing with a 50-ml portion is $\frac{1}{50}$. If 50 ml of wash liquid is added in 10-ml portions and each portion allowed to drain before addition of the next, the first washing will leave $\frac{1}{10}$ the total impurity, the second washing $\frac{1}{10}$ the amount left in the first, and so on. Washing with five 10-ml portions will, therefore, reduce the amount of impurity to $(\frac{1}{10})^5$ or to 1/100,000 the original amount.

Technique. The precipitate is partially washed by decantation before it is transferred to the filter. After the bulk of the mother liquor is removed and no more liquid can be poured off from the precipitate, a small portion (10–30 ml) of wash liquid is added, the precipitate is thoroughly stirred up in the wash liquid and then allowed to settle, and the supernatant liquid is poured through the filter. Generally, the receiver is replaced by an empty flask or beaker before washing is begun. When no more liquid can be poured from the precipitate, another portion of wash liquid is added and the process is repeated. After two or three washings by decantation, the precipitate is transferred to the filter and washing is completed there. In the transfer as much of the precipitate as possible is poured into the filter along with the last portions of liquid. While the beaker is still above the filter, a stream of water from the wash bottle (see Fig. 4–6, page 61) is used to dislodge additional precipitate and wash it into the filter. Finally, a policeman is used to rub adhering particles from the walls of the beaker. Particles are washed from the policeman directly into the filter by a stream of water from the wash bottle. Sometimes the last portions of precipitate adhere so strongly to the walls of the beaker that they must be wiped out with a piece of ashless filter paper, which is then ignited with the precipitate.

In filtration with paper the wash liquid must always be added from a wash bottle fitted with a tip of small bore. The stream of liquid is directed never at the precipitate but around the upper edge of the filter paper so that, as it runs down, it will wash the precipitate to the apex of the cone.

Completeness of washing must be tested in every analysis. The test is performed by collecting 5–10 ml of the filtrate in a test tube and making a simple qualitative test for some readily identified ion, such as chloride or sulfate, which is present in the solution at the time of filtration. It is seldom advisable to make tests before six to eight portions of wash liquid have been used. When suction is used for filtration, a test tube is placed inside the filter flask to catch the test portion, as shown in Fig. 4–10. The tube is suspended by a wire which runs beneath the stopper.

Ignition

The last stage in a gravimetric analysis is the conversion of the precipitate into a form suitable for weighing. This conversion may require only the removal of water, or it may require ignition at a high temperature to effect a partial decomposition of the precipitate.

In all analyses it is necessary to bring the crucible to constant weight *at the same temperature* as that employed for ignition of the precipitate. If this is not done, there may be loss in weight of the crucible itself during ignition. To bring a crucible to constant weight, clean it and ignite for 5–10 minutes. Cool in a dessicator, and weigh accurately. Reignite for a 5–10-minute period, cool, and reweigh. Continue the process until the weight is constant within 0.3 mg after successive heating periods.

The technique of ignition varies widely with the different types of filtration employed. Gooch crucibles may be dried in an oven, or they may require ignition at 600–1000°C. Glass filtering crucibles are not used at temperatures above 200°. Porcelain filtering crucibles may be ignited to 1000°, provided a muffle furnace is used. Filter paper may be ignited in porcelain, silica, or platinum crucibles. Porcelain crucibles can be heated to 1200°C without danger of breaking but are subject to attack by alkaline carbonates, alkalies, and hydrofluoric acid. Platinum is inert to attack by many reagents and can be heated as high as 1700°C. In any ignition (except for substances which attack it), platinum is to be preferred because its high thermal conductivity permits rapid attainment of temperature equilibrium. In flames, precipitates in platinum attain a much higher temperature than in porcelain, but in muffle furnaces both types of crucibles attain the same temperature.

Choice of Burner. Commonly available in all analytical laboratories are Tirrill (double-regulator Bunsen-type burners), Meker, and blast burners. The temperatures that can be attained with these various sources may be seen in Table 4–4. These temperatures are, of

TABLE 4–4.   Approximate Burner Temperatures

Burner	Covered Platinum Crucible	Covered Porcelain Crucible
Tirrill	1050–1150	600– 700
Meker	1100–1250	725– 900
Blast	1100–1300	900–1000

course, only approximate, for there are many variable factors, such as thickness of the crucibles and composition of the fuel gas. In general, Meker burners are preferred to blast lamps because the flame of the former is free of reducing gases, which may permeate the platinum at high temperatures and thus cause damage to the crucible.

A rough measure of the temperature obtained by a crucible may be made by observation of the color. A platinum crucible appears dull red at temperatures of 600–850°C. The color is cherry red in the range 850–1000°C, orange in the range 1000–1100°C, and yellow in the range 1100–1200°C. Above 1200°C the color becomes a very bright yellow, which gradually shifts to white with rising temperature. The white color is beyond the range of the burners employed in the laboratory.

Ignition When Filter Paper Is Used. There are many correct methods for performing an ignition. The following directions will serve as a guide, but they may be modified as circumstances require:

1. Bring the crucible to constant weight. See above.

2. Transfer the precipitate to the crucible. After the filter paper is well drained in the funnel, fold over the top portion so as to enclose the precipitate completely. Lift the bundle from the funnel and place it point down in the crucible. Press it well down to the bottom of the crucible to facilitate drying and charring.

3. Dry the paper and precipitate. Time may be saved by doing this in an oven overnight. If a wet paper is to be ignited immediately, proceed as follows. Put the crucible in a triangle as shown in Fig. 4–11, but put the cover on the crucible instead of slanting it as shown. Place a *small* flame beneath the crucible at a point about midway along the lower side. The tip of the flame must not reach the crucible. Heat until the paper and precipitate are entirely dry.

4. Char the paper. Move the flame until it is beneath the bottom of the crucible, as shown in Fig. 4–11. Slightly increase the size of flame, and continue to heat until all organic matter is distilled from the paper. The cover should be slanted during this operation. Take care that the flame does not strike the cover and deflect reducing gases

onto the precipitate. If the escaping gas bursts into flame, reduce the rate of heating.

5. Oxidize carbon. Increase the size of the flame until the bottom of the crucible is heated to redness. Continue to heat until the carbon residue is burned away. From time to time the crucible should be turned, so as to expose a fresh portion to the heat.

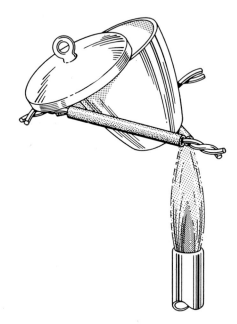

FIGURE 4–11. Ignition of precipitate with access of air.

6. Ignition. Place the crucible upright, remove the cover so as to permit free access of air, and ignite at the prescribed temperature. With some precipitates it is advisable to crush the solid lump before ignition, so as to obtain better heat transfer. This is done with a blunt stirring rod. Any adhering powder is wiped from the rod with a small piece of ashless filter paper which is then placed in the crucible and burned. Do not crush the precipitate unless this is approved by the instructor.

Ignition in a Muffle Furnace. A muffle furnace is an enclosed chamber which can be heated by electricity to the high temperatures (800–1000°C) needed in ashing samples and igniting precipitates. Temperature control is provided by means of a rheostat, and often a pyrometer is attached to indicate the temperature. If such furnaces are available, they greatly increase the efficiency and convenience of the ignition process.

If a precipitate that has been collected in a filtering crucible is to be ignited in a muffle furnace, the crucible containing the precipitate is first dried in an oven at 100°C; it is then set inside an ignition thimble, and the two are placed in a muffle furnace which has been preheated to the proper ignition temperature. The filtering crucible should previously have been brought to constant weight by ignition in the furnace. If the precipitate is collected on paper, the paper is charred; then the ignition is completed, without a cover, in the muffle furnace.

Care of Platinum. Platinum crucibles are expensive and should have special care. The following rules are to be observed:

1. Protect from mechanical damage. Platinum is very soft and pliable. Crucibles may be reshaped on wooden or metal forms if dented or bent.

2. Never use with oxidizing agents, such as aqua regia, chlorine, bromine, and ferric chloride, which react with the metal.

3. Do not ignite compounds of sulfur, phosphorus, arsenic, selenium, mercury, lead, bismuth, silver, tin, and antimony in platinum, since these elements (sometimes formed by reduction of the precipitates containing their compounds) alloy with it.

4. Strong bases and salts that decompose into strong bases, for example, barium or lithium carbonates, must not be heated in platinum. It is perfectly safe to fuse the carbonates of sodium and potassium.

5. Never heat platinum ware in a reducing flame, since this causes the formation of a crystalline deposit on the metal. In time such a deposit penetrates the wall of the crucible, causing it to become brittle.

6. Hot platinum ware should always be handled with platinum-tipped or nichrome forceps. Iron forceps must never be used for crucibles that are at a temperature above 500°C because iron will alloy with the platinum. Platinum or silica triangles should be used. Ordinary clay triangles often contain enough iron to damage platinum if they are used for ignitions.

7. The surface of platinum must always be kept shiny by scouring with talc or sea sand. Hydrochloric acid is the best cleansing agent. Certain metallic stains, especially iron, can be removed by low-temperature fusion with potassium acid sulfate. The molten solid should not be allowed to solidify in the crucible but should be poured onto a dry stone or iron slab.

Platinum iridium alloys (such as are often used for crucibles) undergo an appreciable loss in weight when heated at temperatures of 1000°C or above. The magnitude of this loss is shown in Table 4–5.

TABLE 4–5.   Approximate Loss in Weight of Crucibles

Mg/100 Cm^2/Hour, at Temperature Indicated*

Temperature, °C	Pure Pt	1% Ir– 99% Pt	2.5% Ir– 97.5% Pt
900	0	0	0
1000	0.08	0.30	0.57
1200	0.81	1.2	2.5

* G. K. Burgess and R. G. Waltenburg, *Bur. Standards Sci. Paper*, 280 (1916).

Since a 25-ml platinum crucible has an area of about 80–100 sq cm, the error due to volatility may be appreciable if the crucible is made of an alloy high in iridium.

Not all crucibles will change appreciably in weight, but the student should at all times be aware of this possibility.

Cooling. All precipitates that have been heated should be cooled in a desiccator. Very hot crucibles are not put into desiccators. The crucible is allowed to cool, in the air, for a short time. The cover should be in place during this period. When the temperature has fallen to below 100°C, the crucible is placed in the desiccator. The cover of the desiccator is left slightly open for a few moments, so that heated air may escape from the interior; then the cover is closed and the crucible allowed to stand until it is completely cool. Platinum crucibles will cool in 10–15 minutes, if the desiccator itself is thoroughly cool, but porcelain crucibles require a somewhat longer time.

Care must be exercised in opening the desiccator. The expansion of heated air and subsequent cooling may leave a partial vacuum within the chamber, and, if the lid is opened suddenly, a rush of air may blow a portion of the contents from the crucibles. Always slide the cover to one side very gradually, so that the first opening is on the side opposite the crucible. It is not advisable to cool several crucibles at once because the repeated opening of the desiccator, to remove the crucibles for weighing, permits the crucibles to stand in a not thoroughly dried atmosphere. Rather, several desiccators should be used if many ignitions are in progress simultaneously.

Weighing

Ignited precipitates should be weighed in covered crucibles to prevent contact of the precipitate with moist air. This precaution is particularly to be observed with precipitates that can take up carbon dioxide. Even with covered crucibles, it is sometimes found that a precipitate will show an appreciable gain in weight while on the balance pan. In this event the precipitate should be reignited. For the next weighing, the approximate weight should be placed on the balance pan before the crucible is removed from the desiccator. Observation of the first swing or set of swings will then enable the analyst to compute the exact weight.

Sometimes, when the ignition is prolonged, the loss in weight of platinum crucibles becomes appreciable; then it is a question of what is the proper weight for the crucible when the precipitate has attained constant weight. *In this case the weight of the empty crucible should be determined after the ignition*, not before as in the usual procedure. The precipitate is removed from the crucible, and the crucible is washed, dried by brief ignition, cooled, and weighed.

Evaporation

In many analyses the volume of solutions present must at some stage be reduced by evaporation, and frequently the solution must be evaporated to dryness. Evaporations are made on either a hot plate or on a steam bath. Because of the higher temperature, hot-plate evaporations proceed somewhat faster than those on a steam bath, but because of the higher temperature there is danger that an evaporation on a hot plate may cause the solution to spatter as it approaches dryness.

Evaporations are usually carried out in beakers. Conical flasks are undesirable because steam condenses on the upper part of the wall and the condensate runs back into the liquid. Occasionally it is desirable to replace the beaker by a porcelain evaporating dish or a casserole, but usually the sample to be evaporated is in a beaker; it is better to make the evaporation in the same container rather than to risk the danger of loss during a transfer to another vessel.

During an evaporation the beaker must at all times be covered to prevent foreign matter from falling into the solution. At the same time it is necessary to allow space for the ready escape of steam. Both these objectives can be attained by covering with a watch glass elevated on glass supports, as shown in Fig. 4–12. The glass must be large enough to cover the beaker but should not be unduly large, for then it would require too much space.

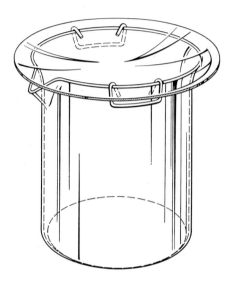

FIGURE 4–12.   Supports for cover glass.

3. Preparation of the Sample for Analysis

Before any quantitative analysis can be made, it is necessary (1) to procure a test sample that will represent the average composition of the material to be analyzed, and (2) to prepare a solution of the test sample.   In quantitative analysis courses, these operations often seem inconsequential, because the student is given a small homogeneous well-ground sample whose analysis is known and he is provided with explicit directions for bringing this sample into solution.   In professional analyses, however, it is often more difficult to prepare a suitable solution than it is to perform the actual analysis.   The material of this section is designed to acquaint the student with the problems involved in the preparation of a sample for analysis and with the general methods for the solution of these problems.   For more detailed information reference books[3] should be consulted.

---

[3] See in particular W. F. Hillebrand, G. E. F. Lundell, H. A. Bright, and J. I. Hoffman, *Applied Inorganic Analysis,* second edition, Wiley, 1953; N. H. Furman, *Scott's Standard Methods of Chemical Analysis,* Van Nostrand, 1939; H. H. Willard and H. Diehl, *Advanced Quantitative Analysis,* Van Nostrand, 1944.

Sampling

Few of the natural or commercial products for which analyses are required are homogeneous in nature. A shipment of ore, for example, will show large variations in composition, even though all of it is taken from the same deposit. Even alloys, such as brass or steel, are inhomogeneous within a single piece, owing to segregations that occur during the solidification of the liquid material. Since, in general, the portion used for the analysis does not weigh more than a gram, it is obviously of paramount importance to exercise great care in the selection of this portion so that it will represent the composition of the whole material. An excellent analysis may be completely invalidated by poor sampling.

The method of sampling is first to select a *gross* sample, which, though itself inhomogeneous, will represent the average composition of the whole, and by systematic mixing and reduction in size of the gross sample to obtain a small homogeneous portion that has the same composition as the entire material.

The size of the gross sample depends on the size of the individual particles of material. If the particles are large or show wide variation in size, as in a carload of coal, the gross sample is necessarily larger than would be needed for a well-ground material of uniform particle size. In general, the gross sample varies from $\frac{1}{20}$ to $\frac{1}{100}$ of the total material, but under special conditions a much smaller fraction may be used satisfactorily.

In taking the gross sample a standardized procedure is followed to insure that the sample shall represent the composition of the whole. Small portions of uniform size, known as sample units, are taken at regular intervals from the entire bulk of material. If the material is in a large bulk, as a boatload of ore, sample units are taken along lines drawn lengthwise of the bin. When the material is in small containers, a portion is taken from every nth container. The size of the sample unit, like that of the gross sample, varies with the size of particles of the material; the larger the individual particles, the larger the sample unit needed.

From the gross sample, which may weigh hundreds of pounds, there is obtained a small sample, weighing 5–10 lb, which is called the laboratory sample. The process of reducing the gross to the laboratory sample consists of systematically mixing, dividing the mixture into two parts, mixing one part, dividing, and so on. Before the first division is made, the sample is crushed to particles 1 in. or less in size, and each fraction is again crushed to smaller particle size before it is subdivided.

When the mixing is done by hand, one of the following methods is used.

1. Long-Pile and Alternate-Shovel Method. This method is used for large samples of 100–1000 lb. The particles should not be larger than 1 in. The sample is mixed by piling into a cone, each shovelful being deposited at the apex so that the material runs down to all sides. The material is shoveled from the cone into a long pile about the width of the shovel; each shovelful is spread along the entire length of the pile, alternate shovels being started at opposite ends. The long pile is divided into two equal portions by taking alternate shovels of material and discarding the odd ones. The portion retained is recrushed, again piled into a cone, and spread into a long pile.

2. Coning. When the size of sample is between 10 and 100 lb, the method of coning is useful. The material, crushed to a proper size in relation to the amount present, is piled into a cone. The cone is then flattened, and the flat pile is divided into four equal portions by passing a straight board through twice, so as to make two cuts at a 90° angle to one another. Alternate quarters are rejected, and the remaining two quarters are crushed to suitable particle size, repiled, and again quartered.

3. Tabling. This method is suitable for finely ground samples of weight 10 lb or less. The material is spread on a canvas cloth and mixed by drawing in turn each corner of the cloth to the diagonal corner, in such a manner as to impart a rolling motion to the particles. After being mixed in this manner, the material is spread in a flat pile and quartered. The two opposite quarters are discarded, as in the coning procedure.

4. Riffling. A sample may be mixed and divided into two portions by a machine known as a riffle, if the particles are small. In this operation the material is fed from a hopper into a row of small chutes arranged so that the even-numbered ones deliver to one side and the odd-numbered ones to the other side. This method is applicable only when there is sufficient sample to fill all the chutes.

Grinding and Crushing

It has been pointed out that thorough mixing requires that the sample be ground or crushed to particles of uniform size. Nearly always, except in the analysis of metals, the test portion must be in the form of a very fine powder. Usually, when the original material is in large lumps, the preliminary breaking down of the particles is done by means of a *jaw crusher*, in which lumps of material are squeezed between massive metal jaws. After the particle size is made small, further reduction is accomplished by grinding.

In the school laboratory, crushing is conveniently done in a hardened-steel mortar, such as shown in Fig. 4–13. Single lumps of the sample are placed in the cylinder and crushed by striking the plunger with a hammer. After the material is crushed in the mortar, it is ground (in small portions) in an agate mortar until no gritty particles can be detected. After grinding, the material is thoroughly mixed, by a tabling process, on a sheet of glazed paper. When large portions of material are to be ground and mixed, a *ball mill* is advantageous. The sample is placed in a closed jar with an equal volume of hard balls, and the jar is continuously rotated on its side for several hours. Rubbing of the balls with one another and with the wall of the container serves to give very effective grinding and mixing. Care must be taken not to have too large a sample at a single charge.

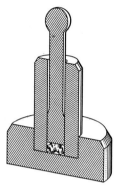

FIGURE 4–13. Steel mortar for crushing sample.

Sieving is an essential part of the grinding operation. No single grinding will pulverize all the particles, but some of the harder ones become coated with powdered material and remain unchanged. These particles must be separated, by passing the material through a sieve. They are then reground. Conventionally, the size of particle to which a material must be ground is designated by the *mesh* of the sieve which the material must pass, the mesh of a sieve denoting the number of wires to the linear inch. Since part of the space is occupied by the wires themselves, an *n*-mesh sieve has openings slightly smaller than 1 *n*th of an inch. The usual sample for analysis should pass a screen of 80–100 mesh or smaller.

Errors in Grinding. The process of grinding a sample may materially alter the composition. Among the changes that may occur are the following:

1. *Contamination by the Apparatus Used for Grinding or Crushing.* In the abrasion necessary to break up particles of the sample, there occurs to some degree an abrasion of the apparatus itself, with consequent contamination. Crushing causes less contamination than

grinding because there is less rubbing of the apparatus with particles of sample. Agate mortars are so hard that few samples will cause an abrasion sufficient to give a detectable amount of the mortar material in the sample. Porcelain mortars, such as are used in elementary chemistry courses, are not at all suitable for grinding samples for quantitative analysis. Porcelain ball mills likewise are unsuitable for quantitative samples of any but soft materials, such as crystalline salts.

2. *Contamination in Sieving.* Sieves are usually made of brass screen because of the resistance to corrosion. There may be enough abrasion of the brass during the operation of sieving to cause appreciable contamination. In many mineral analyses it is preferable to use silk bolting cloth for screening, rather than brass sieves, to eliminate the possibility of introducing copper, zinc, and lead into the sample.

3. *Oxidation of the Sample.* Considerable heat is generated locally in the grinding of a mineral, and in the presence of air an appreciable amount of oxidation may occur. This is particularly noticeable with ferrous compounds, which are readily oxidized to the ferric state. At times it may be desirable to grind samples under absolute alcohol, to prevent changes in composition.

4. *Loss or Gain of Moisture.* The moisture content of a sample may be altered in the grinding operation. A finely divided sample will absorb much more surface moisture than the same material before grinding. An opposite effect may be found in some minerals that contain combined water; the heat generated locally in grinding may expel some of this water.

Moisture

In many analyses it is customary to dry the material at 105–110°C before beginning the determination and to report the result of the analysis on a dry basis. This practice is followed because the moisture content may vary widely with differing atmospheric humidity, and analyses made under different humidity conditions may not be in agreement. Drying at 110°C does not remove all the adsorbed water, particularly in samples which contain oxides of iron or aluminum, nor does it remove all the combined water which may be present as a hydrate. For a given type of sample, however, it may be assumed that drying will give reproducible results.

In many analyses no drying at elevated temperature can be employed. For example, analyses of vegetable or animal tissue must be based on the weight of air-dried material; that is, the material is allowed to stand in air until free water has evaporated and an equilibrium is attained.

Some crystalline materials cannot be dried at 110°C without loss of combined water, and certain analyses may be based on the weight of material as dried at a definite humidity. Such drying is done by placing the sample in a desiccator with a substance of definite known vapor pressure, such as a salt hydrate. For example, borax if dried over hydrated sodium bromide will contain 10 molecules of water of crystallization, whereas if air-dried it will partially decompose to form the pentahydrate.

The determination of the moisture in a sample is a convenient preliminary exercise in quantitative technique. A mixture of barium chloride dihydrate and sodium chloride is an excellent sample. The water of hydration is completely expelled from the barium chloride on drying at a temperature of 105–110°C. The process is accelerated at higher temperatures.

Procedure. Place two weighing bottles in labeled beakers and dry in an electric oven at 105–130°C for half an hour. During this period the stopper of each bottle should be placed in a slanting position in the top of the bottle to permit escape of water vapor. Place the hot bottles in a desiccator, with the stoppers still open, and allow to cool to room temperature. When they have cooled, stopper the bottles, and weigh accurately. Repeat the heating, cooling, and weighing. Successive weighings should agree within 0.3 mg.

Obtain the sample in a stoppered weighing bottle. Transfer about 1–1.5 g of the sample to each weighing bottle, stopper, and reweigh. The difference in weight of the empty and filled bottles is the weight of sample. Place the bottles in the oven, with the stoppers open, and dry for 2 hours (or overnight). Cool in a desiccator, stopper tightly, and weigh. Heat for another 1-hour period, cool, and reweigh. The second weighing should agree with the first within 0.3 mg. If it does not, repeat the heating until constant weight is obtained. From the loss in weight, and the weight of the sample, compute the percentage of water in the sample. The duplicate determinations should agree within 0.2 per cent. Record all the data in the notebook, following the form shown in Table 4–6. Report the result of the analysis as directed by the instructor.

Calculations. It is often convenient to make analyses on undried portions of sample, to determine separately the moisture on another

TABLE 4–6. Determination of Moisture

Sample Notebook Record

Sample number	I	II
Initial weights of bottles	32.5675	34.3246
	32.5676	34.3241
		34.3240
Weight of sample	33.6684	35.2562
	32.5676	34.3240
	1.1008	0.9322
Final weights of bottles with dried sample	33.5307	35.1402
	33.5309	35.1400
Loss in weight	33.6684	35.2560
	33.5309	35.1400
	0.1375	0.1160
Per cent moisture	12.49	12.44

portion and to report the results of the analysis on a dry basis. The method may be illustrated by a sample calculation.

Problem. One portion of a sample was found to contain 5.72 per cent moisture. Analysis of another portion gave 14.56 per cent copper, on an "as received" basis. What is the percentage of copper, expressed on a dry basis?

SOLUTION: Since the sample contains 5.72 per cent moisture, a 1.000-g portion will have 0.9428 g of dry material. This contains 0.1456 g of copper. Therefore the percentage of copper, on a dry basis, is

$$\frac{0.1456}{0.9428} \times 100 = 15.44\%$$

Preparation of Solution

No general directions can be given for bringing samples into solution; each type of material demands specific treatment. The analytic methods of frequent occurrence are given in Table 4–7. It may be noted that the most common methods are: (1) treatment with hydrochloric acid, (2) treatment with nitric acid, (3) fusion with sodium carbonate, and (4) fusion with potassium acid sulfate.

Hydrochloric acid is generally used when a non-oxidizing acid

TABLE 4–7. Methods Frequently Used for Bringing Sample into Solution

Sample	Treatment
Ferrous alloys	Acids
Non-ferrous alloys	HNO_3
Iron ores	HCl (fusion with $KHSO_4$)
Carbonate rocks	HCl or fusion with Na_2CO_3
Silicate rocks	Fusion with Na_2CO_3
Chromium ores	Fusion with $Na_2CO_3 + Na_2O_2$
Pyrites (FeS_2)	Fusion with $Na_2CO_3 + Na_2O_2$
SnO_2	Fusion with $Na_2CO_3 + S$

medium is required and nitric acid when oxidation of a metal is required. In some analyses it is advantageous to employ a mixture of nitric and hydrochloric acids (aqua regia) so that oxidation can be carried out in a hydrochloric acid medium.

Many materials that are not soluble in acids can be brought into a soluble condition by means of reactions with fused salts. The fused salts most frequently employed, known as fluxes, are sodium carbonate and potassium acid sulfate. Sodium carbonate is the most widely used of these. Oxides of an acidic nature, such as SiO_2, are converted into soluble sodium salts by sodium carbonate fusion. In addition, many metal salts are converted into the metal carbonates, which readily dissolve in acids. When oxidation of an acidic oxide is needed, the material may be fused with sodium carbonate plus an oxidizing agent, such as sodium peroxide. Potassium acid sulfate is an acid and is consequently used to render basic oxides soluble by conversion into the sulfates. This flux is most frequently used to bring insoluble iron oxides into solution.

Questions

A. Volumetric Analysis

1. Why is it necessary to calibrate volumetric glassware? Why can the calibration of a buret be omitted if the buret is to be used for both a standardization and an analysis with the same solution? In general, what is the minimum amount of calibration that will permit accurate work?

2. Why is water chosen as the calibrating liquid for most apparatus? What special precautions must be exercised if a buret that has been calibrated with water is used for measuring a viscous liquid, such as a concentrated solution?

3. What is the advantage of calibrating in intervals measured from zero rather than withdrawing successive 5- or 10-ml portions and weighing each portion?

4. Why is it unnecessary to dry the flask used for weighing water in the buret calibration? Why is it important to see that no drops of water adhere to the inside of the flask, near the top?

5. Describe the procedure for withdrawing a measured aliquot portion by means of a pipet. What calibrations are needed if this method is followed for aliquot portions?

6. Describe the procedure for withdrawing aliquot portions by means of a buret. What calibrations are needed?

7. Two calibration marks are sometimes used for a volumetric flask, one for the absolute volume contained and the other for the volume relative to a pipet. When should the flask be filled to each mark?

8. Why, in the buret calibration, should a buoyancy correction not be applied for the weight of the container for weighing water?

9. If a solution has been diluted to volume and mixed at 25°C, what change will be noted if the temperature increases to 28°C? How may an accurate aliquot portion be withdrawn if the room temperature has changed after the solution has been diluted to volume?

10. In the use of a buret, what advantage is there in adjusting the volume exactly to zero before beginning a titration? What disadvantage is there in this procedure?

11. Why are burets cleaned by inverting them in cleaning solution rather than filling the upright burets with the solution?

12. Why is it desirable to have graduation marks on a buret extend at least halfway around the tube?

13. Why is ±0.04 ml set as the outside limit within which buret calibrations must check?

14. Why are the weighings in a buret calibration made to the closest milligram rather than to the closest 0.01 or 0.0001 g?

15. Why is the calibration of the 250-ml volumetric flask not done on the analytical balance? Why is the method of weighing by substitution used?

16. A flask contains 250.1 ml of water at 21°C. If the expansion of glass is neglected, what volume will the flask contain at 30°C?

17. What is meant by a "blank determination"? Should every volumetric determination be accompanied by a blank? Under what conditions may exceptions be made?

18. Why is it often desirable to weigh out a large sample and use aliquot portions for an analysis rather than to weigh out individual portions of the sample for each titration?

19. The following values were obtained in the standardization of a freshly prepared solution:

$$0.1037, \ 0.1045, \ 0.1053, \ 0.1060, \ 0.1065.$$

Would you consider the results reliable? Why? Suggest a cause for these results.

20. Might a buret be calibrated by successive additions of water from a calibrated pipet? Why? Why, then, is this procedure valid for the calibration of a flask?

21. Describe the various methods given for elimination of parallax.

22. A buret calibrator is a modified pipet which is attached to the tip of a buret so that water withdrawn from the buret is forced up into the pipet. The volume delivered by the calibrator is first determined, and then successive portions are withdrawn from the buret to fill the calibrator. For each buret interval the apparent volume withdrawn is noted. Is it necessary to know the temperature when this method is used? Why? Are the errors in buret reading cumulative or not? Is the error in knowing the volume of the calibrator cumulative or not? Explain. Must the calibrator be wet or dry at the beginning of the calibration? Explain. How might drainage errors affect accuracy in using a calibrator?

B. Gravimetric Analysis

23. List the various common methods for filtration. Tell when each is desirable. Which method would you choose for each of the following precipitates: barium sulfate, ferric hydroxide, silver chloride, magnesium ammonium phosphate, zinc sulfide?

24. What is digestion of a precipitate? Is this procedure always employed? Explain.

25. When is it advantageous to employ suction with filter paper? Describe the arrangement necessary to use suction with paper.

26. Describe the preparation of a Gooch crucible.

27. Why should washing be done by the addition of several small portions of liquid rather than by a smaller number of larger portions, provided the same total volume of liquid is used in each method?

28. According to the color scale of temperature given, what temperature do you estimate is attainable with the burner available in your desk?

29. Describe the various steps employed in the ignition of a precipitate in filter paper.

30. Why should ignited precipitates be cooled in a desiccator?

31. What may occur if the desiccator cover is suddenly removed after an ignited precipitate has been cooled?

C. Preparation of Sample

32. Why is not the entire gross sample ground to the fineness of the laboratory sample?

33. Why should samples be ground in small portions when an agate mortar is used?

34. What is the advantage of crushing lumps in the steel mortar described, rather than directly in an agate mortar?

35. In the following analyses should the drying be done in air, in an oven at 110°C, or in a desiccator over a substance of very low vapor pressure?

A plant leaf.	Iodine.
An animal tissue.	An iron ore.
A commercial fertilizer.	

36. If a hydrate is to be brought to constant weight in a desiccator, what must be true about the vapor pressures of the hydrate and the desiccant?

37. What changes might occur in the mineral hercynite, $FeO \cdot Al_2O_3$, during grinding? Name a suitable flux to bring this mineral into solution.

38. How would poor mixing of a laboratory sample be detected during the course of an analysis?

39. Why is it important to record the temperature at which the sample was dried in giving the results of an analysis?

40. Why are most samples dried in an oven before the analyses are started?

41. What methods might be used to dissolve each of the following?

> Arsenious oxide for arsenic analysis.
> Barium sulfate for barium analysis.
> Barium sulfate for sulfate analysis.
> Silver iodide for silver analysis.
> Metallic silver for silver analysis.
> Metallic antimony for antimony analysis.
> An insoluble chromium ore for chromium analysis.
> A silicate rock for silica analysis.
> An ignited aluminum oxide for aluminum analysis.

Problems

1. A 2000-ml volumetric flask is to be calibrated with water, and results must check within 1/1000. If the position of the meniscus can be read within ±1 ml, how sensitive should the balance used for the weighings be?

2. The internal diameter of the neck of a 100-ml volumetric flask is 14 mm. The position of the meniscus can be read to within ±0.2 mm. If the only error in the calibration is in the estimation of the meniscus position, how closely should duplicate calibrations agree?

3. A 10-ml buret is calibrated in 0.01-ml divisions and can be read to ±0.001 ml. If water is to be used as the calibrating liquid and the only error to be considered in setting the tolerance limit for checks is to be the meniscus reading error, how carefully should the water be weighed?

Ans. Nearest 0.2 mg.

4. Repeat Problem 3 with mercury (density = 13.5 g/ml) used as the calibrating liquid.

5. From the following data find the correction for the 0–10-ml interval of a buret:

Initial reading	0.17 ml
Final reading	10.25 ml
Weight of flask	50.632 g
Weight of flask + water	60.649 g
Temperature	24°C

Ans. −0.03 ml.

6. The thermometer used in the previous calibration reads 3° too high. What is the true correction for the 0–10-ml interval? What error in parts per thousand is introduced by the temperature error?

7. At 23°C a 50-ml pipet delivers 49.754 g of water weighed in air with brass weights. What is the calibration correction for the pipet?

Ans. −0.08 ml.

8. If the expansion of glass is neglected, what weight of water will the pipet deliver at 30°C? *Ans.* 49.670 g.

9. Repeat Problem 8, but consider the expansion of the glass.

10. Determine the correction for a 250-ml volumetric flask from the following:

Weight of water contained	249.00 g
Temperature	25°C

11. What is the volume of this flask at 20°C? *Ans.* 249.92 ml.

12. An unmarked volumetric flask is to be calibrated to contain exactly 100.0 ml. The empty flask is weighed on the balance. What additional weights should be placed on the pan for the correct amount of water weighed in air with brass weights at 20°C? *Ans.* 99.721 g.

13. Repeat Problem 12, but use mercury whose density in vacuo at 20°C is 13.5461 g per milliliter. *Ans.* 1354.691 g.

14. A flask is to contain 200.0 ml at 20°C. It is to be calibrated with water at 25°C; if the expansion of glass is neglected, what weight of water determined in air with brass weights should be put in the flask?

15. Repeat Problem 14, considering the expansion of the glass.

16. The neck of a 250-ml volumetric flask is 18 mm in internal diameter. At 20°C the correction for the flask is −0.1 ml. If an amount of water at 20°C corresponding to a mass of 250.0 g in vacuo is placed in the flask, what is the position of the meniscus with respect to the mark? *Ans.* 0.20 cm.

17. Repeat Problem 16 for an amount of water corresponding to 250.0 g of water weighed in air with brass weights. *Ans.* 0.31 cm.

18. From the factor for 25°C in Table 4–2, compute the density of water in vacuo at this temperature. *Ans.* 0.99726 g/ml.

19. Derive the factor similar to those of Table 4–2 for water at 33°C where its density in vacuo is 0.99473 g per milliliter.

20. Derive a factor similar to those of Table 4–2 which could be used for mercury and brass weights at 20°C. At this temperature the density of mercury in vacuo is 13.5461 g per milliliter.

21. After suitable rinsing a 50-ml pipet full of solution is withdrawn from a 250-ml volumetric flask. This aliquot portion is shown to contain 0.0502 g CaO. How many grams of CaO are present in the whole sample if the true volumes of the pipet and flask are 50.05 and 249.7 ml, respectively? How might some of the calculation of this problem have been eliminated by use of different methods of calibration?

22. What is the percentage increase in volume as 1.000 liter of water at 18°C is heated to 28°C?   *Ans. 0.2%.*

23. A buret calibrator (see question 22) delivers 9.95 ml. It is attached to a buret, and the following readings are taken for successive fillings of the calibrator. Compute the correction for each interval in the buret and the total correction to be applied at volumes of 10, 20, 30, 40, 50 ml. Plot the buret corrections.

Buret readings: 0.02, 9.98, 19.93, 29.92, 39.90, 49.91.

Preparation of Sample

24. A sample of $CuSO_4 \cdot 5H_2O$ was ignited until all water was expelled. If the sample weight was 1.035 g, what was the weight of the residue?
                                                                    Ans. 0.662 g.

25. A sample of coal which weighs 2.000 g after air drying loses 0.1575 g of moisture on drying in an oven; 0.3000 g of an oven-dried sample of the same coal yields an ash residue of 0.0500 g. What is the percentage of ash in the coal on an air-dried basis?   *Ans. 15.36%.*

26. A sample of iron ore contains 1.78 per cent moisture and 71.65 per cent iron on a dry basis. Compute the percentage of iron in the original material.
                                                                    Ans. 70.40%.

27. A sample of clay contains 18.3 per cent moisture and 15.8 per cent potassium on an "as-received" basis. What is the percentage of potassium on a completely dry basis? What would be the percentage of potassium if the moisture were reduced to 5.0 per cent of the "as-received" sample?

28. A sample of organic tissue showed a 2.00 per cent loss in weight on air drying and a 57.5 per cent loss of the air-dried weight on oven drying. Colorimetric analysis demonstrated the presence of 12.0 parts per million by weight of copper in the oven-dried sample. Obtain the parts per million of copper on an air-dried and "as-received" basis.   *Ans. 5.10 ppm; 5.00 ppm.*

29. An oven-dried limestone sample analyzes 45.55 per cent loss in weight on ignition to red heat and 33.26 per cent CaO. What is the percentage of CaO in the ignited residue? 15.000 g of an air-dried sample of this limestone was dried in an oven at 120°C to a constant weight of 14.923 g. What would be the percentages of loss on ignition and CaO in an air-dried sample?

PART TWO—THEORY AND

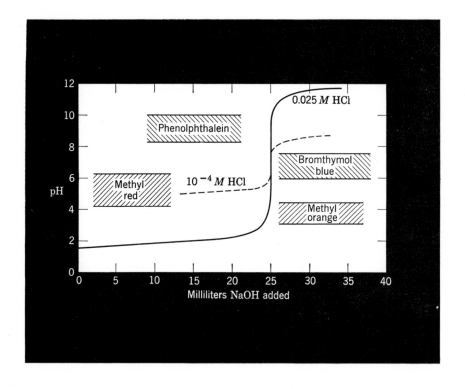

CALCULATIONS
OF ANALYTICAL CHEMISTRY

CHAPTER 5

Stoichiometric Calculations
of Chemical Analyses

Those who take this course have already had extensive training in stoichiometric calculations, i.e., calculations based on the weight relations of chemical formulas and equations. The work of this course will, however, require much greater use of such calculations than was needed in general chemistry. The reason is that every analysis demands some type of calculation. The sought substance is seldom isolated and weighed directly, but instead we compute its weight indirectly, from the weight of a precipitate or from the volume of a solution that reacts with the sample. It is appropriate, therefore, to begin our study of quantitative analysis with a review of stoichiometric calculations, with special emphasis on analytical applications.

Weight Relations in Formulas and Equations

A few illustrative examples will serve to recall how the relations of a chemical equation are used to compute the weights or volumes of the substances involved in a reaction.

Problem.   What weight of HNO_3 (mol. wt. 63.0) is needed to react with 250 g Cu (63.5)?   The reaction is

$$3Cu + 8HNO_3 = 3Cu(NO_3)_2 + 2NO + 4H_2O$$

SOLUTION: The chemical equation shows that 8 moles HNO_3 are required to react with 3 g-atoms of copper.   Since the weight of Cu is given, it is first converted into the number of gram-atoms.

$$\text{Gram-atoms Cu} = \frac{250 \text{ g Cu}}{63.5 \text{ g/g-atom}} = 3.94$$

The number of moles of HNO_3 is $\frac{8}{3}$ the number of gram-atoms of Cu.

$$\text{Moles } HNO_3 = \frac{8 \text{ moles } HNO_3}{3 \text{ g-atoms Cu}} \times 3.94 \text{ g-atoms Cu}$$

$$= 10.51 \text{ moles}$$

$$\text{Weight } HNO_3 = 10.51 \text{ moles} \times 63.0 \text{ g/mole}$$

$$= 662 \text{ g}$$

Problem.   What volume of NO gas, measured at standard conditions, is obtained in the reaction of the preceding problem?

SOLUTION: The equation shows that 2 moles NO is obtained from 3 g-atoms Cu. Since the volume of a mole of gas at STP is 22.4 liters, 3 g-atoms Cu gives 2×22.4 liters or 44.8 liters NO.   We have in 250 g Cu, 3.94 g-atoms.   Therefore,

$$\text{Volume NO} = \frac{44.8 \text{ liters NO}}{3 \text{ g-atoms Cu}} \times 3.94 \text{ g-atoms Cu} = 58.8 \text{ liters}$$

Problem.   A sample of impure magnetite, Fe_3O_4, weighing 1.542 g is dissolved; the iron is oxidized to Fe^{+3} and precipitated as $Fe(OH)_3$.   The precipitate is ignited to Fe_2O_3, giving a weight of 1.485 g.   Calculate the percentage of Fe_3O_4 in the sample.

SOLUTION: Since we are given the weight of pure Fe_2O_3, this is used to compute the number of moles of Fe_2O_3

$$\text{Moles } Fe_2O_3 = \frac{1.485 \text{ g } Fe_2O_3}{159.7 \text{ g } Fe_2O_3/\text{mole } Fe_2O_3}$$

$$= 0.00930 \text{ moles}$$

This is equivalent to $\frac{2}{3}$ as many moles of Fe_3O_4, since 2 moles Fe_3O_4 will yield 3 moles Fe_2O_3 when oxidized.

$$\text{Moles } Fe_3O_4 = \frac{2 \text{ moles } Fe_3O_4}{3 \text{ moles } Fe_2O_3} \times 0.00930 \text{ moles } Fe_2O_3$$

$$= 0.00620 \text{ moles } Fe_3O_4$$

$$\text{Weight Fe}_3\text{O}_4 \text{ in sample} = 0.00620 \text{ moles Fe}_3\text{O}_4 \times \frac{231.55 \text{ g Fe}_3\text{O}_4}{\text{mole Fe}_3\text{O}_4}$$

$$= 1.437 \text{ g Fe}_3\text{O}_4$$

$$\text{Percentage Fe}_3\text{O}_4 = \frac{1.437 \text{ g}}{1.542 \text{ g}} \times 100 = 93.1\%$$

Chemical Factor. In all the preceding problems the final computation involves multiplying the given amount of some substance by a numerical value to obtain the equivalent amount of another substance. In preceding courses we have gone through the detailed steps in such calculations. The experienced chemist condenses the computation by using a factor that relates the weight of the substance sought to the weight of given substance. For example, the preceding three problems may be solved by factors, thereby eliminating some steps of the calculation.

1. *Weight of HNO₃ to react with 250 g Cu.* The factor is 8HNO₃/3Cu, since these are the relative numbers of gram-atoms reacting. Substituting the relative weights gives 504.0/190.5 or 2.645 as the factor. This number is the weight of HNO₃ required to react with 1.000 g Cu. Then,

$$\text{Weight HNO}_3 \text{ to react with 250 g Cu} = 250 \times 2.645$$

$$= 662 \text{ g}$$

2. *Volume of NO gas obtained from reaction of 250 g Cu with HNO₃.* The factor is 2NO/3Cu. Since the answer is desired in volume units, the factor is

$$\frac{2 \times 22.4 \text{ liters NO}}{3 \times 63.5 \text{ g Cu}} = 0.235 \text{ liters, NO/g Cu}$$

Using this value gives

$$\text{Volume NO} = \frac{0.235 \text{ liters NO}}{1.000 \text{ g Cu}} \times 250 \text{ g Cu}$$

$$= 58.8 \text{ liters}$$

3. *Weight of Fe₃O₄ equivalent to 1.485 g Fe₂O₃.* The factor is 2Fe₃O₄/3Fe₂O₃. The number of atoms of Fe must be the same in the numerator and denominator.

$$\text{Weight Fe}_3\text{O}_4 = 1.485 \text{ g Fe}_2\text{O}_3 \times \frac{2\text{Fe}_3\text{O}_4}{3\text{Fe}_2\text{O}_3}$$

$$= 1.485 \text{ g} \times \frac{2 \times 231.55}{3 \times 159.7}$$

$$= 1.437 \text{ g}$$

The three factors used in these calculations may be summarized in tabular form.

Given	Sought	Mole Ratio in Factor	Numerical Factor
Cu	HNO_3	$\dfrac{8HNO_3}{3Cu}$	$\dfrac{8 \times 63.0 \text{ g}}{3 \times 63.5 \text{ g}} = 2.645$
Cu	NO volume	$\dfrac{2NO}{3Cu}$	$\dfrac{2 \times 22.4 \text{ liters}}{3 \times 63.5 \text{ g}} = 0.235 \text{ liters/g Cu}$
Fe_2O_3	Fe_3O_4	$\dfrac{2Fe_3O_4}{3Fe_2O_3}$	$\dfrac{2 \times 231.55 \text{ g}}{3 \times 159.7 \text{ g}} = 0.967$

In using factors note that:

1. The substance sought appears in the numerator of the factor.
2. If the given and sought substances are in the same units, the factor is dimensionless. If the two are in different units, as in the computation of NO gas in problem **2**, show the dimensions of the factor.

Units in Calculations. In the illustrative problems just examined, all quantities are labeled with the units used for the measurements. This usage is probably already known to you from work in general chemistry. If not, a brief study of the preceding examples will prove helpful. Consistent use of units in computations aids in formulating the proper statement of a problem and may avoid such errors as multiplying by a factor when we should divide, or vice versa.

To use units, label each quantity involved in a calculation and then multiply the units just as the numerical values. Thus, if six men work 5 hours each, the total work is 6 men $\times$ 5 hours = 30 man-hours. In additions or subtractions all numbers must have the same units.

Calculations Involving Concentrations of Solutions

> It is desirable to study the material of Chapter 12 before reading the following sections, in order to have a clear understanding of the laboratory operations involved in the computations.

We deal with solutions in almost every analytical determination, and consequently we must employ stoichiometric computations based on their volumes and concentrations. The concentrations of solutions may be expressed in a variety of ways. For convenience we shall classify these under the general headings of physical and chemical methods.

Physical Methods. The simplest ways of expressing the strength of a solution are in terms of the amount of solute present per unit amount of solvent or solution. Such methods are known as physical methods because they are based only on physical measurements of weight or volume and do not take into account the chemical reactions of the solute. The more widely used physical methods are:

1. Grams solute per liter (or 100 ml) solution.
2. Grams solute per liter (or 100 ml) solvent.
3. Grams solute per unit weight of solution. Unit weight is most often chosen as 100 g or 1000 g.
4. Grams solute per unit weight of solvent.
5. Percentage methods. The concentration of solutions is frequently stated in percentage, particularly when only an approximate value is indicated. Unless further defined, the term might mean either the weight or volume of solute per 100 g or 100 ml of solution or of solvent, but by general usage the term has come to mean the weight of solid or the volume of liquid in 100 ml of the final solution. Thus, a 10 per cent salt solution contains 10 g salt per 100 ml solution, and a 10 per cent alcohol solution contains 10 ml liquid alcohol per 100 ml solution. For exact expression of concentrations, the term "per cent by weight" is common. This means the grams of solute per 100 g of solution; it is a designation that is independent of the effect of temperature on volume. All handbooks give density tables for common reagents, in which the per cent by weight is expressed as a function of the density of the solution.

None of the physical methods provides a direct measure of the chemical strength of the solution. For example, a given volume of a 5 per cent by weight solution of hydrochloric acid will react with a greater quantity of a base than will an equal volume of a 7 per cent by weight nitric acid solution. In order to make chemical calculations involving solutions, it is first necessary to transform the concentrations into chemical units, such as the mole. It is convenient, therefore, to state the concentration in chemical units in order to eliminate the necessity for the conversion. Two chemical units are widely employed, the mole and the equivalent.

Molar Methods. A solution containing 1 mole of solute per liter of solution is defined as a *molar solution*. The *molarity* of a solution denotes the number of moles of solute per liter of solution. It is designated by a numerical value[1] preceding the abbreviation "M." Thus,

---

[1] Less often used is the designation $M/10$ to indicate a 0.1 M solution.

0.1 M HCl is a hydrochloric acid solution which contains $\frac{1}{10}$ mole of hydrogen chloride per liter. This terminology does not in any way indicate that a liter of solution is present but merely states that the amount of hydrogen chloride is such that a liter of the solution contains $\frac{1}{10}$ mole.

The total number of moles of solute in a given volume of solution is the product of the volume by the molarity. For example, there are, in 2 liters of 1 M solution, 2 moles of solute, and in 500 ml of 1 M solution there is $\frac{1}{2}$ mole of solute.

It follows from the definition of the mole and of molar solutions that equal volumes of equimolar solutions contain an equal number of molecules. *This statement is the basis for all calculations involving molarity and should be thoroughly understood before the reader proceeds further.* The volumes of equimolar solutions which enter into a reaction are numerically related to one another as are the respective number of molecules in the chemical equation for the reaction. Thus 2 liters of a molar solution of sodium hydroxide are needed for reaction with 1 liter of a molar solution of sulfuric acid.

The preceding definitions may be concisely summarized in the form of equations:

$$\text{Molarity} = \frac{\text{number of moles solute}}{\text{liters of solution}}$$

$$\text{Molarity} = \frac{\text{grams solute}}{\text{molecular weight}} \times \frac{1}{\text{liters of solution}}$$

Moles of solute = molarity $\times$ liters of solution

Grams of solute = molarity $\times$ liters of solution $\times$ molecular weight

In many physicochemical calculations, it is convenient to employ solutions whose strengths are stated in terms of the weight of solvent rather than the volume of solution. A solution which contains 1 mole of solute per kilogram of solvent is known as a *molal* solution. This method of expressing the concentration is not used in quantitative analysis, for in most quantitative procedures the analyst measures the volume of solution.

The use of the mole as a concentration unit has certain disadvantages, and many prefer to use the term "formal" instead of "molar," where a formal solution (1 F) contains a formula weight (in grams) of solute per liter of solution. Thus, 1 F H_2SO_4 contains 98 g H_2SO_4 per liter.

Equivalent Methods. The use of the mole as a concentration unit simplifies chemical calculations, but does not provide a basis for direct comparison of the strength of all solutions. For example, a liter of $1\,M$ sodium hydroxide will neutralize a liter of $1\,M$ hydrochloric acid but will neutralize only $\frac{1}{2}$ liter of $1\,M$ sulfuric acid. It is readily seen that a mole of hydrochloric acid furnishes a mole of hydrogen ion, or 1.0080 g, whereas a mole of sulfuric acid furnishes 2 moles, or 2.0160 g, of hydrogen ion. In neutralization reactions the fundamental change is the disappearance of hydrogen or hydroxide ion from the solution, usually according to the equation

$$H^+ + OH^- = H_2O$$

It is convenient, therefore, to choose as the unit the weight of a substance that will furnish a mole of hydrogen or hydroxide ion. The number of units of all substances involved in a reaction is thus made numerically equal. The *equivalent weight* (eq) provides such a chemical unit. The equivalent weight is defined, for neutralization reactions, as the weight in grams that will furnish or react with one gram-atomic weight of hydrogen ion. For example, an equivalent weight of hydrochloric acid is 1 mole, or sulfuric acid is $\frac{1}{2}$ mole, and of sodium hydroxide is 1 mole. The definition is further illustrated in Table 5–1.

TABLE 5–1. Equivalent Weights of Some Acids and Bases

Substance	Weight of Mole in Grams	Equivalents per Mole	Equivalent Weight in Grams
NaOH	40.00	1	40.00
Ca(OH)$_2$	74.10	2	37.05
MgO	40.32	2	20.16
HCl	36.46	1	36.46
H$_2$SO$_4$	98.08	2	49.04
Na$_2$CO$_3$ (complete neutralization)	106.0	2	53.0
NaHCO$_3$	84.0	1	84.0
NH$_4$OH	35.05	1	35.05

The equivalent weight of a substance depends on the reaction involved in the titration. If H_3PO_4 is titrated with NaOH to the end point with one indicator, the reaction

$$H_3PO_4 + NaOH = NaH_2PO_4 + H_2O$$

takes place, and the equivalent weight of H_3PO_4 is the same as its molecular weight, because each mole of H_3PO_4 has given up 1 mole of H^+. If another indicator is used, the reaction is

$$H_3PO_4 + 2NaOH = Na_2HPO_4 + 2H_2O$$

In oxidation-reduction reactions the equivalent weight is the number of grams of reagent needed to oxidize a gram-atom of hydrogen or reduce a gram-atom of hydrogen ion. It is not appropriate to consider such reactions in detail at this time, but we should note that calculations dealing with solutions of oxidizing or reducing agents are made in exactly the same manner as the ones we now consider for neutralization reactions.

Since an equivalent weight of any substance will furnish or react with a mole of hydrogen ion, the number of equivalents is numerically the same for all substances involved in a reaction. That is to say, one equivalent of any acid will neutralize one equivalent of any base, since each will furnish or react with the same amount of hydrogen ion. Likewise, an equivalent weight of any oxidizing agent will react with an equivalent weight of any reducing agent. Because of this relation, the equivalent weight furnishes a convenient method for expressing the concentrations of solutions on a comparable basis. A *normal* solution is one that contains one equivalent weight of solute per liter of solution. The normality of a solution is denoted by a numerical value followed by the symbol "N." Thus the expression 0.1 N HCl is read as one-tenth normal hydrochloric acid. Such a solution is one containing one-tenth equivalent of hydrogen chloride per liter. (Also used is the symbol $N/10$ for a 0.1 N solution.)

The definitions may be summarized as was done for molar solutions.

$$\text{Normality} = \frac{\text{equivalents solute}}{\text{liters of solution}}$$

$$\text{Normality} = \frac{\text{grams solute}}{\dfrac{\text{molecular weight}}{\text{hydrogen equivalents per mole}}} \times \frac{1}{\text{liters of solution}}$$

Liters of solution $\times$ normality = equivalents of solute
Liters of solution $\times$ normality $\times$ equivalent weight = grams solute

There is a simple relation between the normality and the molarity of a given solution, since the equivalent weight is the molecular weight divided by the number of available hydrogen ions per molecule. The number of equivalents in a mole is equal to one, two, three, or more; consequently the normality is always some factor times the molarity —*never less than the molarity.*

Use of Milliequivalents and Millimoles. In the preceding discussion we have taken the liter as the unit of volume. Practically, in laboratory work, we generally use the milliliter as the unit of volume.

Thus, the volume required in a titration is stated as **44.57** ml and not as 0.04457 liter. This dual selection of units is comparable to the everyday usage in regard to currency. We select the most convenient unit for the purpose at hand. Thus, the price of a car is stated in dollars but the cost of groceries is stated in terms of cents per pound, rather than as dollars per pound. In keeping with this usage, it is convenient to select as the unit of concentration the amount of solute per milliliter of solution, which is $\frac{1}{1000}$ the amount per liter of solution. We can then define a molar solution as one containing 0.001 mole per milliliter of solution, or 1 millimole (mM) per milliliter. Likewise a normal solution is one that contains 0.001 equivalent, or 1 milliequivalent (meq), per milliliter.

Since a mole is the molecular weight in grams, a millimole is a molecular weight in milligrams. Similarly, since an equivalent weight is the weight in grams to furnish or to react with one gram-atom of hydrogen, a milliequivalent is the weight in milligrams necessary to furnish or to react with one milligram-atom of hydrogen. Numerically the weight of solute in grams per liter of solution is the same as the weight in milligrams per milliliter of solution. We have a good analogy in everyday usage; if a commodity costs $5 per 100 pounds, it obviously costs 5 cents per pound.

The defining equations given for molar and normal solutions may be restated in terms of millimoles and milliequivalents by substitution of the prefix "milli" before each term which denotes a quantity. For example,

Milligrams solute
$\quad$ = milliliters solution $\times$ normality $\times$ milliequivalent weight

In subsequent usage in this book, the choice of equivalent or milliequivalent is dictated by the unit of volume chosen. If the volume is given in liters, calculations are made on the basis of equivalents and grams. If the volume is in milliliters, the calculation uses milliequivalents and milligrams.

Titer. It frequently happens that a solution is used repeatedly for a specific analysis, and in making the computations it is convenient to express the strength of the reagent solution as the *titer* or weight of the sample material that is equivalent to 1 ml of the reagent. After a titration the volume of reagent is multiplied by this titer to obtain the weight of sought substance in the titrated sample.

For example, if we are routinely titrating samples of impure Na_2CO_3 with a standard HCl solution, it is convenient to express the

Na_2CO_3 titer of the HCl. As shown later, the titer can be simply related to the normality of the solution.

Illustrative Problems

For typical problems dealing with normalities of solutions, we shall consider now only metathesis reactions. Similar computations for redox reactions are given in Chapter 10.

The basis for all computations dealing with normalities of solutions is the simple relation that the number of equivalents or milliequivalents of one reactant is equal to the number for the other reactant. Thus, if A reacts with B

$$\text{Equivalents A} = \text{Equivalents B}$$

$$\text{Milliequivalents A} = \text{Milliequivalents B}$$

As we solve typical problems, you will note the constant use of this relation.

Preparation of Solutions. Most solutions are prepared by dissolving a weighed amount of solid and adding sufficient water to make the desired volume. The amount of solute in a given solution is computed as in the following problems.

Problem. How many grams of pure sodium hydroxide, mol. wt 40.0, are needed for preparation of 500.0 ml 0.1000 N solution?

SOLUTION: To illustrate the use of equivalents and milliequivalents, we shall work this problem in two ways. To use equivalents we express the volume in liters. According to the definitions previously given,

$$\text{Equivalents solute} = \text{volume in liters} \times \text{normality}$$
$$= 0.5000 \text{ liter} \times 0.1000 \text{ eq/liter} = 0.05000 \text{ eq}$$

To get the weight in grams we need only multiply the equivalents by the weight of 1 eq, or

$$\text{Weight in grams} = 0.05000 \text{ eq} \times 40.00 \text{ g/eq}$$
$$= 2.000 \text{ g}$$

To work the same problem in milliequivalent terminology, we express the volume in milliliters. Then, by the previous definitions,

$$\text{Milliequivalents solute} = \text{volume in milliliters} \times \text{normality}$$
$$= 500.0 \text{ ml} \times 0.1000 \text{ meq/ml} = 50.00 \text{ meq}$$

The weight in milligrams is given by the relation:

$$\text{Weight} = \text{milliequivalents} \times \text{milligrams per milliequivalent}$$
$$= 50.00 \text{ meq} \times 40.0 \text{ mg/meq} = 2000 \text{ mg} = 2.000 \text{ g}$$

Note that in this problem it is equally convenient to express the volume in liters or in milliliters and therefore that either the equivalent or the milliequivalent method can be conveniently used. This is not always true. If, for example, the problem had asked for preparation of 150 liters, very large numbers would be necessary to compute on the milliequivalent basis. Conversely, if the volume were 15 ml, very small numbers would be necessary to work with equivalents. In every problem a choice should be made as to the most convenient unit to employ.

Problem. What is the normality of a solution prepared by dissolving 25.20 g oxalic acid ($H_2C_2O_4 \cdot 2H_2O$, mol. wt. 126.1) in sufficient water to give 1.200 liters of solution? What is the molarity of the solution?

SOLUTION: First we determine the equivalents. The equivalent weight is 126.1/2 or 63.05 g, since oxalic acid has two replaceable hydrogen atoms.

Therefore,

$$\text{Number of equivalents} = \frac{25.20 \text{ g}}{63.05 \dfrac{\text{g}}{\text{eq}}} = 0.3996 \text{ eq}$$

The normality is

$$N = \frac{0.3996 \text{ eq}}{1.200 \text{ liters}} = 0.3330 \text{ eq/liter}$$

To find the molarity, we note that, since a mole furnishes 2 eq, the number of moles per liter is one-half (*never twice*) the number of equivalents. Therefore,

$$\text{Molarity} = \frac{0.3330 \dfrac{\text{eq}}{\text{liter}}}{2 \dfrac{\text{eq}}{\text{mole}}} = 0.1665 \text{ mole/liter}$$

Dilution of Solutions. A solution of known normality is frequently prepared from a more concentrated one of known strength by quantitative dilution.

Problem. What volume of 0.1500 N reagent is needed for the preparation of 500.0 ml 0.1000 N solution? (Note that in this problem it is not necessary to know what the reagent is. Since the normality is given, the computation does not in any way depend on whether the solution contains an acid, base, or salt.)

SOLUTION: We note that both the volume and normality of the final solution are given. Therefore the milliequivalents required are given by the product

$$\text{Milliequivalents required} = 500.0 \text{ ml} \times 0.1000 \text{ meq/ml}$$
$$= 50.00 \text{ meq}$$

To complete the problem we need only determine what volume of 0.1500 N solution is needed to provide 50.00 meq:

$$\text{Volume required} = \frac{50.00 \text{ meq}}{0.1500 \dfrac{\text{meq}}{\text{ml}}} = 333.3 \text{ ml}$$

Remarks. To prepare the desired solution we may add 166.7 ml of water to 333.3 ml of the 0.1500 *N* solution, or, more accurately, place 333.3 ml of the 0.1500 *N* solution in a 500-ml volumetric flask and dilute to the mark. When solutions of a predetermined normality are required, the computation is like that of this problem. A solution is made somewhat stronger than the desired concentration and is accurately standardized. Then, by quantitative dilution, the desired strength is obtained.

Determination of Normality of Solution by Comparison with a Standard.
In volumetric analyses the usual practice is to prepare solutions of the approximate desired strength and then determine the exact strength by titration against a solid primary standard. The following problems illustrate the computations needed. First we shall consider a titration made just to the end point, without any back titration.

Problem. A 0.9324-g sample of pure potassium acid phthalate is dissolved and titrated with 41.73 ml of a base solution, using phenolphthalein indicator. What is the normality of the base solution? (Note that in this problem we have not stated what base is used or what volume of water is added to the potassium acid phthalate sample. It is not necessary to know either. Any base, if strong, will neutralize the potassium acid phthalate completely. The amount of water added to the KHP does not in any way affect the titration since we add enough base to neutralize it completely. In practice we add some 50 ml of water, to dissolve the weighed sample readily, but we do not measure the water.)

SOLUTION: Here the data given are the weight of standard and its equivalent weight (look up in table on back cover of book), 204.2. We compute the milliequivalents of standard by dividing the weight in milligrams by the milliequivalent weight.

$$\text{Milliequivalents} = \frac{932.4 \text{ mg}}{204.2 \text{ mg/meq}} = 4.566 \text{ meq}$$

The normality of the base is obtained by dividing the number of milliequivalents by the volume, in milliliters.

$$\text{Normality} = \frac{4.566 \text{ meq}}{41.73 \text{ ml}} = 0.1094 \frac{\text{meq}}{\text{ml}}$$

When experience has been gained with this type of problem, it should be set up so that it may be solved by a single operation, with logarithms or slide rule.

$$N = \frac{932.4 \text{ mg}}{204.2 \text{ mg/meq} \times 41.73 \text{ ml}} = 0.1094 \frac{\text{meq}}{\text{ml}}$$

In many titrations it is preferable to overrun the end point and then to back-titrate until the end point is just reached. This procedure is illustrated in the following problem.

Problem. In the standardization of an acid and base solution, sodium carbonate of assay value 99.5 per cent was used. A sample weighing 0.2200 g was dissolved, 43.50 ml of acid solution was added, and 2.75 ml of base solution was required for back titration. In another titration it was found that 38.63 ml of acid was needed to react with 33.27 ml of the same base. Calculate the normality of the acid and base solutions.

SOLUTION: First we compute the milliequivalents of sodium carbonate, by dividing the weight of pure substance by the milliequivalent weight. The weight of pure substance is the gross weight times the purity factor.

$$\text{Weight sodium carbonate} = 220.0 \text{ mg} \times 0.995 = 218.9 \text{ mg}$$

$$\text{Milliequivalents sodium carbonate} = \frac{218.9 \text{ mg}}{53.00 \text{ mg/meq}} = 4.130 \text{ meq}$$

Next we compute the *net* volume of acid required to react with 4.130 meq of standard. The total volume of acid, 43.50 ml, is that required for titration of the standard plus that needed to react with the base used in the back titration. We note that the base is stronger than the acid. Therefore,

$$2.75 \text{ ml base} \times \frac{38.63 \text{ ml acid}}{33.27 \text{ ml base}} = 3.19 \text{ ml acid}$$

Here the ratio 38.63/33.27 is taken from the values found in a comparison titration of the acid and base.

The net volume of acid is $43.50 - 3.19 = 40.31$ ml. Hence, the normality of the acid is computed by the equation

$$\frac{4.130 \text{ meq}}{40.31 \text{ ml}} = 0.1025 \frac{\text{meq}}{\text{ml}}$$

The normality of the base is obtained from the results of the comparison of the acid and base.

$$\text{Normality of base} = \frac{0.1025 \times 38.63}{33.27} = 0.1190$$

A very common error in this type of computation is to reverse the acid-base ratio in the last step. This can be avoided by a little thought. Since the comparison requires 38.63 ml of acid for 33.27 ml of base, it is obvious that the base is stronger than the acid. Therefore the normality of acid is multiplied by a ratio that is greater than unity, namely 38.63/33.27, rather than the reciprocal of this ratio.

Analysis of Samples by Titration with Standard Solutions. Acid and base samples are analyzed by titration with a standard solution. A weighed portion of sample is dissolved in water and standard acid or base is added to the proper end point. From the volume of reagent used and the weight of sample, the percentage purity of the sample is computed.

Problem. A sample of impure oxalic acid ($H_2C_2O_4 \cdot 2H_2O$, mol. wt. 126.1) which weighs 0.4750 g requires 35.60 ml 0.2000 N sodium hydroxide for its titration. Calculate the percentage of oxalic acid in the sample.

SOLUTION:

Milliequivalents base = milliequivalents oxalic acid = 35.60 ml $\times$ 0.2000 meq/ml = 7.120 meq

Milligrams oxalic acid in sample = milliequivalents $\times$ weight of 1 meq = 7.120 meq $\times$ 126.1/2 mg/meq = 448.9 mg = 0.4489 g

$$\text{Percentage} = \frac{\text{grams oxalic acid} \times 100}{\text{weight of sample}}$$

$$= \frac{0.4489}{0.4750} \times 100 = 94.51\%$$

Problem. A sample of impure calcite ($CaCO_3$, mol. wt. 100.1) which weighs 0.4950 g is dissolved in 50.00 ml of standard acid, and the excess acid is titrated with 5.25 ml standard base; 1.000 ml of acid is equivalent to 0.005300 g sodium carbonate; 1.050 ml acid = 1.000 ml base. Calculate the percentage of calcium carbonate in the sample.

SOLUTION:

$$\text{Normality of acid} = \frac{5.300 \text{ mg/ml}}{53.00 \text{ mg/meq}} = 0.1000 \text{ meq/ml}$$

$$\text{Net volume acid required for titration of sample} = 50.00 - \left(5.25 \times \frac{1.050}{1.000} \right)$$

$$= 44.49 \text{ ml}$$

Milliequivalents acid = Milliequivalents calcium carbonate

$$44.49 \text{ ml} \times 0.1000 \frac{\text{meq}}{\text{ml}} = 4.449 \text{ meq}$$

Weight calcium carbonate = 4.449 meq $\times$ 50.05 mg/meq = 222.7 mg

$$\text{Percentage calcium carbonate in sample} = \frac{222.7 \times 100}{495.0} = 44.99\%$$

Density and Composition of Aqueous Solutions.   All handbooks of chemistry give tables showing the per cent by weight of various solutions as a function of the density.[2]   Such tables provide a ready means of determining the strength of a solution. We merely measure the density, by means of a hydrometer or pycnometer, and by reference to the table determine the composition. We then compute from the percentage by weight the normality and molarity of the solution.

----

[2] Students often confuse density and specific gravity. Density is the weight of 1 ml of a substance. Specific gravity is the relative weight as compared with an equal volume of water. If the specific gravity is given in relation to water at 4°C, then it is also the density, but, if the water temperature is not 4°C, then the specific gravity must be multiplied by the density of water at the temperature of comparison to convert to density.

Problem. Sulfuric acid of density 1.3028 g per milliliter contains 40.00 per cent H_2SO_4 by weight. Compute the normality and molarity of the solution.

SOLUTION: Weight of 1 ml acid solution = 1302.8 mg

$$\text{Weight of acid in 1 ml} = 1302.8 \text{ mg} \times 0.4000 = 521.1 \text{ mg}$$

$$\text{Normality of acid} = \frac{521.1 \text{ mg/me}}{49.04 \text{ mg/meq}} = 10.63_6 \text{meq/ml}$$

Molarity = one-half normality, since 1 mole furnished two equivalent weights. Therefore molarity = 5.31_8.

Titer and Normality. What is the normality of a hydrochloric acid solution having a sodium carbonate titer of 5.00 mg per milliliter?

SOLUTION: According to the data, 1 ml of the acid will react with 5.00 mg sodium carbonate, whose milliequivalent weight is 53.0 mg. Therefore the normality of the acid is

$$5.00 \text{ mg/ml} \div 53.0 \text{ mg/meq} = 0.0943 \text{ meq/ml}$$

Factor Weight Solutions. It is often convenient to prepare solutions of such strength that, when a predetermined weight of sample is used, the volume of the standard solution required for a titration shall represent the percentage of the constituent sought in the analysis.

Problem. What must be the normality of a sodium hydroxide solution if the volume in milliliters used for the titration of a 0.5000-g sample represents the percentage of acetic acid in the sample?

SOLUTION: According to the data 1.00 ml of the base solution is equivalent to 1.00 per cent or 5.00 mg acetic acid. Therefore the normality of the solution is

$$\frac{5.00 \text{ mg/ml}}{60.05 \text{ mg/meq}} = 0.0833 \text{ meq/ml}$$

It is sometimes convenient to weigh out exact weight samples for use with a solution of a certain normality so that a simple relation exists between the volume of solution used and the percentage of the sought constituent.

Problem. What size sample should be taken for analysis so that each milliliter of 0.1200 N HCl used for titration represents 0.500 per cent Na_2O in a sample of soda ash?

SOLUTION:

$$1.000 \text{ ml acid} = 0.1200 \text{ meq} \quad \text{or} \quad 0.1200 \times \frac{61.98}{2} = 3.720 \text{ mg } Na_2O$$

This weight of Na_2O is 0.500 per cent of the sample, or the sample weighs

$$3.720 \text{ mg} \times \frac{100.0}{0.500} = 744 \text{ mg}$$

Computations Based on Molarities of Solutions

In the preceding problems we have stressed the use of normalities for calculations dealing with volumes and concentrations of solutions. This was done because of the convenience of using equivalents, thereby having the simple relation that the equivalents of A and of B are numerically equal. But we must, in using this method, keep in mind that the value of an equivalent may be ambiguous if not precisely defined. For example, H_3PO_4 may be titrated under conditions in which the equivalent weight is a mole and under other conditions in which a mole furnishes two equivalents. There are cogent arguments against using normalities as concentration units, but the usage is so widespread that every chemist must be familiar with it. He should not, however, be blindly bound to this concept but should be able to make his computations on a molar basis just as readily as on the normal basis. A few examples will illustrate computations by molarities. Here we can also introduce computations of oxidation-reduction reactions, since now the computation does not depend on a definition of the equivalent weight.

Problem. An impure sample of $CaCl_2$ is dissolved and titrated with a solution of $AgNO_3$. The reaction is

$$CaCl_2 + 2AgNO_3 = Ca(NO_3)_2 + 2AgCl$$

It is found that 46.35 ml 0.1034 M $AgNO_3$ titrates a 0.2843-g sample of $CaCl_2$. Compute the percentage of $CaCl_2$ in the sample.

SOLUTION: In solving a problem involving molarities of solutions, we must know the relative numbers of moles reacting. We therefore write the balanced equation before making the computation.

The number of moles of $AgNO_3$ is the volume-molarity product. Since the volume is in milliliters, it is convenient to use millimoles. This gives

$$\text{Millimoles } AgNO_3 = 46.35 \text{ ml} \times 0.1034 \text{ ml} = 4.793 \text{ mM}$$

The equation shows that 2 moles $AgNO_3$ reacts with 1 mole $CaCl_2$. Therefore

$$\text{Millimoles } CaCl_2 = \tfrac{1}{2} \text{ mM } AgNO_3 = \tfrac{1}{2} \times 4.793$$

$$= 2.397$$

$$\text{Weight } CaCl_2 = 2.397 \text{ mM} \times \frac{111.0 \text{ mg}}{\text{mM}} = 266.1 \text{ mg}$$

$$\text{Percentage } CaCl_2 = \frac{266.1 \text{ mg}}{284.3 \text{ mg}} \times 100 = 93.6\%$$

Problem. A solution of $KMnO_4$ is standardized by titration of a weighed sample of $Na_2C_2O_4$. The reaction is

$$2KMnO_4 + 5Na_2C_2O_4 + 8H_2SO_4 =$$
$$2MnSO_4 + 10CO_2 + K_2SO_4 + 5Na_2SO_4 + 8H_2O$$

(In writing an equation for use in stoichiometric calculations, it is preferable to use the molecular form rather than the ionic.)

The following data were obtained in the standardization:

$$\text{Volume } KMnO_4 \quad = 40.41 \text{ ml}$$

$$\text{Weight } Na_2C_2O_4 \quad = 0.2538 \text{ g}$$

$$\text{Purity of } Na_2C_2O_4 = 99.60\%$$

Compute the molarity of the $KMnO_4$ solution.

SOLUTION: Weight pure $Na_2C_2O_4 = 0.2538 \text{ g} \times 0.996$

$$= 0.2528 \text{ g} = 252.8 \text{ mg}$$

$$\text{Millimoles } Na_2C_2O_4 = \frac{252.8 \text{ mg}}{134.0 \text{ mg/mM}}$$

$$= 1.887 \text{ mM}$$

$$\text{Millimoles } KMnO_4 = \tfrac{2}{5} \times \text{millimoles } Na_2C_2O_4$$

$$= \tfrac{2}{5} \times 1.887 \text{ mM} = 0.7548 \text{ mM}$$

$$\text{Molarity } KMnO_4 = \frac{0.7548 \text{ mM}}{40.41 \text{ ml}} = 0.01868 \ M$$

Problem. An iron ore sample is dissolved, the iron reduced to Fe^{+2}, and the resulting solution titrated with standard $KMnO_4$ solution. The reaction is

$$2KMnO_4 + 10FeSO_4 + 8H_2SO_4 = K_2SO_4 + 2MnSO_4 + 5Fe_2(SO_4)_3 + 8H_2O$$

From the following data compute the percentage of iron as Fe_2O_3 in the ore.

$$\text{Weight of sample } = 1.097 \text{ g}$$

$$\text{Volume } KMnO_4 \quad = 37.63 \text{ ml}$$

$$\text{Molarity } KMnO_4 = 0.0210 \ M$$

SOLUTION: The equation shows that 2 moles $KMnO_4$ reacts with 10 moles Fe^{+2}, which is equivalent to 5 moles Fe_2O_3. Therefore, moles of $Fe_2O_3 = \tfrac{5}{2}$ moles $KMnO_4$.

$$\text{Millimoles } KMnO_4 = \text{molarity} \times \text{volume}$$

$$= 0.0210 \text{ mM/ml} \times 37.63 \text{ ml}$$

$$= 0.790 \text{ mM}$$

$$\text{Millimoles } Fe_2O_3 = \tfrac{5}{2} \times 0.790 \text{ mM} = 1.975 \text{ mM}$$

$$\text{Milligrams } Fe_2O_3 = 159.7 \text{ mg/mM} \times 1.975 \text{ mM} = 315.4 \text{ mg}$$

$$\text{Percentage } Fe_2O_3 = \frac{315.4 \text{ mg}}{1097 \text{ mg}} \times 100 = 28.75\%$$

Problem. A solution of $Na_2S_2O_3$ is indirectly standardized by use of a standard $KMnO_4$ solution. A measured volume of the $KMnO_4$ solution is added to a solution containing KI in excess and the liberated I_2 is titrated by the $Na_2S_2O_3$ solution. The reactions are

$$2KMnO_4 + 10KI + 8H_2SO_4 = 6K_2SO_4 + MnSO_4 + 5I_2 + 8H_2O$$

$$5I_2 + 10Na_2S_2O_3 = 10NaI + 5Na_2S_4O_6$$

The molarity of the $Na_2S_2O_3$ is computed from the following data.

$$\text{Volume } KMnO_4 = 40.00 \text{ ml}$$
$$\text{Molarity } KMnO_4 = 0.0200 \, M$$
$$\text{Volume } Na_2S_2O_3 = 35.00 \text{ ml}$$

SOLUTION: Since 2 moles $KMnO_4$ liberates 5 moles I_2 and this in turn reacts with 10 moles $Na_2S_2O_3$, it follows that 1 mole $KMnO_4$ is equivalent to 5 moles $Na_2S_2O_3$.

$$\text{Millimoles } KMnO_4 = 40.00 \text{ ml} \times \frac{0.0200 \text{ mM}}{\text{ml}}$$
$$= 0.800 \text{ mM}$$
$$\text{Millimoles } Na_2S_2O_3 = 5 \times 0.800 \text{ mM} = 4.000 \text{ mM}$$
$$\text{Molarity } Na_2S_2O_3 = \frac{4.000 \text{ mM}}{35.00 \text{ ml}} = 0.1143 \text{ mM/ml}$$

Problem. What volume of 0.1500 M $BaCl_2$ solution is needed to precipitate the SO_4^{-2} ion from a 0.500-g sample of $K_2SO_4 \cdot Al_2(SO_4)_3 \cdot 24H_2O$ (mol. wt. = 948.8), assuming that a 10 per cent excess of $BaCl_2$ solution is used? The equation is

$$K_2SO_4 \cdot Al_2(SO_4)_3 + 4BaCl_2 = 4BaSO_4 + 2KCl + 2AlCl_3$$

SOLUTION: The equation shows that each mole of $K_2SO_4 \cdot Al_2(SO_4)_3 \cdot 24H_2O$ requires 4 moles $BaCl_2$.

$$\text{Millimoles } K_2SO_4 \cdot Al_2(SO_4)_3 \cdot 24H_2O = \frac{500 \text{ mg}}{948.8 \text{ mg/mM}}$$
$$= 0.5275 \text{ mM}$$
$$\text{Millimoles } BaCl_2 = 0.5275 \text{ mM} \times 4 = 2.110 \text{ mM}$$
$$\text{Volume } BaCl_2 = \frac{2.110 \text{ mM}}{0.1500 \text{ mM/ml}} = 14.10 \text{ ml}$$

$$\text{Excess } BaCl_2 \text{ solution} = 0.10 \times 14.10 \text{ ml} = 1.41 \text{ ml}$$
$$\text{Total } BaCl_2 \text{ solution used} = 14.10 + 1.41 \text{ ml} = 15.51 \text{ ml}$$

Indirect Analyses

When a sample contains two components that are difficult to separate, it is sometimes possible to make an analysis by indirect means. The amounts of the combined components are determined in two different forms. These data are used to set up a simultaneous equation which is solved to give the amount of each substance in the original sample.

For example, we have a sample known to contain only NaCl and KCl. The weight of combined NaCl and KCl is used to set up one of the necessary equations. The sample is treated with an excess of $AgNO_3$, which gives a precipitate of AgCl. The weight of the dried precipitate is used to set up the second equation needed for a simultaneous solution.

Problem.   A 0.500-g sample of a mixture of NaCl and KCl gives a precipitate of AgCl weighing 0.9815 g. Calculate the percentages of NaCl and KCl in the sample.

SOLUTION: Let x equal the grams of KCl present, and, because of the composition of the sample, $0.5000 - x = $ grams NaCl

$$\text{Grams AgCl furnished by KCl} = x \times \frac{AgCl}{KCl} = x \times 1.923$$

$$\text{Grams AgCl furnished by NaCl} = (0.5000 - x) \times \frac{AgCl}{NaCl} = (0.5000 - x) \times 2.452$$

The equation to be solved is

$$1.923x + 2.452\,(0.5000 - x) = 0.9815$$

$$x = 0.4621$$

$$\text{Percentage KCl} = \frac{0.4621}{0.5000} \times 100 = 92.4\%$$

$$\text{Percentage NaCl} = 100 - 92.4\%$$

$$= 7.6\%$$

Problem.   A 1.000-g sample of an impure mixture containing both bromides and chlorides is dissolved and treated with an excess of $AgNO_3$ solution, which precipitates a mixture of AgBr and AgCl. The dried precipitate weighs 0.8055 g. A separate 1.000-g portion of the sample is dissolved and titrated with 50.00 ml 0.1000 M $AgNO_3$ solution. Compute the percentage of Br^- and Cl^- in the sample.

SOLUTION: Let x represent the millimoles of Cl^- and y the millimoles of Br^- in the sample. Since 5 mM $AgNO_3$ is used to titrate the combined bromide and chloride ions of the sample (50.00 ml $\times$ 0.1000 mM/ml), we have the relation

$$x + y = 5.000$$

The weight of AgCl is $143.3x$ and the weight of AgBr is $187.8y$. The combined weight is 0.8055 g. This gives the equation

$$143.3x + 187.8y = 805.5 \text{ mg}$$

Note that, since x and y are in millimoles, the combined weight is given in milligrams. A simultaneous solution of the equations gives

$$x = 3.000 \text{ mM Cl}^- \text{ ion}$$

$$y = 2.000 \text{ mM Br}^- \text{ ion}$$

$$\text{Percentage Cl}^- = \frac{3.000 \text{ mM} \times 35.46 \text{ mg/mM}}{1000 \text{ mg}} \times 100$$

$$= 10.64\%$$

$$\text{Percentage Br}^- = \frac{2.000 \text{ mM} \times 79.92 \text{ mg/mM}}{1000 \text{ mg}} \times 100$$

$$= 15.98\%$$

Problems

[The problems of this chapter are grouped by types, to facilitate correlation of assignments with the laboratory work of the course.]

A. Calculations Based on Weight Relations of Formulas and Equations

(Use chemical factors whenever possible in solving
the problems of this section)

1. Calculate the chemical factors:

Weighed	Sought	Factor
$(NH_4)_2PtCl_6$	NH_3	_____
$AgCl$	$Mg(ClO_4)_2$	_____
K_2PtCl_6	K_2O	_____
Cu_2S	CuO	_____
K_2PtCl_6	K	_____
Pb_3O_4	Pb	_____
$BaSO_4$	FeS_2	_____
CaO	$H_2C_2O_4 \cdot 2H_2O$	_____

2. Calculate by factors the following:
(a) The tons of limestone $(CaCO_3)$ needed for the preparation of 5 tons of lime (CaO).
(b) The pounds of $BaCl_2$ needed to furnish 1.50 lb of chlorine by an electrolytic process.
(c) The weight of oxalic acid $(H_2C_2O_4 \cdot 2H_2O)$ in milligrams needed to precipitate the calcium in a 0.5000-g sample of phosphate rock of the composition $Ca_3(PO_4)_2$.
(d) The weight of pyrites, FeS_2, in kilograms that must be burned to produce 3.60 kg sulfuric acid.
(e) The milligrams of Mn_3O_4 that can be obtained by strongly heating 0.2500 g MnO_2.   *Ans.* (a) 9 tons; (b) 4.40 lb.

3. A 0.8000-g sample of pyrolusite (MnO_2) yielded 1.282 g $Mn_2P_2O_7$. What is the percentage purity of the ore?   *Ans.* 98.16%.

4. What is the percentage purity of a sample of $Al_2(SO_4)_3$ if a 0.5000-g sample produced a precipitate of $BaSO_4$ weighing 1.000 g?

5. A 2.350-g sample of K_2CrO_4 yielded an ignited precipitate of Cr_2O_3 which weighed 0.9055 g. Calculate the percentage Cr and the percentage purity of the salt.

6. A 0.8050-g sample of brass, on analysis, yields 0.2537 g SnO_2, 1.0752 g $Zn_2P_2O_7$, and 0.1571 g Cu. Calculate the percentage of copper, tin, and zinc in the sample.

7. An impure sample of Na_2SO_4 weighed 1.562 g. A $BaSO_4$ precipitate from this sample weighed 2.496 g. Calculate the percentage of sulfur and the percentage purity of the salt.   *Ans.* 21.97% S; 97.25% Na_2SO_4.

8. What is the percentage of Fe_3O_4 in an ore sample which weighed 0.5350 g? The sample was dissolved by fusion with $KHSO_4$, oxidized by Br_2 water, and the iron was precipitated as hydrous ferric oxide. The ignited ferric oxide weighed 0.5178 g.   *Ans.* 93.55%.

9. A 0.8500-g sample of pyrites, FeS_2, was fused and the sulfur oxidized to SO_4^{-2}. The precipitate of $BaSO_4$ weighed 1.4300 g. What is the percentage of pyrites in the ore?

10. A chromium ore was brought into solution by an oxidizing flux. 0.5000 g of the ore produced a precipitate of $BaCrO_4$ weighing 0.5000 g. What is the percentage of Cr_2O_3 in the ore?   *Ans.* 30.00%.

11. A sample of rock taken for analysis weighs 1.0000 g on an air-dried basis. After drying for one hour at 110°C the sample weighs 0.9437 g. The calcium is precipitated as oxalate but weighed as $CaSO_4$; this weighs 0.5000 g. The magnesium is precipitated as $MgNH_4PO_4$ which ignited to 0.5000 g $Mg_2P_2O_7$. Find the percentages of CaO and MgO on an oven-dried basis and the percentages of these as well as of H_2O on an air-dried basis.

12. In a steel analysis the CO_2 from the combustion of carbon is absorbed in "ascarite." If the gain in weight of the absorbent for a 0.5000-g sample of steel is 15.6 mg, what is the percentage of C in the steel? *Ans.* 0.851%.

13. A 0.5000-g sample of FeO is ignited to Fe_2O_3; what is the percentage gain in weight?   *Ans.* 11.14%.

14. A 0.7650-g sample of clay, containing 20.50 per cent moisture, gave a precipitate of potassium perchlorate weighing 0.3822 g. What is the percentage of K_2O in the clay on a dry basis?   *Ans.* 21.36%.

15. A limestone sample weighing 0.7735 g gave a precipitate of CaC_2O_4 which was ignited to CaO and weighed 0.3135 g.

(*a*) Express the percentages of the following in the dolomite: CaO, $CaCO_3$, $CaSiO_3$, and $Ca_2Si_2O_7$, on the assumption that all the calcium in the rock is present as the specified constituent.

(*b*) What would the ignited precipitate have weighed had it been ignited to: $CaCO_3$, $CaSO_4$?

16. What weight of sample which contains 56.15 per cent Cl was taken for analysis if the precipitated AgCl weighed 0.5017 g?   *Ans.* 0.2210 g.

B. Calculations of Volumetric Analysis

(Normality of solutions)

17. Construct a table showing for HCl, NaOH, H_2SO_4, Na_2CO_3, and BaO:
(*a*) Equivalent weight.
(*b*) Milliequivalent weight.
(*c*) Weight of solute in grams per liter of 0.15 *N* solution.
(*d*) Weight of solute in milligrams per milliliter of 0.15 *N* solution.
(*e*) Weight of solute in milligrams per milliliter of 0.15 *M* solution.

18. Use molecular weights rounded to the nearest whole number, and solve the following problems mentally to establish the fundamental concepts of stoichiometric calculations and the selection of units.

(a) How many moles of NaOH are present in 1700 ml 0.2 M solution?

(b) How many millequivalents of H_2SO_4 are present in 50 ml 0.4 N solution?

(c) How many milligrams of KOH are present in 25 ml 0.04 N solution?

(d) How many grams of HNO_3 are present in 1500 ml 0.2 F solution?

(e) 85.5 mg of $Ba(OH)_2$ is dissolved in 500 ml of solution. What is the normality of the solution?

(f) How many milligrams of Na_2CO_3 will react with 50 ml 0.2 N HCl?

(g) How many milliliters of 0.1 N NaOH solution will react with 30 ml 0.15 N HCl solution?

(h) How many milliliters of 0.025 M $Ba(OH)_2$ solution will react with 50 ml 0.1 N H_2SO_4?

(i) 25 ml 0.2 N KOH reacts with 50 ml H_3PO_4 solution. What is the normality of the acid solution?

(j) If the titration reaction in (i) is: $H_3PO_4 + 2OH^- = HPO_4^{-2} + 2H_2O$, what is the molarity of the phosphoric acid solution?

Ans. (a) 0.34 mole; (b) 20 meq.

19. What is the equivalent weight of:

(a) Potassium tetraoxalate, $KHC_2O_4 \cdot H_2C_2O_4 \cdot 2H_2O$, if complete neutralization is assumed?

(b) Sodium carbonate if it is titrated to the phenolphthalein end point with HCl? The reaction is $Na_2CO_3 + HCl = NaCl + NaHCO_3$.

(c) Sodium tetraborate decahydrate if the titration reaction is

$$Na_2B_4O_7 \cdot 10H_2O + 2H^+ + H_2O \rightarrow 4HBO_2 + 2Na^+?$$

(d) $Na_3PO_4 \cdot 12H_2O$ if it is titrated with phenolphthalein as an indicator to form HPO_4^{-2}?   *Ans.* (a) 84.73; (b) 106.0; (c) 190.71; (d) 380.14.

20. How many grams of solute are in each of the following solutions?

Solution	Concentration	
(a) 2 liters HCl	2.5 N	
(b) 3 liters H_2SO_4	1.5 N	
(c) 150 ml HCl	0.25 N	
(d) 125 ml NaOH	1.5 N	
(e) 1.5 liters Na_2CO_3	2.0 N	*Ans.* (a) 182.3 g.

21. How may each of the following solutions be prepared?

(a) 150 ml 0.1 N NaOH, from solid NaOH.

(b) 100 ml 0.1 N NH_3, from a 1.1 N solution.

(c) 1.5 liters 0.05 N $Ba(OH)_2$, from solid BaO.

(d) 150 ml 0.15 N HCl, from a 12 N solution.

22. How much of substance B is needed to react with the given amount of A?

A	B
(a) 40.00 ml 0.5000 N H_2SO_4	_____ mg MgO
(b) 106.0 mg Na_2CO_3	_____ ml 0.1000 N HCl for complete neutralization
(c) 30.00 ml 0.2000 M H_2SO_4	_____ ml 0.2000 N NaOH

(d) 25.00 ml 0.1500 N KOH

(e) 204.2 mg potassium acid
    phthalate

(f) 4.000 meq H_2SO_4

(g) 0.5000 g 85.0% H_3PO_4
    solution

_____ mg benzoic acid

_____ ml 0.04000 N KOH

_____ ml 0.0250 M $Ba(OH)_2$

_____ ml 0.1200 N NaOH if reaction is: $H_3PO_4 + OH^- = H_2PO_4^- + H_2O$

Ans. (a) 403.2 mg; *(g)* 80.0 mg.

23. What size sample (to the nearest 0.01 g) should be taken of each of the following, in order that 40.00 ml 0.2000 N acid or base shall be needed for the titration of the sample?

(a) $KH(IO_3)_2$.

(b) Na_2CO_3.   *Ans. (b)* 0.42 g.

24. How many milligrams of calcite, $CaCO_3$, are needed to neutralize 40.15 ml of 0.0982 N H_2SO_4? of 0.0982 M H_2SO_4?

25. What are the normality and molarity of a solution prepared by dissolving 8.050 g $Ba(OH)_2 \cdot 8H_2O$ in 1500 ml of solution?

Ans. 0.03401 N; 0.01701 M.

26. To what volume should 50.00 ml 1.250 N H_2SO_4 solution be diluted to prepare a 0.8000 N solution?   *Ans.* 78.13 ml.

27. 35.00 ml H_2SO_4 solution yields a precipitate of $BaSO_4$ which weighs 0.8000 g. What is the normality of the acid?   *Ans.* 0.1959 N.

28. 25.00 ml HCl solution is required to react with 0.1854 g of pure Na_2CO_3. What is the normality of the acid? 32.16 ml of the acid reacts with 29.65 ml NaOH. What is the normality of the NaOH?

Ans. 0.1399 N acid; 0.1517 N base.

29. Find the normality of acid and base solutions from the following:

     Weight Na_2CO_3 (99.5% pure)   0.2027 g

     Volume HCl   45.50 ml

     Volume NaOH (back titration)   3.57 ml

     35.05 ml NaOH = 31.03 ml HCl

30. 45.00 ml 0.1163 N H_2SO_4 was added to 0.4000 g of a sample of soda ash which is 67.72 per cent Na_2CO_3. What volume of 0.1053 N NaOH is required for back titration?   *Ans.* 1.17 ml.

31. A soda-lime sample is 90 per cent NaOH and 10 per cent CaO. If 3.00 g is dissolved in 250 ml, what is the total normality of the solution as a base? How many milliliters of 0.5100 N H_2SO_4 would be required to titrate 100 ml of the solution?   *Ans.* 0.3126 N; 61.29 ml.

32. Compute the normality and molarity of each of the following solutions:

(a) HCl, of density 1.12, containing 24.0 per cent HCl by weight.

(b) HNO_3, of density 1.42, containing 72.0 per cent HNO_3 by weight.

(c) H_2SO_4, of density 1.83, containing 95 per cent H_2SO_4 by weight.

(d) H_2SO_4, of density 1.8022, containing 71.84 per cent SO_3 by weight.

(e) NH_3, of density 0.89, containing 30 per cent NH_3 by weight.

(f) NaOH, of density 1.10, containing 10 per cent NaOH by weight.

Ans. (a) 7.37 N and M.

33. How many milliliters of an HNO_3 solution whose density is 1.389 g per milliliter and which contains 63.00 per cent acid by weight are required to prepare 1500 ml 0.3000 N solution? *Ans.* 32.41 ml.

34. A solution of sulfuric acid has a density of 1.250 g per milliliter and contains 49.00 per cent H_2SO_4 by weight.

(a) How many milliliters of this acid are needed to prepare 250.0 ml 0.2000 N solution?

(b) 250.0 ml of the concentrated acid is diluted to 2.000 liters. What is the normality of the diluted solution?

35. What is the normality of a sulfuric acid solution of density 1.2185 g per milliliter at 20°C? (Refer to a handbook for the necessary data to solve this problem.)

36. Express titers as follows:

(a) 0.1500 N HCl in terms of: NaOH, Na_2O, BaO, K_2CO_3.

(b) 0.1200 N NaOH in terms of: HCl, H_2SO_4, potassium acid phthalate.

37. What is the normality of a H_2SO_4 solution whose $Ba(OH)_2$ titer is 1.714 mg? *Ans.* 0.02000 N.

C. Calculations Based on Molarity of Solutions

38. How many milliliters of NH_3 (density = 0.91 g/ml, 25% NH_3) are needed to precipitate the iron as hydrated ferric oxide from a 0.1262-g sample of $FeCl_2$?

39. How many milliliters of 0.1 M NH_3 solution are needed to precipitate the iron as ferric hydroxide from a 2.000-g sample of ore which is approximately 80 per cent Fe_3O_4? *Ans.* 622 ml.

40. If the iron in a 0.1500-g sample of ore is reduced and subsequently requires 15.03 ml 0.02000 M $KMnO_4$ for titration, what is the purity of the ore expressed as percentage of: Fe, FeO, Fe_2O_3? *Ans.* 55.96% Fe; 72.00% FeO; 80.00% Fe_2O_3.

41. How many milliliters of a solution which contains 1.0 g $AgNO_3$ per 20.0 ml are needed to precipitate the AgCl from a 0.5000-g sample of $BaCl_2 \cdot 2H_2O$ if a 5 per cent excess is to be used?

42. If a 10 per cent excess is to be used, how many milliliters of 0.15 M $AgNO_3$ are needed to precipitate the AgCl from a sample which weighs 0.5000 g and contains 3.00 mM $ZnCl_2$?

43. How much 0.15 M Na_2SO_4 solution is needed to precipitate the $BaSO_4$ from the sample of problem 41 if no excess is to be used?

44. Find the number of milliliters of a solution which contains 0.2000 mole $BaCl_2$ in 500-ml solution that is needed to precipitate SO_4^{-2} as $BaSO_4$ from 3.00 mM $Al_2(SO_4)_3$. *Ans.* 22.50 ml.

45. 25.00 ml $AgNO_3$ reacts with 5.000 mM NaCl; 35.00 ml of the silver solution is required to titrate a sample of $ZnCl_2 \cdot 6H_2O$. How many milligrams of the hydrate are present in the sample? *Ans.* 855.4 mg.

Adding these three, we see that if they all occur in the same direction the total uncertainty may be of the order 1.6/1000. Since another determination might have an uncertainty of the same magnitude but in the opposite direction, we anticipate that the range might be twice this value or about 3.2/1000. Actually we do not expect the uncertainty to be quite as large as the estimated amount, for it is likely that some of the random errors will occur by chance in the opposite direction to others and the total effect be made smaller. But if the range is wider than we have computed, it indicates either that our computation is too optimistic or that the experimenter is careless. Experience shows that many students obtain results showing a smaller range than we have computed, for operations such as that for which the computation is made. We conclude then that a range of about 3/1000 is a reasonable target at which to aim in making determinations of this type.

TABLE 6–4. Student Results for Percentage of Chlorine in Soluble Sample

Each student has a separate sample

	A	B	C	D	E
	51.30	55.69	55.22	54.47	53.65
	51.25	55.76	55.08	54.48	53.62
	51.29	55.76	55.06	54.36	53.75
Mean	51.28	55.74	55.12	54.44	53.67
Absolute range	0.05	0.07	0.16	0.12	0.13
Relative range	1/1000	1.2/1000	2.9/1000	2.2/1000	2.4/1000

The data of Table 6–4 verify our conclusion that a range of 3/1000 is a reasonable figure to expect in student titration results. They give student results for the percentage of chloride ion in a soluble sample, which is determined by titrating the sample with a solution of silver nitrate.

It is noted in the preceding section that the size of the likely error is in absolute measurement units, such as milligrams, milliliters, etc. We can minimize the relative error by proper choice of the size of sample and the volume of solution used. An error of 0.02 ml is 2/1000 if it is made in measuring a 10-ml sample but only 0.5/1000 if the sample is 40 ml. We try therefore to work with as large samples and as large volumes of solutions as is convenient. Our burets are 50-ml capacity. We try to select the size of samples so that a titration volume of 35–45 ml is used. It is not desirable to exceed 50 ml, for this would require refilling the buret and introduce the error of two additional readings.

In this connection we should point out that it may be a waste of time to do overly exact work in one operation of an analysis if some other operation limits the over-all precision of the final result. For example, if we are titrating one-gram samples and the titration itself has a likely range of 2/1000, it is necessary to weigh the sample only to the nearest milligram (show that this is 1/1000). Nothing is gained by weighing the sample to 0.0001 g (express this in parts per 1000). On the other hand, if the sample weighs 0.2 g, it is necessary to weigh to 0.1 or 0.2 mg to obtain the desired precision. The following problem illustrates the evaluation of the precision required in an operation of an analysis.

Problem. Copper is deposited from a sample and weighed as Cu. A semi-microbalance sensitive to 0.01 mg is used to weigh the deposited copper. If the sample contains approximately 5 per cent copper, what weight should be used for the analysis in order that the error in weighing the deposit not exceed 1/1000?

SOLUTION: Since each weighing is reliable to 0.01 mg and two weighings are required to get the weight of the deposit, the total error may be as large as 0.02 mg. Since this is to be 1/1000, we have the relation

$$\frac{0.02 \text{ mg}}{\text{weight Cu}} = \frac{1}{1000}$$

$$\text{Weight Cu} = 0.02 \text{ mg} \times 1000 = 20 \text{ mg}$$

Since the weight of copper is 5 per cent of the weight of the total sample, the sample weight is 400 mg. An ordinary analytical balance, sensitive to 0.1 or 0.2 mg, can be used to weigh out this amount. The semi-microbalance is required only for weighing the final deposit.

Rejection of Extraneous Observations. Most students will find that at some time or other the three results of a triplicate analysis will not agree as closely as is expected from the degree of uncertainty in the various measurements required. It is inevitable that some gross errors will occur at times, and the three results obtained may show such a wide range that a report based on them would constitute more of a "guess" than an analysis. For example, one student, making a chloride analysis in the class with those students whose results have been given in Table 6–4, found the three values 52.63, 52.71, and 53.11 per cent chlorine. The range is 0.48 per cent, which is about 9/1000, far beyond what we expect to find in this determination. The immediate reaction of a student when a result such as this is obtained is to discard the outlying value and report the average of the two results that are in fair agreement. This, however, is not a sound procedure, for we have no assurance that the divergent value is wrong and the other two right. When such a situation arises, as it probably will for everyone, we recommend the following procedure:

1. Check over the data and computations carefully, to see whether an error has been made in the figures. Usually this check will yield nothing, but occasionally we find an error that brings the extraneous value into line with the others.

2. If another portion of sample is available, carry out a fourth determination. If this is within a reasonable range of the two concordant values, apply the Q test, as directed in the next section, and use this statistical method to determine whether the divergent result should be discarded or not. If it is discarded, report the mean of the remaining three results. If it is not sound to discard the divergent result, report the mean of the four trials as the best value.

3. If it is not possible to obtain another portion of sample and make a fourth determination, report the *median* value of the original results rather than the mean. When this is done, the report must point out that the value given is the median of three. For an odd number of determinations the median is the middle value. For an even number the median is the average of the two results nearest the middle, i.e., the average of the second and third values when N is four. The median has the advantage of not emphasizing the wide spread in range caused by a single discordant result, but certainly it is not a highly reliable value.

We have spoken of results in which there are two concordant values and a third that lies at some distance from these two. It occasionally happens that we obtain three results with a wide range, no two of them being in close agreement. This constitutes no problem; the analysis is worthless and should be repeated with a new sample.

The Q Test for Extraneous Results. Dean and Dixon[1] have described a simple statistical test that can be applied to doubtful observations. It is made with the aid of a statistical table of Q values, some of which are given in the last column of Table 6–3. The rejection quotient Q (0.90) is the "range quotient" above which there is only 10 per cent probability that the discarded value should be retained. The range quotient is obtained by dividing the distance of a doubtful observation from its nearest neighbor by the total range of the series. The quotient obtained is compared with the Q (0.90) value of Table 6–3 for the given number of samples. If the observed Q is as large as the one of the table, the value is discarded and the mean of the remaining values is reported.

To illustrate use of the Q test we shall apply it to the chloride results mentioned above. On the first set of three the values found are 52.63, 52.71, and 53.11 per cent. Another sample is run, giving a value of 52.59. For the four samples the range is $53.11 - 52.59 = 0.52$. The distance of the suspected value (53.11) from its nearest neighbor is $53.11 - 52.71 = 0.40$. The quotient is $0.40/0.52 = 0.77$. We find in Table 6–3 that the rejection quotient for four samples is 0.76. Since the observed quotient is approximately equal to this, we reject the value 53.11 and report the average of the remaining three, 52.64 per cent. The range for these three is 0.12 per cent, which now compares favorably with the ranges given by other students for three samples.

Students are cautioned not to have as much confidence in values from which an extraneous result has been dropped as in a series that had no discordant values. In application of the Q test there is fair probability that the value dropped is discordant because of a careless error not present in other values, but this is not certain. It is always possible that some unknown factor has brought about the closer agreement among the three or more values retained and that the dropped value is actually nearer the true one than are the others.

Significant Figures in the Final Result. We have previously stated that the significant figures of a number are the digits whose values are known with reasonable certainty. All of us tend to lose sight of the degree of certainty in the original data when we use these data to obtain derived numbers (in multiplications, divisions, roots, powers, etc.). Use of too many significant figures is a widespread error that pervades even textbooks and current scientific publications. The quantitative analysis course is a good place to practice proper use of significant figures and to develop good habits along these lines.

It is quite proper to use one significant figure beyond the last digit we are absolutely certain of if the precision of the measurement goes beyond that indicated by the certain figure. It is never proper, however, to use two figures beyond the last one that is absolutely certain. The range or precision of data usually determine the proper number of significant figures to use. Consider, for example, the copper analyses reported by five students in Table 6–2. Students C and D have analyses with ranges respectively of 0.03 and 0.04 per cent. Their values are properly reported to the second decimal place, since the only deviations occur in this place. In the reports of students A, B, and E, it is incorrect to use two decimal places. Their ranges are respectively 0.30, 0.26, and 0.40 per cent, and the values of the second decimal place are not known with any certainty. The mean values for these analyses are properly reported to the first decimal place only, and even the first decimal place digits are not known with certainty.

It is helpful to keep the proper number of significant figures in mind at each stage of a computation. This not only insures that the answer contains the proper number but also saves time. The following simple rules should be followed.

1. In *addition* and *subtraction* retain in the answer only as many figures to the right of the decimal point as there are in the number which contains the fewest of such digits.

In the addition of

$$203.1$$
$$7.21$$
$$0.3734$$

the numbers should be rounded to

$$203.1$$
$$7.2$$
$$0.4$$

and the result written as 210.7. Note how meaningless it would be to write the result as 210.6834 when the first number in the sum might just as well have been 203.0 or 203.3 if it had been remeasured.

2. In *multiplication* and *division* retain in the answer sufficient digits to express the precision of the least precise result, i.e., the result with the fewest number of significant figures.

In the multiplication of

$$7654.1 \times 0.002234 \times 0.131$$

the least precise factor is 0.131, which has a precision of 1/131 or 7/1000 if the last digit is precise. We therefore round the other numbers to this precision, which gives

$$7.65 \times 10^3 \times 2.23 \times 10^{-3} \times 1.31 \times 10^1 = 2.24$$

The answer, 2.24, implies a precision of 1/224 or 5/1000.

Computation Tools. A slide rule is satisfactory for computation only when a precision of not greater than 2–3 parts per thousand is required. For higher precision longhand arithmetic, electrical computers, or logarithm tables have to be used. For a precision of 1/1000, that commonly sought in analytical work, a four-place logarithm table would be satisfactory, but a five-place table, such as is included in this book, is handier because interpolation is unnecessary.

In the process of computation no factor or intermediate result should be rounded off to a precision less than that to be expressed in the final answer. It is often desirable to carry an extra significant figure in each step and to do the final rounding in the answer. At times the final answer may contain a doubtful figure which may be so indicated by writing it as a subfigure. Thus 3.0_4 indicates that the digit 4 is doubtful, but that the precision is greater than 1/30, as would be indicated if the digit 4 were omitted.

Questions

1. Would you consider all the zeros significant figures in each of the following statements? Rewrite each with the proper number of significant figures.

The population of a city is 100,000.
A factory sold for $100,000.
The average life of a car is 100,000 miles.
The age of a geological specimen is 40,000,000 years.

2. Which of the following can be concluded from the fact that three results of an analysis check closely: (a) random errors have been minimized, (b) the average result is accurate, (c) the reagents used were absolutely pure, (d) systematic errors have been minimized, (e) a blank determination is unnecessary.

3. Which of the following errors may have occurred in an analysis which has good precision but poor accuracy:

Mechanical loss of solution during titration.
Use of uncalibrated weights.
An error in recording the weight of one sample.
Use of the wrong indicator for the titration.

4. Criticize and correct the following statements:

(a) "A series of results cannot have high precision without also having high accuracy."

(b) "To express a precision in a measurement to one part in a thousand always requires the use of four significant figures."

5. List the methods used to minimize systematic errors.

6. Why is it desirable in this course, and with the equipment provided, to use samples which weigh at least 0.2000 g and to use volumes of at least 40.00 ml in titration?

Problems

1. How many significant figures are in each of the following numbers: (a) 410, (b) 410.0, (c) 0.041, (d) 0.04100, (e) 0.41×10^{-8}?

2. Express the results of the following calculations to the proper number of significant figures:

(a) $1.060 + 0.05974 - 0.0013$.

(b) $14.356 + 0.015 + 12.6$.

(c) $101.65 + 1.283 - (3 \times 10^{-5})$.

(d) $2.5 \times 101{,}562$.

(e) $15.75 \times (300 \times 10^4) \times 0.000035$.

(f) $\dfrac{29.74 \times 400}{0.080}$.

(g) $\dfrac{12.3456 \times 2.0}{0.002}$.

Ans. (a) 1.119; (e) 1.66 × 10³.

3. The factor for the conversion of U. S. gallons into liters is 3.78533.

(a) Express 0.55 gallons as liters.

(b) Express 11.99 ml as gallons.   Ans. (a) 2.08 liters.

4. If the convention of significant figures is properly used, what is the smallest graduation interval on a thermometer for which a reading of 96.54°C is recorded?

5. Express to the proper number of significant figures:

(a) The density of an object whose volume is 13.25 ml and which weighs 62.5013 g.

(b) The percentage of silica in a mineral which weighs 1.0065 g and which yields a precipitate of SiO_2 which weighs 0.0142 g.

(c) The result of an analysis which calculates to 0.00296457 and which involves in the calculation a weight of sample of 0.9572 ± 0.0002 g and a volume of standard solution of 56.13 ± 0.08 ml.   Ans. (b) 1.41%.

6. By use of a table of atomic weights express the molecular weights of the following compounds to the proper number of significant figures to agree with the least precisely known atomic weight involved: (a) HCl, (b) Au_2O_3, (c) $Eu(NO_3)_3$.

7. Express the following measurements with the proper number of significant figures to indicate the highest accuracy obtainable with the instrument used:

(a) 5 g measured on a laboratory trip scale.

(b) 30 g measured on a balance sensitive to 10 mg.

(c) 10 mg weighed on an analytical balance.

(d) 10 mg weighed on a microbalance sensitive to 10^{-6} g.

(e) 9 ml measured in a 100-ml graduated cylinder.

(f) 9 ml measured in a 10-ml graduated cylinder.

(g) 9 ml measured from a 50-ml buret.

Ans. (a) 5.0 g; (d) 0.010000 g; (f) 9.0 ml.

8. For the results 0.0519, 0.0521, 0.0522, and 0.0520 find:

(a) The average deviation in parts per thousand.

(b) The standard deviation in parts per thousand.

(c) The range in parts per thousand.

(d) The 95 per cent confidence interval.

Ans. (a) 1.9/1000; (b) 2.5/1000; (c) 6/1000; (d) ±0.0002.

9. An analyst obtained the following results for the standardization of HCl solution: 0.1135, 0.1134, 0.1136, 0.1144, 0.1132. May any one of these be discarded?   Ans. Discard 0.1144.

10. The following results were obtained for the percentage of iron in a sample of pure FeO: 76.95, 77.02, 76.90, 77.25.

(a) Compute the standard deviation in parts per thousand.

(*b*) Express the range in parts per thousand.

(*c*) What can be said about the presence of systematic errors in these results?

11. Given the results 24.65, 24.78, and 24.55:

(*a*) Express the range in parts per thousand.

(*b*) Express the average deviation in parts per thousand.

(*c*) What is the 95 per cent confidence interval?

12. May any of the results 40.02, 40.12, 40.16, and 40.18 be discarded?

13. By how many parts per thousand do the following pairs of results deviate from each other:

(*a*) 9.5 and 9.7.

(*b*) 0.507 and 0.512.

(*c*) 40.06 and 40.10.

(*d*) 0.09719 and 0.09717. *Ans.* (*a*) 20/1000.

14. The weight of a 10-g sample is known to 1/100,000. How carefully was the sample weighed?

15. If a microburet can be read to the nearest 0.001 ml, what total volume should be withdrawn from the buret so that the volume will be known to a precision of 2/1000? *Ans.* 1 ml.

16. How carefully should a 5-g sample be weighed to achieve a precision of: (*a*) 1 per cent, (*b*) 0.5 per cent, (*c*) 0.1 per cent?

17. A sample of an alloy is to be analyzed for silver by electrodeposition. If the sample is approximately 20 per cent Ag, what size sample should be taken for analysis so that the error in determining the weight of the deposited silver does not exceed 1/1000, provided a balance sensitive to 0.1 mg is used? *Ans.* 1 g.

18. Copper is to be determined by electrodeposition of the metal from a solution which is approximately 0.5 per cent Cu. If the error in determining the weight of the deposited copper is not to exceed 1/100 and a balance which is sensitive to 0.05 mg is to be used, what size sample should be taken for analysis? How precisely should the sample be weighed?

CHAPTER 7

Acid-Base Equilibria
and pH of Solutions

An acid is a substance that furnishes protons or H^+ ions and a base a substance that accepts protons. In classical terminology the acid HA dissociates in aqueous solution according to the equation

$$HA = H^+ + A^-$$

and a base MOH dissociates to give $M^+ + OH^-$ ions. The reaction of HA with MOH is

$$HA + MOH = MA + H_2O$$

This reaction goes more or less to completion because the H_2O molecule is only slightly dissociated. To understand the limitations of acid-base titrations and the selection of proper indicators for such titrations, the student must understand the equilibrium relations involved in neutralization reactions.

Equilibrium Constant

A chemical reaction is at equilibrium when the rates of the forward and the reverse reactions are equal. When this condition is attained, the concentrations of the reacting components are related by the equilibrium constant equation. Thus, for the reaction

$$mA + nB = pD + qE$$

we have the relation

$$\frac{[D]^p [E]^q}{[A]^m [B]^n} = K_e$$

in which the square brackets indicate the molar concentrations of the reacting species, and each molar concentration is raised to a power that is the coefficient of that substance in the chemical equation. The constant K_e is known as the equilibrium constant. Conventionally the products of the reaction are written in the numerator of the expression.

The equilibrium constant is experimentally determined for every reaction by measuring the concentrations of the reacting species after a condition of equilibrium has been attained. The constant is valid only at the temperature at which the determination is made; a change in temperature will change the equilibrium concentrations, thereby changing the numerical value of K_e.

Ionization of Acids

In our study of neutralization reactions we are concerned with the extent to which acids are dissociated in solution. This dissociation is due to attraction of the water molecules for the ions formed. For a proton from an acid, the attraction of water is explained by the electronic structure of the water molecule, which can add a proton[1] at a free electron pair

$$H^+ + H:\overset{..}{\underset{..}{O}}:H = H:\overset{..}{\underset{..}{O}}:H^+$$
$$H$$

to form the hydronium ion, H_3O^+.

In the general treatment of acids and bases the classical definitions of the older Arrhenius theory have been broadened by Brönsted and others, to better explain the role of the solvent in ionization. In Brönsted terminology an acid in aqueous solution is a substance that furnishes the H_3O^+ ion and a base is a substance that can accept or unite with a proton. Thus, in the reaction that occurs when HCN is

---

[1] A proton in aqueous solution holds more than one molecule of water, but one of these is thought to be held more tightly than the others and consequently it is customary to write H_3O^+.

dissolved in water, we have the equation

$$H:C:::N: + H:\overset{\cdot\cdot}{\underset{\cdot\cdot}{O}}:H = H:\overset{\cdot\cdot}{\underset{\underset{H}{\cdot\cdot}}{O}}:H^+ + :C:::N:^-$$

or, as the equation is usually written, without indicating the electronic structure,

$$\underset{\text{acid 1}}{HCN} + \underset{\text{base 2}}{H_2O} = \underset{\text{acid 2}}{H_3O^+} + \underset{\text{base 1}}{CN^-}$$

In conformity with the newer definitions, both HCN and H_3O^+ can furnish protons and are acids; similarly H_2O and CN^- can accept a proton and are called bases. The equilibrium conditions for a given reaction are determined by the relative strengths with which the two bases in a system hold the proton. In the HCN example H_2O is a much weaker base than CN^- ion. The HCN molecule is dissociated only slightly to H_3O^+ and CN^-.

Electrolytes are described as strong or weak, depending on the degree to which they dissociate in aqueous solutions. The strongest of the strong electrolytes dissociate almost completely; i.e., all or nearly all the molecules break up into ions.[2] Weak electrolytes are those that dissociate only to a small extent. An example is acetic acid, $HC_2H_3O_2$, which for convenience we abbreviate as HOAc. In aqueous solution we have the equilibrium

$$\underset{\text{acid 1}}{HOAc} + \underset{\text{base 2}}{H_2O} = \underset{\text{acid 2}}{H_3O^+} + \underset{\text{base 1}}{OAc^-}$$

In $0.1\ M$ solution HOAc is slightly more than 1 per cent dissociated. This means that OAc^- ion has more affinity for a proton than H_2O does. Most salts, the common inorganic acids, and the inorganic bases are strong electrolytes. The organic acids, ammonia solutions, and substituted ammonia compounds are weak electrolytes. A list of the more common weak electrolytes is given in Tables A–1 and A–2 of the Appendix.

Ionic Equations. In writing equations for chemical reactions in solution, it is helpful to show each substance in the state representing the majority of the molecules of that species. Thus, if acetic acid is

[2] It is not possible to make a hard and fast classification of electrolytes as strong or weak, since there are many that fall into an intermediate region. For example, we think of HNO_3 as a strong electrolyte, but in $10\ M$ solution the degree of dissociation is only about 50 per cent. Sulfuric acid, commonly thought of as a strong electrolyte, dissociates almost completely into H_3O^+ and HSO_4^- ions in dilute solution, but the HSO_4^- ion is only about 12 per cent dissociated into H_3O^+ and SO_4^{-2} ions in $0.1\ M\ H_2SO_4$.

a reactant, it is expressed as a molecule, HOAc, since almost 99 per cent of the total molecules present are undissociated. When NaOH is a reactant, it is shown as the ions Na^+ and OH^- because it is almost completely dissociated. Thus the equation for the neutralization of acetic acid by sodium hydroxide is written as

$$HOAc + Na^+ + OH^- = Na^+ + OAc^- + H_2O$$

The advantage of writing equations in the ionic form is that in metathesis reactions (double decompositions) the reaction goes in the direction to remove ions from the solution. In the previous reaction, H_2O furnishes fewer protons than acetic acid—the reaction therefore goes to the right—or HOAc is converted almost completely into OAc^- ion and water.

Similarly, in precipitation reactions the formation of the precipitate removes ions from solution. Thus, if $AgNO_3$ is treated with NaCl solution, the equation is

$$Ag^+ + NO_3^- + Na^+ + Cl^- = AgCl \downarrow + Na^+ + NO_3^-$$

The reaction goes to the right because precipitation of AgCl removes Ag^+ and Cl^- ions from the solution. When no ions of a solution can combine with one another to form a precipitate or an undissociated substance, then no reaction occurs. Thus, if we mix solutions of $NaNO_3$ and KCl, the equation shows that no reaction is to be expected

$$Na^+ + NO_3^- + K^+ + Cl^- = \text{no reaction}$$

for neither of the possible products, NaCl or KNO_3, would remove any of the ions from the solution.

Ionization Constants for Weak Electrolytes. When a weak electrolyte is put into an aqueous solution, an equilibrium is established. For HOAc the reaction is

$$HOAc + H_2O = H_3O^+ + OAc^-$$

The mathematical equation for this equilibrium is

$$\frac{[H_3O^+][OAc^-]}{[HOAc][H_2O]} = K_e$$

The concentration of H_2O is constant, and we therefore combine the two constants to give

$$\frac{[H_3O^+][OAc^-]}{[HOAc]} = K_i$$

The constant K_i is known as the ionization constant of the weak electrolyte (the ionization constant of a weak acid may also be denoted by K_a or for a weak base by K_b). Numerical values of K_i have been determined for most of the known weak electrolytes. Those for the more common weak acids and bases are shown in Tables A–1 and A–2, Appendix.

Determination of K_i. The numerical value of ionization constants is computed from experimental measurements of the concentrations of the reactants present at equilibrium. For example, it is found experimentally that the concentration of the H_3O^+ ion is 0.00134 M in a solution prepared by dissolving 0.1 mole acetic acid and diluting to 1 liter. Since each molecule of acid that dissociates gives a hydronium ion and an acetate ion, we have the following concentrations in the solution.

$$[H_3O^+] = 0.00134 \text{ moles/liter}$$

$$[OAc^-] = 0.00134 \text{ moles/liter}$$

$$[HOAc] = 0.10000 - 0.00134 = 0.09866 \text{ moles/liter}$$

Substituting these numerical values into the equilibrium constant equation gives

$$\frac{(0.00134)(0.00134)}{(0.09866)} = K_i = 0.000018 = 1.8 \times 10^{-5}$$

If the measurements are repeated for other initial concentrations of acetic acid, say for 0.05 M, 0.01 M, etc., we find that essentially the same value of K_i is obtained for each initial concentration of acid. Thus, the value of K_i is a measure of the strength or tendency to dissociate.

Activity and Activity Coefficient. When very precise measurements of ionization constants are made, it is found that the values of K_i for different concentrations are not absolutely identical. The previously given theory is overly simplified. It is tacitly assumed in the statement of the Law of Chemical Equilibrium that the individual ions or molecules of a solution are not affected by other ions present. This is not rigorously true, except in very dilute solutions. Ions of one type of charge are attracted by ions of opposite charge, and because of this attraction the effective concentrations of ions are somewhat different from the actual concentrations. The Law of Chemical Equilibrium is strictly obeyed only when the effective concentrations are substituted for the actual ones. These effective concentrations are known as the *activities* of solutes. The relation of activity to concentration is given by the equation

$$a = \gamma C$$

where a is the activity, C the molar concentration, and γ a number known as the activity coefficient. Usually γ has a numerical value of less than 1.00, and as solutions are made increasingly dilute it approaches 1.00.

In the work of this course we shall assume that in all cases $\gamma = 1.00$, or that the activity is equal to the molar concentration. Although not strictly rigorous, the assumption is valid enough for our purposes. We should, however, keep in mind that the assumption does introduce a degree of uncertainty into our computations.

Ionization of Strong Electrolytes. The behavior of strong electrolytes in solution is best interpreted on the assumption that most of them have a high degree of dissociation when in dilute aqueous solutions. That is, if one-tenth of a mole of $NaCl$ is dissolved, we have in solution not $NaCl$ molecules but a tenth of a mole each of Na^+ and Cl^- ion. Since there is no equilibrium between ions and undissociated molecules, an equilibrium constant is not applicable to this system. We apply our equilibrium constants only to systems in which weak electrolytes are involved.

pH and Hydrogen Ion Concentration. Pure water dissociates slightly, forming equivalent amounts of hydronium and hydroxide ions, according to the equation

$$H_2O + H_2O = H_3O^+ + OH^-$$

The H^+ ion formed by dissociation of one H_2O molecule unites with another H_2O molecule to form an hydronium ion. At room temperature the concentrations of H_3O^+ and OH^- ions in pure water are each 1×10^{-7} molar.

The equilibrium expression for dissociation of water is

$$\frac{[H_3O^+][OH^-]}{[H_2O]^2} = K_e$$

Since $[H_2O]$ is constant, we have

$$[H_3O^+][OH^-] = K_w$$

The constant K_w is evaluated by substituting known numerical values for the concentrations of the ions and solving. This gives

$$K_w = (1 \times 10^{-7})(1 \times 10^{-7}) = 1 \times 10^{-14}$$

This is the value for the constant at 25°C. It rises rapidly with increase in temperature, and at 100° the value is about 60 times the 25° value.

If the concentration of one of the ions is increased, by addition of acid or base, the concentration of the other ion is correspondingly decreased. It is the product of the concentrations of the two ions that remains constant. Thus, in a 0.1 M HCl solution the concentration of the H_3O^+ ion is 0.1 M and the concentration of the OH^- ion is 10^{-13} M. Similarly, in 0.1 M NaOH the concentration of the hydroxide ion is 10^{-1} M and of the hydronium ion 10^{-13} M.

We are, in titrations, concerned with hydronium ion concentrations ranging from those of strongly acid solutions to those of base solutions, or a range from about 10^{-1} M to 10^{-13} M. In titration curves we need plots of some function of the hydronium ion. It is awkward to construct a plot covering such wide ranges as 10^{-1} to 10^{-13}, but this can be avoided by using the logarithms of the numbers instead of the numbers themselves. However, the logs of these numbers are negative, and the obvious simple solution is to use a related function, known as pH, to express hydronium ion concentrations. The function pH is defined by the equation

$$pH = -\log_{10}[H_3O^+]$$

The relation between pH and the concentrations of hydronium and hydroxide ions is shown in Table 7–1, which also gives corresponding

TABLE 7–I.   Relation of pH and Hydronium Ion Concentration

H_3O^+	10^{-1}	10^{-3}	10^{-5}	10^{-7}	10^{-9}	10^{-11}	10^{-13}
OH^-	10^{-13}	10^{-11}	10^{-9}	10^{-7}	10^{-5}	10^{-3}	10^{-1}
pH	1	3	5	7	9	11	13
pOH	13	11	9	7	5	3	1

values for the concentration of OH^- ion and pOH, defined by

$$pOH = -\log_{10}[OH^-]$$

It will be noted that the sum of pH + pOH is always 14 when the solution is at room temperature.

The conversion of hydronium ion concentration to pH and the reverse process are illustrated in the following problems.

Problem.   What is the pH of a solution in which the hydronium ion concentration is 0.0002 M?

SOLUTION: Express the concentration in exponential form, with the numerical portion a number between 1 and 10. This is to facilitate taking the log of the number, since the characteristic for the log of a number between 1 and 10 is zero.

$$[H_3O^+] = 0.0002 = 2 \times 10^{-4}$$

Look up the log of the number, and add to this the log of the power of 10 used to express the concentration.

$$\log (2 \times 10^{-4}) = \log 2 + \log 10^{-4}$$
$$= 0.30 - 4 = -3.70$$

Change the sign of the log to obtain the pH

$$pH = -\log [H_3O^+] = -(-3.70) = 3.70$$

In such calculations the logs are usually given to only two decimal places, since pH measurements are usually not more precise than this.

Problem. The pH of a solution is 5.40. What is the concentration of hydronium ion?

SOLUTION: Since pH $= -\log [H_3O^+]$, we have the inverse relation that $[H_3O^+] = 10^{-pH}$.

$$[H_3O^+] = 10^{-5.40}$$

To evaluate the number we must express the decimal portion of the exponent as a positive number. We rewrite in the form,

$$10^{-5.40} = 10^{+0.60} \times 10^{-6}$$

Look up the antilog of 0.60 and multiply by 10^{-6}.

$$(\text{Antilog } 0.60) \times 10^{-6} = 4.0 \times 10^{-6} = [H_3O^+]$$

pH of Aqueous Solutions

In considering the feasibility of titrations and in selecting the proper indicator for a given titration, we often need to compute the pH of solutions under various conditions. The following sections summarize the types of solutions we deal with and the methods used to compute the pH of each.

1. *Strong Acid.* Since strong acids are highly dissociated, the concentration of hydronium ion is considered to be approximately the molarity of the strong acid in solution.

Problem. What is the pH of the resulting solution when 50 ml 0.1 M NaOH has been added to 75 ml 0.1 M HCl?

SOLUTION: Each mole of NaOH added neutralizes a mole of HCl. The number of moles HCl remaining after the base is added is the original number minus the number of moles of base added. Since the volumes are given in milliliters, it is convenient to work with millimoles.

$$\text{Millimoles HCl} = 75 \text{ ml} \times 0.1 \text{ mM/ml} = 7.5 \text{ mM}$$
$$\text{Millimoles NaOH} = 50 \text{ ml} \times 0.1 \text{ mM/ml} = 5.0 \text{ mM}$$
$$\text{Millimoles HCl remaining unneutralized} = 7.5 - 5.0$$
$$= 2.5 \text{ mM}$$

The total volume is 75 ml + 50 ml = 125 ml. The concentration of HCl, which is the concentration of hydronium ion, is then 2.5 mM/125 ml = 0.02 M.

$$pH = -\log 0.02 = -\log (2 \times 10^{-2}) = 1.70$$

2. *Strong Base.* The value of pOH is computed from the molarity of the base solution, and this value is subtracted from 14 to obtain the pH.

Problem. What is the pH of a solution obtained by adding 85 ml 0.1 M NaOH to 75 ml 0.1 M HCl?

SOLUTION: The total amount of NaOH is 8.5 mM; since the solution originally contained 7.5 mM HCl, there is an excess of 1.0 mM NaOH in a volume of 160 ml, a concentration of 6.25×10^{-3} M. The pOH is 2.21 and pH is 11.79.

3. *Salt of Strong Acid and Strong Base.*[3] Since neither ion can combine with an ion of water to form a weak electrolyte, the pH is 7.0, that of water, provided CO_2 is removed from the solution. When the solution is in equilibrium with air of normal CO_2 content, the pH is approximately 5.7, due to the reaction

$$CO_2 + 2H_2O = H_3O^+ + HCO_3^-$$

4. *Weak Acid (in absence of salts of the acid).* The hydronium ion comes from dissociation of the acid. In general the amount of this ion is so large in comparison to that furnished by dissociation of water that the latter can be neglected.

Problem. What is the pH of a 0.05 M solution of acetic acid?

SOLUTION: The acid dissociates according to the equation

$$HOAc + H_2O = H_3O^+ + OAc^-$$

for which we have the ionization constant

$$\frac{[H_3O^+][OAc^-]}{[HOAc]} = 1.8 \times 10^{-5}$$

Let x be the concentration of H_3O^+. This is also the concentration of OAc^-. The concentration of undissociated HOAc is $0.05 - x$. In future calculations we shall indicate these values by writing them above the respective symbols in the equation,

$$\overset{0.05-x}{HOAc} + H_2O = \overset{x}{H_3O^+} + \overset{x}{OH^-}$$

Substituting the concentrations into the ionization constant equation gives

$$\frac{x^2}{0.05 - x} = 1.8 \times 10^{-5}$$

[3] According to Brönsted's treatment the fundamental process involved in acid-base reactions is

$$acid\ 1 + base\ 2 = base\ 1 + acid\ 2$$

and the term "salt" has no meaning. In speaking of NaCl as the salt of a strong acid and strong base, we are reverting to the earlier terminology wherein NaOH, rather than its OH^- ion, is called a base.

An exact solution can be obtained by use of the quadratic formula, but usually it is sufficiently accurate to make an approximate solution by assuming that $0.05 - x \cong 0.05$ (that x is negligible in comparison with 0.05) and solving.

$$\frac{x^2}{0.05} = 1.8 \times 10^{-5}$$

$$x^2 = 9.0 \times 10^{-7}$$

To find the pH we may solve for x, then $-\log x$. It is more convenient to take $\log x^2$ and divide by 2 to obtain $\log x$.

$$\log x^2 = \log (9.0 \times 10^{-7}) = 0.95 - 7 = -6.05$$

$$\log x = -3.03$$

$$\mathrm{pH} = -\log x = 3.03$$

The approximate solution we have used may be expressed as a general equation

$$[\mathrm{H_3O^+}] = \sqrt{CK_a}$$

where C is the concentration of the weak acid present and K_a is its dissociation constant.

5. *Weak Base (in absence of its salts).* The concentration of $\mathrm{OH^-}$ given by dissociation of the base is computed in the same way as the $\mathrm{H_3O^+}$ computation for a weak acid.

Problem. What is the pH of a 0.1 M $\mathrm{NH_3}$ solution?

SOLUTION: Write the equation for dissociation[4] and indicate the respective concentrations of the various species.

$$\overset{0.1-x}{\mathrm{NH_3}} + \mathrm{H_2O} = \overset{x}{\mathrm{NH_4^+}} + \overset{x}{\mathrm{OH^-}}$$

Substitute the respective concentrations into the equilibrium constant equation and solve for pOH and pH. Use the approximation that $0.1 - x \sim 0.1$.

$$\frac{x^2}{0.1} = 1.8 \times 10^{-5}$$

$$x^2 = 1.8 \times 10^{-6}$$

$$\log x^2 = 0.26 - 6 = -5.74$$

$$\log x = -2.87$$

$$\mathrm{pOH} = 2.87$$

$$\mathrm{pH} = 14 - 2.87 = 11.13$$

---

[4] Strictly speaking, the reaction of $\mathrm{NH_3}$ with $\mathrm{H_2O}$ is not a dissociation, but rather an acid-base reaction as expressed by the equation given, where $\mathrm{NH_3}$ is a base and $\mathrm{H_2O}$ an acid.

The generalized approximate equation for the dissociation of a weak base is

$$[OH^-] = \sqrt{CK_b}$$

6. *Weak Acid Plus Its Salt.* If a salt that contains the same anion is added to a solution of a weak acid, the effect is to decrease the concentration of hydronium ion. The salt, completely ionized, increases the concentration of the anion, thereby displacing the chemical equilibrium.

In the titration of a weak acid by a strong base, each mole of base added gives a mole of salt. The effect of this salt must be considered in computing the pH of the solution.

Problem. What is the pH of an acetic acid solution when 30 ml 0.15 M NaOH have been added to 50 ml 0.1 M HOAc?

SOLUTION: First compute the amounts of the various species present.

Millimoles HOAc originally present $= 50 \times 0.1 = 5.0$ mM

Millimoles salt $=$ millimoles base added $= 30 \times 0.15 = 4.5$ mM

Millimoles HOAc remaining unneutralized $= 0.5$ mM

The $[H_3O^+]$ of the solution is due to dissociation of the remaining HOAc.

$$\overset{\frac{0.5}{80} - x}{} \qquad \overset{x}{} \qquad \overset{\frac{4.5}{80} + x}{}$$
$$HOAc + H_2O = H_3O^+ + OAc^-$$

The concentrations of each species are shown above the symbols. Substituting these concentration values into the ionization constant equation gives

$$\frac{x \left(\dfrac{4.5}{80} + x \right)}{\dfrac{0.5}{80} - x} = 1.8 \times 10^{-5}$$

(Note that we have not divided the number of millimoles by the volume, to evaluate the various concentrations. As we shall see later, it is convenient to leave the concentrations in this form, since the volume terms cancel from the equation.)

Neglecting x when added to or subtracted from larger numbers, we have the approximate equation

$$\frac{(x) \left(\dfrac{4.5}{80} \right)}{\dfrac{0.5}{80}} = 1.8 \times 10^{-5}$$

The solution volume cancels from the numerator and denominator, giving

$$\frac{4.5x}{0.5} = 1.8 \times 10^{-5}$$

Solving, we have

$$9x = 1.8 \times 10^{-5}$$

$$x = 0.2 \times 10^{-5} = 2 \times 10^{-6}$$

$$pH = -(0.30 - 6) = 5.70$$

The general equation for the pH of a solution containing both a weak acid and its salt is

$$\frac{[H_3O^+]n_s}{n_a} = K_a \qquad \text{or} \qquad [H_3O^+] = \frac{n_a K_a}{n_s}$$

where n_s is the number of millimoles of salt present and n_a is the number of millimoles of unneutralized acid. Note that the general equation involves the approximation made when the x term is dropped.

7. *Weak Base Plus Salt with Common Ion.* The treatment is similar to that for the weak acid of the preceding section.

Problem. What is the pH of a solution containing 0.535 g NH$_4$Cl in 50 ml 0.1 M NH$_3$?

SOLUTION: Write the equation for reaction of NH$_3$ with H$_2$O. Above each term show the concentration

$$\underset{\text{NH}_3}{0.1-x} + \text{H}_2\text{O} = \underset{\text{NH}_4^+}{\frac{10}{50}+x} + \underset{\text{OH}^-}{x}$$

Substitute the concentrations into the ionization constant equation

$$\frac{(x)\left(\dfrac{10}{50} + x\right)}{0.1 - x} = 1.8 \times 10^{-5}$$

Approximating by dropping the x when added to or subtracted from larger numbers gives

$$\frac{0.2x}{0.1} = 1.8 \times 10^{-5}$$

$$x = 0.9 \times 10^{-5} = 9 \times 10^{-6}$$

$$pOH = -\log x = -(0.95 - 6) = 5.05$$

$$pH = 14 - 5.05 = 8.95$$

The general equation for a solution containing a weak base plus its salt is

$$\frac{[OH^-]n_s}{n_b} = K_b \qquad \text{or} \qquad [OH^-] = \frac{n_b K_b}{n_s}$$

8. *Salt of Weak Acid and Strong Base.* When an equivalent amount of NaOH has been added to a solution of a weak acid (such as HOAc), the solution is not neutral, as it is when an equivalent amount of strong base has been added to a strong acid. The reason is that

two bases, the OAc^- and the OH^- ions, are competing for the protons. At the equivalence point we have added a mole of OH^- ion for each mole of HOAc originally present. But, since a small fraction of the total number of protons is still held by the OAc^- ion, as undissociated HOAc molecules, we have an excess of OH^- ions present. The pH of the solution is computed from the equilibrium constants of the two competing reactions.

Problem. What is the pH at the equivalence point when 50 ml 0.1 M HOAc is titrated with 0.1 M NaOH?

SOLUTION: The concentrations of the various species in the solution are the same regardless of whether the solution is prepared (1) by adding an equivalent amount of base to a weak acid or (2) by dissolving the salt of the weak acid in water. The reaction of the salt with water

$$OAc^- + H_2O = HOAc + OH^-$$

is known as hydrolysis.[5]

The equilibrium expression is

$$\frac{[HOAc][OH^-]}{[OAc^-][H_2O]} = K_e$$

Or, since $[H_2O]$ is constant,

$$\frac{[HOAc][OH^-]}{[OAc^-]} = K_h \text{ (hydrolysis constant)}$$

The hydrolysis constant is evaluated by combining the constants for the two equilibra in the solution

(1)
$$[H_3O^+][OH^-] = K_w = 1 \times 10^{-14}$$

(2)
$$\frac{[H_3O^+][OAc^-]}{[HOAc]} = K_a = 1.8 \times 10^{-5}$$

Dividing (1) by (2) gives

$$\frac{[HOAc][OH^-]}{[OAc^-]} = \frac{K_w}{K_a} = \frac{1 \times 10^{-14}}{1.8 \times 10^{-5}} = K_h$$

Since practically all the original 5 mM HOAc in the solution is neutralized at the equivalence point, let

$$\frac{5}{100} - x = [OAc^-]$$

$$x = [HOAc] = [OH^-]$$

Substituting the values of the two constants and the concentrations of the various species, we have

$$\frac{x^2}{0.05 - x} = \frac{1 \times 10^{-14}}{1.8 \times 10^{-5}}$$

---

[5] The term "hydrolysis" is not used in the Brönsted treatment. Rather, OAc^- is treated as a base which competes with the OH^- ion for protons.

Solving,

$$x^2 = 2.78 \times 10^{-11}$$

$$\log x^2 = 0.44 - 11 = -10.56$$

$$\log x = -5.28$$

$$pOH = -\log x = 5.28$$

$$pH = 8.72$$

The general expression for the concentration of OH^- ion in a solution of a salt of a weak acid and strong base is

$$[OH^-] = \sqrt{\frac{C_s K_w}{K_a}}$$

where C_s is the salt concentration, neglecting the small amount which reacts.

9. *Salt of Weak Base with Strong Acid.* The equilibrium expression is treated exactly the same as for a weak acid. The following problem illustrates the general method.

Problem. What is the pH of a solution containing 10 mM NH_4Cl in a volume of 100 ml?

SOLUTION: Here the reaction of the salt with water is the reverse of the neutralization reaction

$$NH_4^+ + Cl^- + H_2O = H_3O^+ + NH_3 + Cl^-$$

but since the Cl^- ion does not participate in the reaction, we give only the essential species,

$$\overset{0.1-x}{NH_4^+} + H_2O = \overset{x}{H_3O^+} + \overset{x}{NH_3}$$

Here the bases NH_3 and H_2O are competing for the proton. Treating the equilibrium in the usual manner gives

$$\frac{[H_3O^+][NH_3]}{(NH_4^+)} = \frac{x^2}{0.1-x} = \frac{K_w}{K_b} = \frac{1 \times 10^{-14}}{1.8 \times 10^{-5}} = 0.55 \times 10^{-9}$$

$$x^2 = [H_3O^+]^2 = 0.55 \times 10^{-10} = 5.5 \times 10^{-11}$$

$$\log [H_3O^+]^2 = 0.74 - 11 = -10.26$$

$$\log [H_3O^+] = -5.13$$

$$pH = 5.13$$

Titration Curves

Graphs of pH versus the volume of reagent added in a titration are known as *titration curves*. We shall now consider the various types of titration curves we may encounter in acid-base titrations. These curves can be obtained experimentally by use of a pH meter, as described in Chapter 22, or we can use the methods of the preceding sections to compute the pH of the solutions.

A typical strong-acid versus strong-base titration curve is shown in Fig. 7–1. The values have been computed for the titration of 100 ml $0.025\,M$ HCl by $0.1\,M$ NaOH, using the methods of sections 1, 2, and 3 of the preceding treatment. It is assumed that CO_2 is removed by boiling the solution so that the pH at the equivalence point is that of a NaCl solution, or 7.00.

Exercise.   Compute the pH for additions of 0, 10, 20, and 30 ml $0.1\,M$ NaOH to 100 ml $0.025\,M$ HCl. Use the computed values and the pH of 7.00 at the equivalence point to plot the titration curve. It should agree with Fig. 7–1.

Note the following features of the titration:

1. The pH changes slowly at first, until the equivalence point is approached.

2. In the region of the equivalence point there is a rapid change in pH. There is a nearly vertical rise in the region from pH 4 to 10.

3. After the equivalence point is passed, the curve flattens out, as excess of NaOH is added.

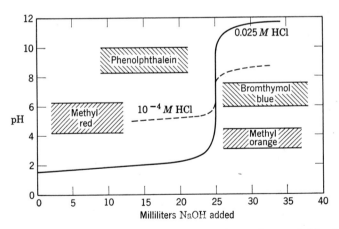

FIGURE 7–1.   Titration of HCl by NaOH. Dotted line for $10^{-4}\,M$ solutions. Transition ranges of indicators are shown in the shaded areas.

The effect of concentration on the shape of a titration curve is shown in the dotted graph of Fig. 7–1. It is for titration of $0.0001\,M$ HCl by $0.0001\,M$ NaOH. Note the short vertical region of this curve as compared with the one for a more concentrated solution.

Titration curves for acetic and boric acids are shown in Fig. 7–2. Computed pH values for the acetic acid curve are given in Table 7–2,

TABLE 7–2.   Construction of Titration Curve for HOAc versus NaOH

25 ml 0.1 M HOAc diluted to 100 ml and titrated with 0.1 M NaOH

NaOH, ml	Volume of Solution	$[H_3O^+]$	pH
0.00	100.0	6.71×10^{-4}	3.17
1.00	101.0	3.21×10^{-4}	3.49
2.50	102.5	1.62×10^{-4}	3.79
5.00	105.0	7.20×10^{-5}	4.14
10.00	110.0	2.70×10^{-5}	4.57
15.00	115.0	1.20×10^{-5}	4.92
20.00	120.0	4.50×10^{-6}	5.35
22.50	122.5	2.00×10^{-6}	5.70
23.00	123.0	1.56×10^{-6}	5.81
24.00	124.0	7.50×10^{-7}	6.12
24.50	124.5	3.67×10^{-7}	6.44
24.75	124.8	1.82×10^{-7}	6.74
24.90	124.9	7.24×10^{-8}	7.14
24.95	125.0	3.66×10^{-8}	7.44
24.98	125.0	1.54×10^{-8}	7.82
24.99	125.0	8.16×10^{-9}	8.08
24.995	125.0	6.43×10^{-9}	8.19
25.00	125.0	3.00×10^{-9}	8.52
25.025	125.0	$5.0 \ \times 10^{-10}$	9.30
25.10	125.1	1.25×10^{-10}	9.90
25.20	125.2	6.25×10^{-11}	10.20
25.50	125.5	2.51×10^{-11}	10.60
26.00	126.0	1.26×10^{-11}	10.90
28.00	128.0	4.27×10^{-12}	11.37
30.00	130.0	2.60×10^{-12}	11.59

which is based on titration of 100 ml 0.025 M acetic acid by 0.1 M NaOH. Four different types of computations are involved.

1. Before any base is added the solution contains acetic acid alone. See section 4, page 149, for the method.

2. In the region below the equivalence point, the solution has acetic acid plus its salt. The computation method is given in section 6, page 151.

3. At the equivalence point, when 25.00 ml NaOH is added, the solution contains the salt, NaOAc. The computation method is given in section 8, page 152.

4. After the equivalence point is passed, the solution contains

Application of the law of chemical equilibrium to the ionization of an acid indicator gives the expression,

$$\frac{[H_3O^+][In^-]}{[HIn]} = K_{ind}$$

which can be rearranged to the form,

$$[H_3O^+] = K_{ind}\frac{[HIn]}{[In^-]}$$

It is evident from this equation that the ratio in the concentrations of the two colored forms of the indicator will vary continuously as the hydrogen ion concentration is changed.

The color changes that the eye observes depend on the colors of the two different forms of the indicator and on their relative concentrations. If, for example, HIn is yellow and In^- is red, the solution will appear red when In^- is in great excess and yellow when HIn is in great excess. When the two are present at about the same concentration, the solution is neither yellow nor red but orange.

If we assume that in the mixture of colors for In^- and HIn the solution appears red when $[In^-]$ is ten times or more $[HIn]$ and that the color is yellow when $[HIn]$ is ten times or more $[IN^-]$, we can readily show that to change the solution from one color to the other will require a pH change of 2 units or more. If the ionization constant has the value 10^{-8}, we have

$$[H_3O^+] = (\tfrac{1}{10}) \cdot 10^{-8} \text{ or } 10^{-9} \text{ when the solution is red}$$

$$[H_3O^+] = (\tfrac{10}{1}) \cdot 10^{-8} \text{ or } 10^{-7} \text{ when the solution is yellow}$$

Thus, at pH above 9 the solution is red, and at pH below 7 it is yellow. In the intermediate region, between pH 7 and 9, the solution is neither yellow nor red, but it goes through various shades of orange as the pH is changed.

Actually, many indicators do have a transition interval of about 2 pH units, although some are considerably smaller and some larger. This indicates that, for the eye to see the change from one color to the other, the excess must be, as previously assumed, at least tenfold in most cases.

It is not necessary, however, to have a pH change of 2 units to see an end point in a titration. Many indicators are of such colors that the eye recognizes a change when the pH varies only a few tenths of a unit. This is particularly true of some of the mixed indicators.

The apparent sharpness of an indicator end point depends on factors other than the contrast afforded by the colored forms of the indicator; most important is the concentration of the solution used for the titration. The more concentrated the solution, the greater the change in pH when one drop of reagent is added, and therefore the sharper the color change. The effect of concentration on the sharpness of the pH change is shown by the titration curves of $0.025\,M$ and $0.0001\,M$ HCl, in Fig. 7–1.

In some titrations, where there is a somewhat continuous change of indicator color near the end point, it is better to titrate to a definite color tint than to a sharp change. For such titrations there is a color standard, prepared by adjusting a comparison solution to the pH that should prevail at the end point and adding indicator to this. Each sample is then titrated to match the color of the standard.

Chemistry of Indicators. Two questions arise concerning the use of indicators: (1) what volume of titration reagent is required for reaction with the titration indicator, and (2) what chemical change in the indicator accompanies the color change?

The amount of indicator required to give the necessary intensity of color is so small that a negligible amount of reagent is required for reaction with the indicator. At most, less than 10^{-3} mM of indicator is employed, and not more than 0.01 ml 0.1 N acid or base is required to neutralize the indicator.

The mechanism of indicator color change has been widely investigated. The generally accepted theory is that the formation of an indicator ion is attended by a molecular rearrangement which gives rise to a chromophoric, or color-forming, group. Paranitrophenol may be cited as an example. In acid solution this is colorless, but in basic solution it is yellow. The structural formulas for the colorless molecule and colored ion are as follows:

The structural changes in phenolphthalein and methyl orange are more complicated.

Phenolphthalein:

Colorless in
acid solution
(molecule)

Red in
base solution
(ion)

Methyl orange:

Yellow in
base solution

Pink in
acid solution

Mixed Indicators. The transformation range of an indicator may often be decreased and the sharpness of the color change increased by the addition of a suitable dye.[7] The added dye may be either a second indicator or an inert substance whose color does not change with pH. The effect of the added substance is to decrease the range of wavelengths transmitted by the solution, so that the light transmitted by the two colored forms of the indicator is not masked by light of other wavelengths.

Selection of Indicator for Titration

We have seen that the pH at the equivalence point may vary over a wide range in the different types of titrations we encounter. Consequently it is necessary to select for each titration that indicator whose end point occurs at or near the equivalence point. The proper indicator for one titration may be useless for another.

For example, we see in Fig. 7–1 that any of the indicators whose transition ranges are shown on the plot will give a good end point in the titration of 0.025 M HCl. The transition ranges all fall on the

[7] See I. M. Kolthoff and C. Rosenblum, *Acid-Base Indicators,* Macmillan, 1937, for a list of mixed indicators and their transition intervals.

steep portion of the titration curve. But in the HOAc-NaOH titration of Fig. 7–2 only one of the indicators shown, phenolphthalein, is suitable. Its color change occurs at the equivalence point when 25.00 ml NaOH is added. The other indicator, methyl red, shows a gradual color change in this titration and gives no sharp end point.

One of the following methods is used to select the indicator for a given titration.

1. If the titration curve is known, we select that indicator whose end point most closely corresponds to the pH of the equivalence point.

2. If the ionization constant for the acid or base we wish to titrate is approximately the same as the constant for the acid or base of a known titration, we choose the indicator that is suitable for this known titration. Thus, when titrating an acid whose K_i is 10^{-4} or 10^{-5}, we can safely use phenolphthalein indicator, for experience has shown it to work for acetic acid whose ionization constant is 1.8×10^{-5}.

3. If the titration is one for which we have no prior experience, we compute the pH at the equivalence point, using the method of section 8, page 152, or section 9, page 154. The computed pH is the basis for selection of the indicator.

Problem. Compute the pH at the equivalence point for titration of 50 ml 0.01 M benzoic acid with 0.01 M NaOH, and select the proper indicator from Table 7–3 for this titration.

SOLUTION: The ionization constant for benzoic acid is 7×10^{-5}. Since this is somewhat stronger than acetic acid, we could by analogy select phenolphthalein as the indicator if the titration were for solutions of usual concentration, 0.05–0.1 M. Since the specified solutions are more dilute than those we have previously used, we compute the pH at the equivalence point.

At the equivalence point the solution contains 0.5 mM sodium benzoate, in a volume of 100 ml, or a concentration of 0.005 M. The titration reaction is

$$\overset{x}{HC_7H_5O_2} + \overset{x}{OH^-} = H_2O + \overset{0.005-x}{C_7H_5O_2^-}$$

Treating the equilibrium by the methods of section 8, page 152, we have

$$[OH^-]^2 = \frac{0.005\, K_w}{K_a}$$

$$= \frac{0.005 \times 1 \times 10^{-14}}{7 \times 10^{-5}}$$

$$pOH = 6.07$$

$$pH = 7.93$$

Reference to Table 7–3 shows that we can use either phenol red (pH range 6.8–8.4) or cresol red (7.2–8.7) for this titration.

Feasibility of Titrations. As previously shown, a pH change of about 2 units is required to cause a color change in an acid-base indicator. This color change must be abrupt (occurring with addition of not more than 0.1–0.2 ml of reagent) if the analyst is to observe a sharp end point. Otherwise he will see a gradual fading of one color and appearance of another, and he will not be able to say at exactly what volume of reagent the titration is completed.

In the titration of a very weak acid or base the pH change at the equivalence point is gradual. This effect is shown in the titration curve for boric acid, Fig. 7–2. The pH at the equivalence point is about 11, but the pH change is so gradual at this point that about 5 ml of base is needed to cause a pH change of one unit. This titration, therefore, is not feasible and should not be attempted with an internal acid-base indicator (it could be done by electrometric titration).

After selecting the proper indicator for a titration, we should evaluate the feasibility before attempting to carry out the operations. The feasibility may be evaluated by one of the following methods.

1. By inspection of the titration curve, if this is available for the concentrations of solutions we plan to use. We see in Fig. 7–2 an example of a feasible titration, acetic acid, and one that is not feasible, boric acid.

2. By analogy to titrations that are known to be feasible. We know, for example, that solutions whose concentrations are of the order of 0.1 M can be titrated when the ionization constant of the weak acid or base is 10^{-5} or greater, since we have had experience with titrations of acetic acid and ammonia. The titrations of other acids or bases of strengths comparable to these two are therefore known to be feasible, provided the concentrations of the solutions are also comparable to those we have employed.

3. By computation of the pH change caused by addition of 0.1 ml of reagent, at the equivalence point. That is, we compute the pH at the equivalence point and the pH for the further addition of 0.1 ml of reagent. If the change is of the order of 1 pH unit or greater, we may expect the end point to be sufficiently sharp to permit accurate observation.

Problem. Is the titration of 0.01 M benzoic acid by 0.01 M NaOH feasible?

SOLUTION: In a previous calculation it was shown that the pH at the equivalence point for this titration is 7.93. We now compute the pH when an additional 0.1 ml NaOH is added. The amount of NaOH added is 0.001 mM (0.1 ml of 0.01 M solution), and the total volume of solution is 100.1 ml. Since NaOH is in excess, we compute the OH^- concentration of this excess.

$$[OH^-] = [NaOH] = 0.001 \text{ mM}/100.1 \text{ ml}$$
$$= 10^{-5} M$$
$$pOH = 5.0$$
$$pH = 9.0$$

Thus, 0.1 ml of reagent causes a pH change from 7.93 to 9.0, or slightly more than 1 pH unit, and we expect the titration to be feasible. In practice it has been found that the end point is sufficiently sharp.

4. As a rough guide to feasibility of titrations, we can generalize that when the solutions are $0.1 M$ or stronger, the titration of a weak acid or base of ionization constant 10^{-5} or greater is always feasible, provided the titration is made with a strong base or acid. It is not feasible to titrate a weak acid with a weak base or vice versa. When acids or bases whose K_i is 10^{-5} or greater are titrated in very dilute solutions, a computation should be made of the rate of change of pH with volume of reagent, as in the preceding section. When K_i approaches 10^{-6}, the titration may or may not be feasible, depending on the particular indicator available. When K_i is less than 10^{-6}, the titration is generally not feasible.

Titration of Salts

We have noted that when the ionization constant for a weak acid is below 10^{-6}, the acid cannot be titrated directly by a strong base. We can, however, titrate the salt of such acids with a strong acid. For example, if sodium borate is titrated by HCl, we have the reaction

$$BO_2^- + H_3O^+ = HBO_2 + H_2O$$

At the equivalence point the solution contains HBO_2, whose pH is readily computed from the ionization constant and the concentration.

The most widely used salt titration is that of carbonate ion, as described in Chapter 13. Carbon dioxide in water is a polyprotic acid, ionizing in two stages

$$CO_2 + 2H_2O = H_3O^+ + HCO_3^- \qquad K_1 = 4 \times 10^{-7}$$
$$HCO_3^- + H_2O = H_3O^+ + CO_3^{-2} \qquad K_2 = 5 \times 10^{-11}$$

When CO_3^{-2} is titrated with HCl, we have the two reactions

$$CO_3^{-2} + H_3O^+ = HCO_3^- + H_2O$$
$$HCO_3^- + H_3O^+ = CO_2 + 2H_2O$$

The titration curve of Na_2CO_3 by HCl is shown in Fig. 7–5. There are two breaks, corresponding to the two stages in the reaction. The first break, corresponding to formation of HCO_3^-, is so gradual that it does not yield a good end point. The second break, corresponding to formation of CO_2, is somewhat sharper than the first but with

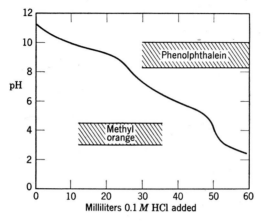

FIGURE 7–5. Titration of Na_2CO_3 by 0.1 M HCl.

acid as dilute as O.1 M the end point is not good. The appropriate indicator, methyl orange, gives a good end point if the titration is made with stronger HCl, say 0.25 M or greater. The sharpness of the end point is improved by boiling off CO_2, as described in Chapter 13. The titration is made, in the cold, until the indicator shows a color change at pH near 5. At this stage most of the HCO_3^- is converted to CO_2. The solution is now boiled, to remove CO_2; after this is expelled, the solution contains only NaCl, so that the pH rises to 7.0. After cooling, the titration is completed by addition of a few drops of the HCl, which neutralizes any bicarbonate ion remaining. The shape of the titration curve after the CO_2 is removed is like the lower portion of the NaOH–HCl curve, of Fig. 7–3, and the end point is sharp.

If carbonate is titrated cold, so that the CO_2 remains in solution, the pH at the equivalence point is readily computed from the dissociation constant.

Problem. A solution containing 5 mM Na_2CO_3 is titrated by HCl. The volume at the equivalence point is 100 ml. Select the proper indicator.

SOLUTION: At the equivalence point the solution contains CO_2 at a concentration of 0.05 M. This reacts with water, according to the equation above. Let $x = [H_3O^+] = [HCO_3^-]$. Then $0.05 - x = [CO_2]$. Substituting these concentrations into the ionization constant equation gives

$$\frac{x^2}{0.05 - x} = 4 \times 10^{-7}$$

The hydronium ion from the first stage of the reaction prevents any appreciable reaction in the second stage, and for our purposes we neglect the further dissociation of the bicarbonate ion. Neglecting x in the denominator and solving, we obtain a value of 3.85 for the pH. We find in Table 7–3 that methyl yellow, bromphenol blue, and methyl orange are all suitable indicators.

Buffers

A buffer solution is one that contains a weak acid and its salt or a weak base and its salt. The name is based on the fact that an acid or base added to a buffer solution causes less change in pH than an acid or base added to pure water or to an unbuffered solution. To illustrate the buffer effect, we shall consider a solution containing acetic acid and a salt, sodium acetate. As shown in section 6, page 151, the hydronium ion concentration is given by the relation

$$[H_3O^+] = \frac{n_a}{n_s} K_a$$

Let us assume a solution containing 5 mM each of HOAc and OAc⁻ ion. Then,

$$[H_3O^+] = \frac{5 \times 1.8 \times 10^{-5}}{5} = 1.8 \times 10^{-5}$$

$$pH = 4.74$$

If a millimole of HCl is added, it reacts with the OAc⁻ ion to form a millimole of HOAc. The solution now contains 6 mM HOAc and 4 mM OAc⁻. We now have

$$[H_3O^+] = \frac{6 \times 1.8 \times 10^{-5}}{4} = 2.7 \times 10^{-5}$$

$$pH = 4.57$$

If a millimole of NaOH is added to another portion of the buffer, it reacts with HOAc to form a millimole of NaOAc. The solution now contains 4 mM HOAc and 6 mM OAc⁻, or

$$[H_3O^+] = \frac{4 \times 1.8 \times 10^{-5}}{6} = 1.2 \times 10^{-5}$$

$$pH = 4.92$$

13. How many milliliters of $0.1000\,M$ HCl must be diluted to 250.0 ml to produce a solution of pH 2.50?   *Ans.* 8.0 ml.

14. How many drops of $0.1\,M$ HCl must be added to 100 ml of water to change the pH from 7 to 4? Assume that a drop is 0.05 ml. Repeat the calculation, using $1\,N$ acid.

15. 200 ml $0.001\,M$ NaOH is added to 300 ml $0.0005\,M$ HCl. What is the pH of the solution?   *Ans.* 10.0.

16. Is it possible at room temperature to prepare a NaOH solution of pH 15?

17. 10 ml $0.2\,M$ HCl is added to 100 ml $0.1\,M$ HCl. What is the pH of the mixture?   *Ans.* 0.96.

18. 25 ml $0.2\,M$ HOAc is diluted to 75 ml and titrated with $0.1\,M$ NaOH. Find the pH when half the acid has been neutralized.

19. What color will each of the following indicators have in the solutions of problem 6? Bromphenol blue, methyl red, phenol red.

20. An indicator which is a weak monobasic acid is blue in a strongly acid solution and yellow in a strongly basic one. If K_{ind} is 10^{-10}, what is the pH when 40 per cent of the indicator is in the yellow form? What color would this indicator have in a solution of pH 7?   *Ans.* 9.82; blue.

21. The K_{ind} for an indicator which is a weak monoacidic base is 10^{-5}. What is the pH when the indicator is equally divided between its two colored forms?

22. What is the concentration of HCl present in solution when methyl yellow shows its intermediate color?

23. Repeat problem 22, but use HOAc.   *Ans.* Around $8.7 \times 10^{-3}\,M$.

24. What is the pH at the equivalence point when 50 ml $0.1\,M$ HCl is diluted to 75 ml and titrated with $0.2\,M$ NaOH?

25. Repeat problem 24, but use HOAc.

26. What is the pH at the equivalence point in the titration of 5 mM KHP (potassium acid phthalate) dissolved in 100 ml of water and titrated with $0.1\,M$ NaOH? Neglect anything but the first step of the hydrolysis.
                                        Ans. 9.05.

27. 0.025 mole Na_2CO_3 is dissolved in 150 ml of water and titrated with $1.0\,M$ HCl, using methyl red as the indicator. Find the approximate pH at the equivalence point.

28. Select an indicator for the titration of 50 ml $0.1\,M$ benzoic acid (in a volume of 225 ml) by $0.2\,M$ KOH.

29. The acid HA has an ionization constant of 10^{-5}. If 50 ml $0.1\,N$ HA is diluted to 200 ml and titrated with $0.1\,N$ NaOH:
    (*a*) Find the pH before any base is added.   *Ans.* 3.30.
    (*b*) Find the pH when 25 ml of base has been added.   *Ans.* 5.00.
    (*c*) Find the pH at the equivalence point.   *Ans.* 8.65.
    (*d*) What indicator should be used?

(e) Draw the approximate titration curve, using the points for which the pH has been computed.

(f) Give reasons for thinking that the titration is feasible, without computing the pH at points near the equivalence point.

30. The base $(CH_3)_3N$ has an ionization constant of 7×10^{-5}. If 50 ml $0.1 N$ $(CH_3)_3N$ is titrated with $0.1 N$ HCl:

(a) Find the pH before any acid is added.

(b) Find the pH at the equivalence point.

(c) Find the pH when the base is two-fifths neutralized.

(d) Draw the approximate titration curve.

(e) What indicator should be used?

31. Repeat the calculations of problem 29 for the neutralization of 50 ml $0.001 N$ HA by $0.001 N$ NaOH, assuming an initial volume of 200 ml.

Ans. (a) 4.35; (b) 5.00; (c) 7.65.

32. Compute pH values and plot on a large scale the titration curve for $0.1 N$ HCl and NaOH.

33. In a titration 25 ml $0.001 N$ HCl is diluted to 100 ml and titrated by $0.001 N$ NaOH. Compute the pH for the addition of 0, 5, 20, 24, 26, 30 ml of base. Plot the results on the same sheet, and compare with the plot of problem 32.

34. 50 ml $0.1 M$ HX ($K_i = 10^{-7}$) is diluted to 75 ml and titrated with $0.2 M$ KOH.

(a) What color will the indicator methyl red exhibit when enough base has been added to neutralize half the acid present?

(b) Select an indicator for the titration.

(c) Draw the approximate titration curve, and discuss the feasibility of the titration.   Ans. (a) Yellow; (b) thymolphthalein.

35. A $0.05 M$ solution of a salt $C_6H_5NH_3Cl$ has a pH of 3.00. What is the ionization constant of $C_6N_5NH_2$?

36. It is found that 50.00 ml of a carbonate-free base is neutralized to the phenolphthalein end point by 50.00 ml $0.1000 N$ HCl.

(a) What volume of HCl is needed to neutralize 50.00 ml of the base to the methyl orange end point, pH 3.80?

(b) What is the normality of the base solution when used with phenolphthalein indicator?

(c) What is the normality of the base solution when used with methyl orange indicator without an indicator blank, assuming an end point at pH 3.80?   Ans. (c) $0.1003 N$.

37. An acid HA has an ionization constant of 5×10^{-6}. If 50 ml of a $0.1 M$ solution of the acid is diluted to 200 ml and titrated with $0.1 N$ NaOH:

(a) Compute the pH when 49.95 ml of base has been added.

(b) Compute the pH when 50.05 ml of base has been added.

38. (a) How many grams of NaOAc must be added to 300 ml $0.01 M$ HOAc to produce a buffer of pH 6.50?

(b) 2.00 mM NaOH is added to the buffer above. What is the pH after the addition of the base?

(c) What would happen to the buffer if 2.00 mM NaOH additional were introduced?   Ans. (a) 14 g.; (b) 7; (c) pH becomes 11.53.

39. 10 ml 0.1 M HCl is added to 100 ml of a solution which is 0.1 M both in HOAc and NaOAc. By how many units does the pH change because of the HCl addition?

40. What is the pH of a solution which produces a slight greenish color when bromthymol blue is added as an indicator? What must be the ratio of $[OAc^-]/[HOAc]$ to produce a solution of this pH?

41. Find the pH of a buffer made by dissolving 3 mM NaCN and 5 mM HCN in 250 ml of solution.

42. What is the pH of a buffer which has been prepared by mixing 250 mg NH_4Cl and 200 ml 0.1 M NH_3 solution? What is the pH of this after 10 mM HCl has been added? The addition of the acid dilutes the mixture to 250 ml.

Ans. (a) 9.85; (b) 9.09.

43. What is the pH of a buffer made by mixing 43.00 ml 0.1000 N NaOH with 50.00 ml 0.1000 M KHP and diluting to a volume of 100 ml?

44. Estimate the pH of a buffer made by adding 10 g $NaHCO_3$ to 250 ml of a solution which is saturated with CO_2 (0.032 M).   Ans. 7.57.

45. Estimate the pH of a buffer made by mixing 50 ml 0.1 M KH_2PO_4 with 25 ml 0.1 M NaOH and diluting to 200 ml.

CHAPTER 8

Equilibria in Precipitation Reactions

The precipitates we deal with in analytical chemistry are composed of ions, packed together to make up the crystal lattice. When such a crystal is placed in water, an equilibrium is set up, ions leaving the surface to go into solution and ions from the solution depositing on the surface. At equilibrium the rates of solution and redeposition are equal, and we have a saturated solution.

The solubility of a given crystal species, at a given temperature, depends on two competing forces. The force tending to pull ions into solution is the attraction of water (solvent) molecules for the ions. Opposed to this is the attraction of the ions in the lattice for one another. All ions are strongly attracted by water, for in some form all ions are soluble. The lattice forces depend on the particular species that make up the given crystal. When the lattice forces are large, the solubility is low; and conversely when the lattice forces are small, there is a large solubility. Some substances are soluble to the extent of several moles per liter, whereas other crystals are so insoluble that most careful measurements are required to show that any of the ions have dissolved.

The reaction that occurs when a solid salt, such as AgCl, is brought in contact with water is given by the equation

$$AgCl(s) = Ag^+ + Cl^-$$

Strictly speaking, this equation does not tell the whole story, for the ions in solution are to some extent hydrated, just as a dissolved proton is hydrated. The bonding of other ions to water is nowhere near as tight as that of the hydrogen ion, however. Moreover, in the case of most ions we do not know the extent of hydration. We therefore represent the dissolved ion by its ionic formula and make no attempt to show the bonded water molecules. The (s) following the AgCl formula indicates that it is present as a solid. Nearly all the dissolved material is in solution as ions.

Solubility Product

An equilibrium constant can be determined for this two-phase reaction, in the same manner as was done for single-phase systems in the preceding chapter. Thus, we have at equilibrium

$$\frac{[Ag^+][Cl^-]}{[AgCl(s)]} = K_e$$

Since the concentration of AgCl, the solid crystalline material, is constant, we combine the two constants of the equation

$$[Ag^+][Cl^-] = [AgCl(s)] \times K_e = K_{sp}$$

The ion product constant, K_{sp}, is known as the solubility product constant or, more briefly, as the "solubility product." The generalized expression for the solubility product of the equilibrium

$$A_m B_n (s) = m(A \text{ ions}) + n(B \text{ ions})$$

is

$$[A \text{ ions}]^m [B \text{ ions}]^n = K_{sp}$$

Thus, for a saturated solution of $Mg(OH)_2$ we have the equilibrium

$$Mg(OH)_2(s) = Mg^{+2} + 2OH^-$$

and the solubility product is

$$[Mg^{+2}][OH^-]^2 = K_{sp}$$

Numerical values of the solubility products are convenient for tabulating solubilities of sparingly soluble substances. Values for the more common precipitates of analytical chemistry are given in Table A–3, Appendix.

Since the K_{sp} gives the products of the molar concentrations of the ions in a saturated solution, it follows that

1. If the numerical value of the product of the concentrations of the ions in a given solution is less than the K_{sp} value, the solution is less than saturated, and more solid will dissolve if placed in the solution.

2. If the numerical value of the product of the concentrations of the ions in a given solution is greater than the K_{sp} value, the solution is supersaturated and ions will precipitate out of solution until the equilibrium concentrations are reached.

Evaluation of K_{sp} from Experimental Measurements. At room temperature the solubility of AgCl is approximately 0.0015 g per liter, or 1.05×10^{-5} moles per liter. Since the salt is completely dissociated in solution, the concentration of Ag^+ equals the concentration of Cl^-, and each is $1.05 \times 10^{-5} M$. This gives

$$(1.05 \times 10^{-5})(1.05 \times 10^{-5}) = 1.1 \times 10^{-10} = K_{sp}$$

Usually the value is rounded to 1×10^{-10} since we are interested in calculations at room temperature, which may vary by a few degrees from $25°C$.

Problem. The solubility of $Mg(OH)_2$ is 0.0009 g per 100 ml. Compute the K_{sp} value.

SOLUTION: The solubility is 0.009 g per liter. Dividing by the molecular weight, 58.3, gives a molar solubility of 1.54×10^{-4} moles per liter. Each mole of $Mg(OH)_2$ in solution yields 1 mole of Mg^{+2} ion and 2 moles of OH^- ion. Thus we have

$$[Mg^{+2}] = 1.54 \times 10^{-4} M$$
$$[OH^-] = 2 \times 1.54 \times 10^{-4} = 3.1 \times 10^{-4} M$$

Substituting these values into the K_{sp} expression gives

$$K_{sp} = [Mg^{+2}][OH^-]^2$$
$$= (1.54 \times 10^{-4})(3.1 \times 10^{-4})^2$$
$$= 1.5 \times 10^{-11}$$

Computation of Solubility from K_{sp}. When the ions in solution are due only to the dissolved substance, we can compute the solubility from the K_{sp} value.

Problem. K_{sp} for $Fe(OH)_2$ is 10^{-15}. What is the solubility in grams per milliliter?

SOLUTION: Let S be the molar solubility. Then S is the concentration of Fe^{+2} and $2S$ the concentration of OH^- [we make the assumption that the contribution of OH^- from dissociation of water is negligible in comparison to the ion furnished by the dissolved $Fe(OH)_2$].

Substituting these values in the K_{sp} expression gives

$$(S)(2S)^2 = 10^{-15}$$

Solving,

$$4S^3 = 10^{-15}$$
$$S = 0.6 \times 10^{-5} M$$

The weight of $Fe(OH)_2$ is 0.5×10^{-3} g per liter or 0.5×10^{-4} g per 100 ml.

Note that the computed concentration of OH^- ion is $3.4 \times 10^{-5} M$. Since this is much larger than the concentration of OH^- ion furnished by water, it is valid to neglect the latter in solving the problem. In many computations of this type we must take the OH^- ion concentration from water into account. For example, K_{sp} for $Fe(OH)_3$ is 10^{-37}. If the method above is used to compute its solubility, we obtain a value of less than 10^{-7} for the concentration of OH^-, which is obviously absurd. Actually, the solubility is so slight that the hydroxide ion concentration of a saturated solution is essentially that of water or $10^{-7} M$.

Effect of Common Ion on Solubility. Since the product of the concentrations of ions in a saturated solution is constant, it follows that if we add an excess of one ion of a precipitate to a saturated solution, the concentration of another ion must be decreased to maintain the equilibrium.

Problem. How many grams of AgCl will dissolve in a liter of 0.01 M NaCl?

SOLUTION: Let S be the moles of AgCl that dissolve. Then S is the concentration of Ag^+ ion and $0.01 + S$ is the concentration of Cl^- ion. Substituting these values into the equilibrium expression gives

$$(S)(0.01 + S) = K_{sp} = 1 \times 10^{-10}$$

Since S is small in comparison to 0.01, we simplify to

$$(S)(0.01) = 10^{-10}$$

$$S = 10^{-8} \text{ moles/liter}$$

The weight of AgCl is 143×10^{-8} g per liter. Note that this is much smaller than the solubility in pure water, 1.5×10^{-3} g per liter. Advantage is taken of this effect in precipitations. A slight excess of the precipitant is added to reduce the amount of the sought ion that remains in the solution, i.e., to make the precipitation more complete.

Precipitation

The K_{sp} relations may be used to determine whether precipitation will occur when given amounts of reagents are mixed.

Problem. Will a precipitate of AgCl be obtained when 3 mg $AgNO_3$ and 2 mg NaCl are added to 250 ml of water?

SOLUTION: Compute molar concentrations of silver and chloride ions, and the product of the two concentrations. If this product exceeds K_{sp}, precipitation will occur. We are given 0.0177 mM $AgNO_3$ in 250 ml, a concentration of $7 \times 10^{-5} M$. Since the salt is completely dissociated, this is the concentration of silver ion. Similarly, the concentration of chloride ion is $1.4 \times 10^{-4} M$. The product of the two molar concentrations is 10^{-8}, which is larger than the K_{sp} of 10^{-10}, and consequently precipitation occurs.

Problem. K_{sp} for $Mg(OH)_2$ is 10^{-11}. If 0.1 g $MgSO_4$ is added to 100 ml 0.1 M NH_3 solution, will a precipitate be obtained?

SOLUTION: Compute the concentrations of Mg^{+2} and OH^- ions and substitute the numerical values into the K_{sp} expression.

The concentration of Mg^{+2} is that of $MgSO_4$. The molar concentration is $0.0083\ M$. The concentration of OH^- is computed from the reaction of NH_3 with water, as shown in Chapter 7. We have

$$\underset{0.1-x}{NH_3} + H_2O = \underset{x}{NH_4^+} + \underset{x}{OH^-}$$

$$\frac{x^2}{0.1-x} = 1.8 \times 10^{-5}$$

$$x = [OH^-] = 1.34 \times 10^{-3}$$

Substituting these concentrations into the K_{sp} expression

$$[Mg^{+2}][OH^-]^2 = (8.3 \times 10^{-3})(1.34 \times 10^{-3})^2 = 1.5 \times 10^{-8}$$

Since the ion product value exceeds K_{sp}, precipitation occurs.

Problem. If 1.0 g NH_4Cl is added to the solution of the preceding problem, will precipitation of $Mg(OH)_2$ occur?

SOLUTION: As we have seen in Chapter 7, the presence of a salt with common ion in the NH_3 solution represses the reaction, and the concentration of OH^- is much less than in a solution of NH_3 alone.

Compute the OH^- concentration for the solution containing 1.0 g NH_4Cl in 100 ml 0.1 M NH_3 solution. We have 1.0 g NH_4Cl in 100 ml or 10 g per liter, which gives a molar concentration of 0.18 M. Treating the equilibrium in the usual manner and solving we have

$$\underset{0.1-x}{NH_3} + H_2O = \underset{0.18+x}{NH_4^+} + \underset{x}{OH^-}$$

$$\frac{(0.18+x)(x)}{(0.1-x)} = 1.8 \times 10^{-5}$$

$$x = [OH^-] = 1 \times 10^{-5}$$

This gives

$$[Mg^{+2}][OH^-]^2 = (8.3 \times 10^{-3})(10^{-5})^2 = 8.3 \times 10^{-13}$$

Since this is less than K_{sp}, precipitation does not occur from the solution. In certain analyses the precipitation of $Mg(OH)_2$ is prevented by adding NH_4Cl to the solution.

Fractional Precipitation

In analytical procedures we frequently have occasion to add a reagent to a solution which contains two or more ions that can form a precipitate with the particular reagent. The K_{sp} values for the possible precipitates can be used to compute the concentration conditions at which precipitation of the various possible substances will occur. When there is insufficient reagent to precipitate everything from the solution, the substance whose K_{sp} is first exceeded will first form a precipitate. Thus, by controlled addition of the reagent we may separate two insoluble substances by fractional precipitation.

A widely used analytical application of fractional precipitation is the Mohr titration of Cl^- ion by Ag^+ ion, using CrO_4^{-2} ion as indicator. The process is described in Chapter 17. Solubility relations are such that most of the Cl^- ion is precipitated as AgCl before any Ag_2CrO_4 precipitates. The appearance of the latter precipitate, which is colored, serves to indicate the end point of the titration. The K_{sp} values of the two precipitates may be used to compute the chloride ion concentration remaining in solution when the end point of the titration is reached.

Consider the titration of 5 mM Cl^- in a volume of 100 ml at the end point. The solution contains K_2CrO_4 at a concentration of 0.002 M. At the end point we have two precipitates, Ag_2CrO_4 and AgCl. The K_{sp} equations for the two are

$$[Ag^+]^2 [CrO_4^{-2}] = 10^{-12}$$

$$[Ag^+] [Cl^-] = 1 \times 10^{-10}$$

Since the end point is the point at which a precipitate of Ag_2CrO_4 first appears, the chromate ion concentration is the original one, $2 \times 10^{-3}\ M$. Substituting this into the K_{sp} for Ag_2CrO_4 and solving for the silver ion concentration at the end point, we have

$$[Ag^+]^2 = \frac{10^{-12}}{2 \times 10^{-3}} = 5 \times 10^{-10}$$

$$[Ag^+] = 2.2 \times 10^{-5} \quad \text{or} \quad 2 \times 10^{-5}\ M$$

Substituting this value of the silver ion concentration into the K_{sp} for AgCl and solving for the chloride ion concentration gives

$$[Cl^-] = \frac{1 \times 10^{-10}}{2 \times 10^{-5}} = 0.5 \times 10^{-5}$$

It is recalled that in a saturated solution of AgCl the chloride ion concentration is $\sqrt{K_{sp}}$ or $1 \times 10^{-5}\ M$. Since the concentration at the end point is lower, 0.5×10^{-5}, we see that the end point does not exactly coincide with the equivalence point. There is sufficient excess silver ion present at the end point to reduce the chloride ion concentration from 1×10^{-5} to 0.5×10^{-5}.

The error due to the excess silver ion required for an end point is small. The additional Ag^+ is that required (1) to increase Ag^+

from 1×10^{-5} to 2×10^{-5} M in the solution and (2) reduce Cl^- from 1×10^{-5} to 0.5×10^{-5} M, by precipitating more AgCl. This amount is computed as follows.

The total silver ion in 100 ml of solution is 1×10^{-3} mM at the equivalence point and 2×10^{-3} mM at the end point. The total chloride ion in 100 ml of solution is 1×10^{-3} mM at the equivalence point and 0.5×10^{-3} mM at the end point. The excess amount of silver ion added after the equivalence point provides 1×10^{-3} mM in solution and precipitates 0.5×10^{-3} mM AgCl, a total of 1.5×10^{-3} mM silver nitrate in excess. If the titration is made with 0.1 M AgNO$_3$ solution, as usual, the excess volume of solution is 1.5×10^{-3} mM/0.1 mM per ml = 0.015 ml. The percentage of error is the amount in excess divided by the total AgNO$_3$ used, or

$$\text{Percentage error} = \frac{0.0015 \text{ mM excess}}{5 \text{ mM total}} \times 100\%$$

$$= 0.03\%$$

Problem. A solution which contains 5 mM each of Br$^-$ and Cl$^-$ is titrated by silver nitrate solution. What concentration of bromide ion remains in solution when precipitation of AgCl commences? Assume a volume of 100 ml at this point.

SOLUTION: The solubility products are

$$[Ag^+][Cl^-] = 10^{-10}$$

$$[Ag^+][Br^-] = 10^{-12}$$

The concentration of chloride ion is 5 mM in 100 ml or 0.05 M. The concentration of silver ion when AgCl begins to precipitate is given by the relation

$$[Ag^+] = \frac{1 \times 10^{-10}}{5 \times 10^{-2}} = 2 \times 10^{-9} \, M$$

Substituting this concentration of silver ion into the K_{sp} for AgBr gives

$$[Br^-] = \frac{1 \times 10^{-12}}{2 \times 10^{-9}} = 5 \times 10^{-4} \, M$$

The amount of bromide ion remaining in 100 ml of solution is the concentration-volume product or 5×10^{-2} mM, which is 3.9 mg.

Questions

1. Is the solubility product for a salt ever equal to the cube of the molar solubility? Explain.

2. What types of precipitates are more soluble in acid than in neutral solutions? Explain, for one example, the nature of the effect of acid on the solubility.

Problems

1. From the solubility-product values in the Appendix, find the solubility expressed in millimoles per milliliter and in grams per 100 ml for:

(a) PbI_2.

(b) Hg_2Cl_2.

(c) $MgCO_3$.

2. How many milligrams of $PbSO_4$ will dissolve in 300 ml H_2O?

Ans. 9.1 mg.

3. Find the number of milligrams of AgCl which are left unprecipitated in 200 ml of solution in which the excess concentration of Ag^+ is 10^{-2} M.

Ans. 2.8×10^{-4} mg.

4. A precipitate of CaC_2O_4 is digested in a volume of 300 ml. Calculate the percentage loss (in terms of a theoretical weight of precipitate of 0.2500 g) caused by solubility in the precipitating solution if:

(a) No excess of precipitant had been added.

(b) The supernatant liquid contains an excess of 5.0 ml 0.10 M $(NH_4)_2C_2O_4$.

5. The solubility of Ag_2CrO_4 is 2×10^{-5} mole per 100 ml.

(a) If the possibility of supersaturation is neglected, what is the minimum volume of water required to dissolve 0.5 mg Ag_2CrO_4?

(b) How many milligrams of Ag_2CrO_4 remain unprecipitated in 200 ml of solution in which there is a stoichiometric excess of 1.00 mM CrO_4^{-2}?

Ans. (a) 7.5 ml; (b) 2.6 mg.

6. The precipitate AB_2 (mol. wt. = 100) is soluble to the extent of 10^{-4} moles per liter.

(a) How many milligrams of AB_2 remain unprecipitated in 250 ml of solution in which the concentration of excess A^{+2} ion is 0.001 M?

(b) What is the maximum number of milligrams of AB_2 which can dissolve in 200 ml of wash water? *Ans.* 2 mg.

(c) 3.00 mM BaB_2 is dissolved in 100 ml of water, and 10 per cent in excess of the equivalent amount of 0.10 M $A(NO_3)_2$ is added. How many milligrams of AB_2 remain unprecipitated? *Ans.* 0.28 mg.

7. The solid M_2N (mol. wt. = 75) is soluble to the extent of 0.00015 g per 100 ml.

(a) Find the solubility product for M_2N. *Ans.* 32×10^{-15}.

(b) Calculate the milligrams of M_2N which remain unprecipitated in 300 ml of solution in which the excess concentration of M^+ is 10^{-3} M.

(c) Calculate the milligrams of M_2N which could be lost in 250 ml of wash water.

(d) Is the actual loss in washing likely to equal the amount found in (c)?

8. The solubility of silver chloride is 10^{-5} moles per liter. How much wash water may be used in an analysis if the loss of chloride ion in the wash water is not permitted to exceed 0.1 mg? *Ans.* 282 ml.

9. A sample which weighs 0.5000 g contains 40.00 per cent Cl^-. This sample

is dissolved in 200 ml of water, and an amount of 0.100 M AgNO$_3$ solution is added which is equivalent to the Cl$^-$ in the sample and 5.00 ml in excess. Find the percentage error due to solubility in the precipitating solution. The precipitate is drained free from supernatant solution and washed with 300 ml 0.05 M HNO$_3$. Find the maximum percentage error due to solubility in the wash water.

10. What will be the weight of the precipitated AgCl from 0.2000 g of a sample which is 50.00 per cent Cl if the losses caused by solubility in the precipitating solution and wash water are taken into account? The conditions of precipitation and washing are:

Final volume of precipitating solution	200 ml
Excess 0.1 M AgNO$_3$ present	1.0 ml
Volume of wash water used	500 ml

Ans. 0.4036 g.

11. 0.5000 g AgNO$_3$ and 0.7000 g KBrO$_3$ are added to 250 ml of water. The precipitated AgBrO$_3$ is collected on a filter and washed with 250 ml of water. After drying in an oven, how much does the precipitated AgBrO$_3$ weigh?

12. Find the milligrams of Cl$^-$ left in solution if 50 ml 0.1 M AgNO$_3$ is added to 200 ml of solution which contains 0.2500 g NaCl. *Ans.* 3×10^{-4} mg.

13. 40.00 ml 0.5000 M AgNO$_3$ is added to 230 ml of water which contains 1.942 g K$_2$CrO$_4$. What percentage of chromium remains unprecipitated?

Ans. 0.2%.

14. M(OH)$_2$ is a slightly soluble strong base which dissolves to the extent of 10^{-6} mole per 100 ml. To 200 ml of 0.1 M MSO$_4$, solid NaOH is added.

(*a*) How many grams of NaOH have been added before a precipitate begins to form?

(*b*) What is the pH when 75 per cent of M^{+2} has precipitated?

Ans. (*b*) 7.60.

15. The solubility of a hydroxide, X(OH)$_2$, is 0.003 g per liter. The molecular weight is 150.

(*a*) Compute the solubility product.

(*b*) At what pH will a precipitate begin to form if an acid solution of 0.01 M X^{+2} is neutralized?

(*c*) If 1.0 g XSO$_4$ and 1.0 g NH$_4$Cl are added to 100 ml 0.1 M NH$_3$, will a precipitate be obtained? Justify the answer by a computation of the product [X^{+2}][OH$^-$]^2 and comparison of the value with the solubility-product value.

16. At what pH will magnesium hydroxide begin to precipitate from a 0.01 M magnesium sulfate solution? *Ans.* 9.5.

17. If 25 mg magnesium chloride is added to 100 ml 0.1 N ammonia, will a precipitate be formed?

18. If 1 g ammonium chloride is added to the solution above, will a precipitate be obtained?

19. How many milliliters of 2 M HNO$_3$ must be diluted to 400 ml so that 2.00 g silver acetate will completely dissolve in this volume?

20. 50 ml 0.4 M MgSO$_4$ and 50 ml 1.0 M NaOAc solutions are mixed. Is there any precipitation of Mg(OH)$_2$?

CHAPTER 9

Systems with Competing Equilibria

In the preceding chapters we have discussed equilibria in acid-base and precipitation reactions. These discussions have been simplified by dealing with systems in which there is only one reaction to consider. Frequently we have systems in which the same ions participate in two different equilibria that must be taken into account in computing the concentrations of the various species in the solution.

Dissolving Precipitates

As shown in Chapter 8, if we have a precipitate MX in contact with its saturated solution, there is the relation

$$[M^+] [X^-] = K_{sp}$$

If a substance is added to the solution that can react with the M^+ ion or the X^- ion so as to reduce the concentration of one of the ions in the solution, this disturbs the equilibrium condition. More of the solid must be dissolved to restore the concentration of the ion to its equilibrium value.

There are a number of ways for bringing precipitates into solution. The more common are:

1. Treatment with H_3O^+ when X^- is the anion of a weak acid.

Ferrous sulfide, for example, will dissolve in acids

$$\text{FeS}(s) + H_3O^+ = HS^- + Fe^{+2} + H_2O$$

The concentration of S^{-2} in solution is reduced by formation of the very slightly dissociated HS^- ion. As this occurs, more FeS dissolves to maintain the relation

$$[Fe^{+2}][S^{-2}] = K_{sp}$$

We recall that some sulfides dissolve in acids but others, such as CuS or PbS, are not soluble. The reason for this insolubility is that the saturated solution of such sulfides has even lower concentration of S^{-2} ion than that furnished by dissociation of HS^- ion. Consequently such sulfides will precipitate in acid solution.

2. Oxides and hydroxides are almost universally soluble in strong acids because of the very slight dissociation of water.

$$MO(s) + 2H_3O^+ = M^{+2} + 3H_2O$$

3. Some hydroxides of low solubility in water are appreciably soluble in ammonium salts, which react with the OH^- ion in equilibrium with the precipitate.

$$MOH(s) + NH_4^+ = M^+ + NH_3 + H_2O$$

4. When M^+ ion can form a complex ion with reagent R, the complex is frequently so slightly dissociated that a precipitate of MX will dissolve in a solution of R.

$$MX(s) + R = MR^+ + X^-$$

An example is the solubility of AgCl in NH_3 solutions.

$$AgCl(s) + 2NH_3 = Ag(NH_3)_2^+ + Cl^-$$

Here the formation of the complex ion reduces the concentration of silver ion to less than that furnished in the saturated solution of AgCl, thereby making the product of the concentrations of silver and chloride ions less than the K_{sp} value. To restore equilibrium, more AgCl must be dissolved.

5. Amphoteric hydroxides dissolve in solutions of strong bases

$$Al(OH)_3(s) + OH^- = Al(OH)_4^-$$

This is a special case of complex-ion formation. The concentration of Al^{+3} ion in solution is reduced by formation of the slightly dissociated tetrahydroxoaluminate ion.

In a quantitative treatment of the equilibria involved when a precipitate is dissolved, we take into account the two competing reactions to develop the equilibrium constant for the over-all reaction.   The following examples will illustrate the general methods.

Problem.   The dissociation constant for diamine silver complex is 5×10^{-8}.   The solubility product constant for AgCl is 1×10^{-10}.   How many grams of AgCl will dissolve in 100 ml 1 M NH$_3$ solution?

SOLUTION: The reaction is

$$AgCl(s) + 2NH_3 = Ag(NH_3)_2^+ + Cl^-$$

The equilibrium constant for this reaction is

$$\frac{[Ag(NH_3)_2^+][Cl^-]}{[NH_3]^2} = K_e$$

The concentration of solid AgCl, which is constant, is included in the value for K_e. The two equilibria in the solution are

$$AgCl(s) = Ag^+ + Cl^-$$

$$Ag(NH_3)_2^+ = Ag^+ + 2NH_3$$

The equilibrium constants are

$$[Ag^+][Cl^-] = K_{sp} = 1 \times 10^{-10}$$

$$\frac{[Ag^+][NH_3]^2}{[Ag(NH_3)_2^+]} = K_{complex} = 5 \times 10^{-8}$$

These two constants can be combined, dividing K_{sp} by $K_{complex}$, to give the equilibrium constant we need for the system.   This gives

$$\frac{[Ag(NH_3)_2^+][Cl^-]}{[NH_3]^2} = \frac{K_{sp}}{K_{complex}} = \frac{1 \times 10^{-10}}{5 \times 10^{-8}} = 0.2 \times 10^{-2}$$

Let x be the moles of AgCl that dissolve.   Since each mole AgCl furnishes a mole of Cl$^-$ ion and a mole of the complex ion, the concentrations of Ag(NH$_3$)$_2^+$ and of Cl$^-$ ions are each x/V moles per liter.   Since the volume, V, is 0.1 liter, the concentrations are $x/0.1$ or $10x$.   In 100 ml 1 M NH$_3$ we have 0.1 mole, but $2x$ moles are used in forming the complex ion.   This leaves $0.1 - 2x$ moles of NH$_3$, and the concentration is $\dfrac{0.1 - 2x \text{ moles}}{0.1 \text{ liters}}$ or $(1 - 20x)$ molar.

Substituting these concentrations into the equilibrium constant expression gives

$$\frac{(10x)(10x)}{(1 - 20x)^2} = 0.2 \times 10^{-2} = 20 \times 10^{-4}$$

Extract the square root of each side,

$$\frac{10x}{1 - 20x} = 4.5 \times 10^{-2}$$

Solving, $x = 0.0041$ moles AgCl that dissolve.   The weight of AgCl is 0.0041 moles $\times$ 143 g per mole, or 0.59 g.

If a similar calculation is made for the amount of silver iodide ($K_{sp} = 10^{-16}$) that dissolves in 100 ml 1 M NH$_3$, we find the amount to be quite small.   Thus, a rather complete separation of AgCl from AgI can be effected by treating the mixture with ammonia solution, filtering out the AgI, and reprecipitating the AgCl by neutralizing the solution with HNO_3.

Problem. Zinc sulfide is precipitated by adding H_2S to a solution of a zinc salt. The reaction is

$$Zn^{+2} + H_2S + 2H_2O = ZnS + 2H_3O^+$$

If 100 ml 0.1 M $ZnSO_4$ solution is saturated with H_2S, to a final concentration of 0.1 M, what fraction of the Zn^{+2} ion originally present is precipitated?

SOLUTION: The equilibrium constant for the reaction is

$$\frac{[H_3O^+]^2}{[Zn^{+2}][H_2S]} = K_e$$

The two equilibria in the solution are

$$ZnS(s) = Zn^{+2} + S^{-2}$$

$$H_2S + 2H_2O = 2H_3O^+ + S^{-2}$$

for which we have the constants

$$[Zn^{+2}][S^{-2}] = K_{sp} = 1 \times 10^{-20}$$

$$\frac{[H_3O^+]^2[S^{-2}]}{[H_2S]} = K_a = 1 \times 10^{-20}$$

The latter is obtained by combining the two ionization constants for H_2S (Table A–1, Appendix). Dividing K_a by K_{sp} gives

$$\frac{[H_3O^+]^2}{[Zn^{+2}][H_2S]} = \frac{1 \times 10^{-20}}{1 \times 10^{-20}} = 1$$

In 100 ml 0.1 M $ZnSO_4$ we have 0.01 moles Zn^{+2}. Let x be the number of moles of ZnS formed. There are $0.01 - x$ moles of Zn^{+2} remaining in solution, giving a concentration of $(0.01 - x)/0.1 = (0.1 - 10x)$ molar. Two moles of hydronium ion are formed for each mole of ZnS precipitated. The concentration of H_3O^+ is $2x/0.1$ liters, or $20x$. The concentration of H_2S in the saturated solution is 0.1 M. Substituting these numerical values for the concentrations gives

$$\frac{(20x)^2}{(0.1 - 10x)(0.1)} = 1$$

$$\frac{(20x)^2}{0.01 - x} = 1$$

$$400x^2 + x - 0.01 = 0$$

$$x = \frac{-1 + \sqrt{1 + 16}}{800} = 0.0039 \text{ moles ZnS}$$

The fraction of the zinc ion precipitated is $0.0039/0.01 = 0.39$, or 39 per cent.

It is recalled that to get more complete precipitation of zinc, as zinc sulfide, it is necessary to neutralize the hydronium ion by adding ammonia.

Problem. Cupric and zinc ions are separated in qualitative analysis by saturating 0.1 M solutions of the two ions with hydrogen sulfide, with the hydronium ion concentration adjusted to about 0.3 M before adding hydrogen sulfide. Initial concentrations of Cu^{+2} and Zn^{+2} are each 0.1 M. The final concentration of H_2S is 0.1 M. What is precipitated?

SOLUTION: The K_{sp} values for ZnS and CuS are respectively 10^{-20} and 10^{-36}. Since the constant for CuS is so very small, we make the initial assumption that practically all the cupric ion is precipitated. Each mole of CuS formed gives 2 moles H_3O^+, or 0.2 M H_3O^+ ion is formed. Adding this to the initial concentration gives a final concentration of 0.5 M for the hydronium ion.

The equilibrium expression for precipitation of CuS, developed as in the preceding section, is

$$\frac{[H_3O^+]^2}{[Cu^{+2}][H_2S]} = \frac{K_a}{K_{sp}} = \frac{10^{-20}}{10^{-36}} = 10^{16}$$

Substituting numerical values for the concentrations of hydronium ion and hydrogen sulfide, we have

$$\frac{(0.5)^2}{[Cu^{+2}](0.1)} = 10^{16}$$

$$[Cu^{+2}] = \frac{0.25}{(10^{16})(10^{-1})} = 0.25 \times 10^{-15} M$$

Thus, the precipitation of CuS is complete.

For precipitation of ZnS we have the relation developed in the preceding problem.

$$\frac{[H_3O^+]^2}{[Zn^{+2}][H_2S]} = 1$$

Substituting numerical values for concentrations of hydronium ion and hydrogen sulfide gives

$$\frac{(0.5)^2}{[Zn^{+2}](0.1)} = 1$$

$$[Zn^{+2}] = \frac{0.25}{(1)(0.1)} = 2.5 M$$

Since this equilibrium concentration is *greater* than the original concentration of zinc ion in the solution, no ZnS precipitates, and the separation of the two is complete.

Problem. Solid magnesium hydroxide dissolves in ammonium salt solutions to some extent. The reaction is

$$Mg(OH)_2(s) + 2NH_4^+ = Mg^{+2} + 2NH_3 + 2H_2O$$

How many grams of $Mg(OH)_2$ will dissolve in 1 liter 1 M NH_4Cl solution?

SOLUTION: The equilibrium constant equation is

$$\frac{[Mg^{+2}][NH_3]^2}{[NH_4^+]^2} = K_e$$

The equilibria in the solution are

$$Mg(OH)_2(s) = Mg^{+2} + 2OH^-$$

$$NH_3 + H_2O = NH_4^+ + OH^-$$

Combining the equilibrium constants for these two reactions in the same manner as in the preceding problems gives

$$\frac{[Mg^{+2}][NH_3]^2}{[NH_4^+]^2} = \frac{K_{sp}}{K_b{}^2} = \frac{1 \times 10^{-11}}{(1.8 \times 10^{-5})^2} = 0.031$$

Let x equal the number of moles Mg^{+2}, or since the volume is 1 liter, $x = [Mg^{+2}]$. Then, $2x = [NH_3]$ and $1 - 2x = [NH_4^+]$. Substituting the numerical values into the equilibrium constant gives

$$\frac{(x)(2x)^2}{(1 - 2x)^2} = 0.031$$

$$\frac{4x^3}{(1 - 2x)^2} = 0.031$$

The simplest solution for this cubic equation is by the method of successive approximations. We start by assuming that $2x$ is negligible in relation to 1. This gives

$$4x^3 = 0.031$$

$$x = 0.198$$

We now use this value for x in the denominator, and $1 - 2x$ becomes 0.604. The equation now is

$$\frac{4x^3}{(0.604)^2} = 0.031$$

$$x = 0.141$$

Substituting this value for x gives us the third approximation

$$\frac{4x^3}{(0.718)^2} = 0.031$$

Solving, $x = 0.158$.

Continue in this manner until successive solutions give a value of x that is within 1 per cent of the previous one. The sixth solution gives a nearly constant value of $x = 0.154$. The weight of $Mg(OH)_2$ dissolving is 0.154 moles $\times$ 58.3 g per mole = 9.0 g.

Problem. Aluminum hydroxide dissolves in a strong base, according to the equation

$$Al(OH)_3(s) + OH^- = Al(OH)_4{}^-$$

The dissociation constant for the tetrahydroxoaluminate (III) complex is

$$\frac{[OH^-]}{[Al(OH)_4{}^-]} = 2.5 \times 10^{-2}$$

If solid $Al(OH)_3$ is treated with 100 ml 0.1 M NaOH, how many grams dissolve?

SOLUTION: Let x equal the $[Al(OH)_4{}^-]$ in solution and $0.1 - x$ the $[OH^-]$. Substituting these values into the dissociation constant gives

$$\frac{0.1 - x}{x} = 0.025$$

$$x = 0.098 \ M$$

In 100 ml there are 0.0098 moles, or 0.76 g.

Equilibria Involving Ions from Water

In the previous treatments of dissociations of weak acids or bases and of the completeness of neutralization reactions, we have neglected the contributions of the ions from water to the concentrations of hydronium or hydroxide ions. This causes our previously developed equations to be only approximations, but under the conditions in which they have been used the approximations are sufficiently exact. In this section we shall set up exact statements of the equilibria and then make approximations as differing conditions may require. By this treatment we are at all times aware of just what the approximations are and can more critically evaluate results of computations than is possible with the simplified previous treatment.

pH of Salt Solutions. When a salt of a weak acid with strong base, a weak base with a strong acid, or a weak acid with a weak base is dissolved in water, there are two competing equilibria to consider. We shall illustrate the method by considering the salt of a weak acid with a strong base. Since the symbols become somewhat complicated, we shall for the remainder of this chapter use H^+ for the hydronium ion instead of H_3O^+. In using this symbol, however, the student is cautioned not to lose sight of the effect of water in combining with the proton.

Consider a solution of a salt, NaA, where A^- is the anion (or conjugate base) of a weak acid, HA. When the salt dissolves, a small portion of the A^- unites with hydronium ion from water, forming undissociated HA molecules. There are two equilibria to consider:

$$H_2O = H^+ + OH^-$$

$$A^- + H^+ = HA$$

and we have the two consonants

(1) $$[H^+][OH^-] = K_w$$

(2) $$\frac{[HA]}{[H^+][A^-]} = \frac{1}{K_a}$$

Dissociation of water gives equal concentrations of H^+ and OH^- ions, but some of the H^+ ions are removed by formation of HA. Consequently we have the relation

(3) $$[OH^-] = [H^+] + [HA]$$

From (1) and (2) we have

$$[OH^-] = \frac{K_w}{[H^+]}$$

$$[HA] = \frac{[H^+][A^-]}{K_a}$$

Substituting these values into (3) gives

$$\frac{K_w}{[H^+]} = [H^+] + \frac{[H^+][A^-]}{K_a}$$

$$K_a K_w = K_a[H^+]^2 + [H^+]^2[A^-]$$

$$[H^+]^2(K_a + [A^-]) = K_a K_w$$

$$[H^+]^2 = \frac{K_a K_w}{K_a + [A^-]}$$

This is an exact expression, involving no assumptions, but to get a simple solution we must make an approximation,[1] since we do not precisely know the concentration of A^- ions in the solution. For this approximation we shall assume that the concentration of A^- ions is C, the concentration of the salt in the solution. Thus we are neglecting the small fraction of the A^- ions that react to form HA. We now have

(4) $$[H^+]^2 = \frac{K_a K_w}{C + K_a}$$

When C is much larger than K_a, this reduces to

$$[H^+]^2 = \frac{K_a K_w}{C}$$

but when C is very small, we use equation (4). The following problems illustrate conditions for using each form of the equation.

Problem. What is the pH of a 0.01 M solution of salt NaA, where A^- is the anion of an acid whose ionization constant is 10^{-6}?

SOLUTION: Here C is much greater than K_a, and we use the equation

$$[H^+] = \sqrt{\frac{K_a K_w}{C}}$$

---

[1] Intelligent use of approximations for solving equations is a useful tool in scientific work. The phenomena of nature are complex, and rigorous mathematical expressions to describe these phenomena are often impossible. Frequently, we can make an approximation, which is an educated guess, to simplify the mathematical solution. When this is done, we should always re-examine the final solution to see that the approximation has not introduced a gross error into the computation.

Substituting numerical values gives

$$[H^+] = \sqrt{\frac{(10^{-6})(10^{-14})}{10^{-2}}}$$

$$= \sqrt{10^{-18}}$$

$$= 10^{-9}$$

$$pH = 9.0$$

Problem. What is the pH of a $10^{-6}\,M$ solution of the salt used in the preceding problem?

SOLUTION: Since the concentration is of the same magnitude as K_a, we must use equation (4). Substituting numerical values gives

$$[H^+]^2 = \frac{(10^{-6})(10^{-14})}{10^{-6} + 10^{-6}} = 0.5 \times 10^{-14}$$

$$[H^+] = 0.71 \times 10^{-7}$$

$$pH = 7.15$$

We can show that the approximation used in developing the equations has not introduced an appreciable error into the computations.

For the $0.01\,M$ solution of NaA we found an hydronium ion concentration of 10^{-9}. Substituting this value for H^+ into equation (2), we have

$$\frac{[HA]}{(10^{-9})(10^{-2})} = \frac{1}{10^{-6}}$$

This gives $[HA] = 10^{-5}\,M$, which is negligible in comparison with $[A^-]$ (about 0.1 per cent).

In the very dilute solution ($10^{-6}\,M$), the computed $[H^+]$ concentration is $0.71 \times 10^{-7}\,M$. Substituting this into equation (2) and solving for $[HA]$,

$$\frac{[HA]}{(0.71 \times 10^{-7})(10^{-6})} = \frac{1}{10^{-6}}$$

$$[HA] = 0.71 \times 10^{-7}$$

Thus, **7** per cent of the A^- ion has reacted to form HA. Even so, the use of the approximate formula has given a true result for $[H^+]$. Substituting the value of $[A^-] = C - [HA] = 0.93 \times 10^{-6}$ in equation (4) and solving, we have

$$[H^+]^2 = \frac{(10^{-6})(10^{-14})}{0.93 \times 10^{-6} + 10^{-6}}$$

$$[H^+] = 0.72 \times 10^{-7}$$

which agrees closely with the value obtained using the approximate formula.

Salts of Polyprotic Acids.   The equilibria involved in neutralization of a polyprotic acid are more complex than those we have seen for the simple monoprotic acids discussed in Chapter 7. To illustrate, we shall discuss the pH of a solution of bicarbonate ion, HCO_3^-. This ion may be formed by adding CO_2 to a base, such as $NaOH$, or by adding an acid to a solution of a carbonate, as was done in the titration curve of Fig. 7–4.

In a solution of $NaHCO_3$ we have the following reactions

(1) $\qquad\qquad H_2O = H^+ + OH^- \qquad\qquad [H^+][OH^-] = K_w$

(5) $\qquad\qquad HCO_3^- + H^+ = H_2O + CO_2 \qquad \dfrac{[CO_2]}{[H^+][HCO_3^-]} = \dfrac{1}{K_1}$

(6) $\qquad\qquad HCO_3^- = H^+ + CO_3^{-2} \qquad\qquad \dfrac{[H^+][CO_3^{-2}]}{[HCO_3^-]} = K_2$

The concentrations of the reactants are related by the equation

(7) $\qquad\qquad [H^+] = [OH^-] + [CO_3^{-2}] - [CO_2]$

The validity of this is seen by inspection of the equations just given. A hydrogen (hydronium) ion is formed for each OH^- and each CO_3^{-2} ion in the solution, and one is removed for each molecule of CO_2 formed.

From the equilibrium constants we have

$$[OH^-] = \frac{K_w}{[H^+]} \qquad\qquad \text{from (1)}$$

$$[CO_3^{-2}] = \frac{K_2[HCO_3^-]}{[H^+]} \qquad\qquad \text{from (6)}$$

$$[CO_2] = \frac{[H^+][HCO_3^-]}{K_1} \qquad\qquad \text{from (5)}$$

Substituting these values into equation (7) gives

$$[H^+] = \frac{K_w}{[H^+]} + \frac{K_2[HCO_3^-]}{[H^+]} - \frac{[H^+][HCO_3^-]}{K_1}$$

Rearranging and solving, we have

$$[H^+]^2 = K_w + K_2[HCO_3^-] - \frac{[H^+]^2[HCO_3^-]}{K_1}$$

$$[H^+]^2 \left(\frac{K_1 + [HCO_3^-]}{K_1} \right) = K_w + K_2[HCO_3^-]$$

$$[H^+]^2 = \frac{K_1 K_w + K_1 K_2 [HCO_3^-]}{K_1 + [HCO_3^-]}$$

The concentration of HCO_3^- in solution is the salt concentration, C, less the concentrations of CO_2 and CO_3^{-2} formed from the salt.

$$[HCO_3^-] = C - [CO_2] - [CO_3^{-2}]$$

Usually we are interested in solutions with bicarbonate ion concentrations of the order of 0.05–0.1 M, which are large in comparison to the concentrations of CO_2 and CO_3^{-2}. Assuming that $[HCO_3^-] = C$, we have

$$[H^+]^2 = \frac{K_1 K_w + K_1 K_2 C}{C + K_1}$$

The numerical value of K_1 is 4×10^{-7}, of K_2 is 5×10^{-11}, and of K_w is 10^{-14}. With solutions of the usual concentration of salt $C \gg K_1$ and $K_2 C \gg K_w$. The equation therefore reduces to

$$[H^+]^2 = K_1 K_2 = 2 \times 10^{-17}$$

$$pH = 8.35$$

Note that when $C \gg K_1$ and $K_2 C \gg K_w$, the pH of the bicarbonate solution is independent of its concentration.

Ionization of Weak Acids. The treatment given in Chapter 7 for the ionization of weak acids is not valid for very weak acids or for very dilute solutions. In both of these the contribution of the hydronium ion from water must be taken into account.

We shall consider dissociation of acid HA. In the solution we have the reactions

(1) $\qquad H_2O = H^+ + OH^- \qquad\qquad [H^+][OH^-] = K_w$

(8) $\qquad HA = H^+ + A^- \qquad\qquad \dfrac{[H^+][A^-]}{[HA]} = K_a$

These reactions show that the concentrations are related by

(9) $\qquad\qquad [H^+] = [OH^-] + [A^-]$

From equation (1) we have

$$[OH^-] = \frac{K_w}{[H^+]}$$

and from (8)

$$[A^-] = \frac{K_a[HA]}{[H^+]}$$

Substituting these values into (9) gives

$$[H^+] = \frac{K_w}{[H^+]} + \frac{K_a[HA]}{[H^+]}$$

(10) $$[H^+]^2 = K_w + K_a[HA]$$

The concentration of HA is the initial concentration of the acid, C, less the concentration of A^- ions formed in the dissociation. That is,

$$[HA] = C - [A^-]$$

As we have seen in Chapter 7, acids with ionization constants of the magnitude of 10^{-5} are but slightly dissociated in solutions of the order of 0.1 M. When this is true, we make the approximaton that

$$C - [A^-] \sim C$$

or that

$$[HA] = C$$

When this is valid and $CK_a \gg K_w$, equation (10) reduces to

$$[H^+]^2 = CK_a$$

which is the expression obtained in Chapter 7.

When $[A^-]$ is not negligible in comparison with C, it is necessary to make approximations for the solution of the cubic equation. The general method is illustrated in the following discussion. To simplify treatment of the mathematics we replace the various quantities by algebraic symbols. Let $x = [H^+]$, $y = [OH^-]$, $z = [A^-]$, and $C - z = [HA]$. From equation (9) we have

(11) $$x = y + z$$

Rewriting (1) and (8) with these symbols gives

(12) $$xy = K_w$$

(13) $$\frac{xz}{C - z} = K_a$$

Solving (12) for y gives

(14) $$y = \frac{K_w}{x}$$

and solving (13) for z gives

(15) $$z = \frac{CK_a}{x + K_a}$$

Substituting the values for y and z into (11), we have

$$x = \frac{K_w}{x} + \frac{CK_a}{x + K_a}$$

Solving gives

(16) $$x^3 + K_a x^2 - x(K_w + CK_a) - K_a K_w = 0$$

The following examples illustrate various types of approximations used in solving (16).

Problem. Compute the degree of dissociation, α, for a 0.1 M solution of acid HA, whose ionization constant is 1×10^{-5}.

SOLUTION: Substitute numerical values of the constants into (16).

$$x^3 + 10^{-5}x^2 - x(10^{-14} + 10^{-1} \cdot 10^{-5}) - 10^{-5} \cdot 10^{-14} = 0$$

We assume that $x \gg 10^{-7}$, since the acid is comparable to acetic acid in strength, and we know from experience that an acetic acid solution of this concentration has a pH near 3. If x is 10^{-5} or greater, each of the first three terms is much greater than 10^{-19}, the last term. We therefore simplify the equation by dropping the last term. This gives

$$x^3 + 10^{-5}x^2 - 10^{-6}x = 0$$

Factoring,

$$x(x^2 + 10^{-5}x - 10^{-6}) = 0$$

Since x is not zero, we have

$$x^2 + 10^{-5}x - 10^{-6} = 0$$

$$x = \frac{-10^{-5} + \sqrt{10^{-10} + 4 \times 10^{-6}}}{2} = 1 \times 10^{-3}$$

Substituting this value of x into (15) and solving for z,

$$z = \frac{10^{-1} \cdot 10^{-5}}{10^{-3} + 10^{-5}} = 10^{-3}$$

$$\alpha = \frac{z}{C} = \frac{10^{-3}}{10^{-1}} = 10^{-2} \qquad \text{or } 1\%$$

Problem. What is the degree of dissociation in a 10^{-5} M solution of acid HA?

SOLUTION: Substitute values of the constants into (16). This gives

$$x^3 + 10^{-5}x^2 - x(10^{-14} + 10^{-10}) - 10^{-19} = 0$$

We note that x cannot exceed 10^{-5}, since this is the initial concentration of HA. Assuming that dissociation is almost complete and substituting this value for x, we see that each of the first three terms is of the order of 10^{-15} and the last is 10^{-19}. This term is dropped, giving

$$x^3 + 10^{-5}x^2 - 10^{-10}x = 0$$

Factoring and solving as in the preceding problem gives

$$x = 0.62 \times 10^{-5}$$

Using this value of x in (15) and solving for z, we have

$$z = 0.62 \times 10^{-5}$$

$$\alpha = \frac{0.62 \times 10^{-5}}{10^{-5}} = 0.62 \qquad \text{or } 62\%$$

Problem. Compute the degree of dissociation in a 10^{-7} M solution of acid HA.
SOLUTION: Substitute numerical values of the constants into (16)

$$x^3 + 10^{-5}x^2 - x(10^{-14} + 10^{-12}) - 10^{-19} = 0$$

At this low concentration we may assume that the acid is essentially all ionized, or that H^+ is about 10^{-7} M. Using this value for x, we see that the first term is of the order of 10^{-21} and the other terms of the order of 10^{-19}.

We, therefore, drop the first term, giving

$$10^{-5}x^2 - 10^{-12}x - 10^{-19} = 0$$

Dividing by 10^{-5},

$$x^2 - 10^{-7}x - 10^{-14} = 0$$

$$x = 1.6 \times 10^{-7}$$

Substituting this value into (15)

$$z = \frac{10^{-12}}{1.6 \times 10^{-7} + 10^{-5}} = \frac{10^{-12}}{1.02 \times 10^{-5}} = 0.98 \times 10^{-7}$$

$$\alpha = \frac{z}{C} = \frac{0.98 \times 10^{-7}}{1 \times 10^{-7}} = 0.98 = 98\%$$

Problems

1. Consider the effect of the ionization of water, and make a reasonable approximation of the solubility of cupric hydroxide from its solubility product.

2. A solution contains hydrogen, copper, and zinc ions at a concentration of 0.1 M. This solution is saturated with hydrogen sulfide, to make a final concentration of 0.1 M. Compute the concentration of copper and zinc ions remaining in solution.

3. Can manganous sulfide, MnS, be quantitatively precipitated by saturating a 0.05 M $MnCl_2$ solution with H_2S?

4. Calculate the pH of a buffer which will permit all but 0.1 mg of Zn^{+2} to be precipitated from 200 ml 0.1 M $ZnSO_4$ solution when it is saturated with H_2S. How many grams of formic acid and sodium formate should be added to the solution to prepare this buffer? (Neglect the effects of dilution; use sufficient sodium formate to have 50 per cent in excess of the amount required to react with the H^+ produced by the precipitation.)

5. How much silver iodide will dissolve if the solid salt is shaken with 250 ml concentrated ammonia (15 M)?   *Ans.* 1.7×10^{-4} mole.

6. A solution contains the following ions at the stated concentrations:
   $[Fe^{+3}] = 0.1\ M$ $[H^+] = 0.1\ M$ $[SO_4^{-2}] = 0.2\ M$ $[Cl^-] = 0.1\ M$
From the point of view of the electrical neutrality of solutions, must any other ions also be present?

7. How many equations would have to be set up to determine precisely the pH of a 0.1 M K_2HPO_4 solution?

8. Find the pH of a solution made by diluting one drop (0.04 ml) of 0.10 N HCl to 10 liters.   *Ans.* 6.37.

9. Repeat problem 8, but use HOAc.

10. Find the pH of 0.1 M solutions of:
    (a) $NaHSO_3$.
    (b) KHS.   *Ans.* (a) 4.61.

11. Compute the degree of hydrolysis for 10^{-1}, 10^{-3}, 10^{-5}, 10^{-7} M solutions of sodium acetate. Construct a curve showing α as a function of pC ($-$ log concentration). The degree of hydrolysis is the ratio of the concentration of acetic acid to the original concentration of salt present.

12. Make a computation similar to that of problem 8 for the hydrolysis of ammonium chloride, and plot a curve showing α as a function of pC.

13. Compute the degree of ionization for HOAc at concentrations of 10^{-1}, 10^{-3}, 10^{-5}, 10^{-7}, 10^{-9} M. Plot α versus $-$ log C.

14. Will $Fe(OH)_3$ precipitate from a solution which is 0.1 M in KOH and 0.2 M in $K_3Fe(CN)_6$? The dissociation constant for $K_3Fe(CN)_6$ is 10^{-44}.

15. What concentrations of Fe^{+3} and Zn^{+2} ions are in equilibrium with their hydroxides in a buffer solution made by adding 20 g NaOAc to 150 ml 0.15 M HOAc?   *Ans.* $[Fe^{+3}] = 5 \times 10^{-11}\ M$.

CHAPTER 10

Oxidation-Reduction Analyses

Reactions in aqueous solutions are of two general types, metathesis, or double decomposition, and oxidation-reduction. We have seen applications of metathesis reactions in neutralization and in precipitation. In this chapter we shall discuss the general methods of oxidation-reduction or redox analyses, which have widespread applications.

Oxidation-reduction reactions are those in which there are changes in the oxidation states of two substances. These changes are due to the transfer of one or more electrons from one atom or ion to another. The substance losing electrons is *oxidized* and the one gaining electrons is *reduced*. Thus, in the reaction

$$Ce^{+4} + Fe^{+2} = Ce^{+3} + Fe^{+3}$$

ferrous ion is oxidized and ceric ion is reduced. The ceric ion is the oxidizing agent and the ferrous ion the reducing agent.

Since an oxidation-reduction involves the transfer of electrons from one substance to another, the two reactions must occur simultaneously. We cannot have an oxidation without an equivalent reduction within the system.

Any oxidation-reduction reaction between ions in solution may be used as the basis for a volumetric analysis method if it fulfills the conditions that (1) there is only one reaction under the given conditions, (2) this reaction goes essentially to completion at the equivalence point, and (3) a suitable indicator (or other means) is available to locate the end point. The number of reactions fulfilling these conditions is very large. Fortunately, however, the general principles involved are common to all, and a limited study of only a few reactions is sufficient to give a good understanding of the entire field of redox analyses. The general methods of redox analyses fall in one of the following three categories.

1. A solution of a susbtance that is readily oxidized is titrated by a standard solution of a strong oxidizing agent. By "strong" oxidizing agent we mean one that has a large attraction for electrons. The most widely used strong oxidizing agents for volumetric analyses are MnO_4^- ion in acid solution, $Cr_2O_7^{-2}$ ion in acid solution, Ce^{+4} ion in acid solution, I_2 in I^- solution, and MnO_4^- ion in basic solution. Applications of all but the last of these are discussed in the following chapters. They suffice for analysis of almost all the ions that are good reducing agents.

In titrations of reducing agents by a strong oxidizing agent, it is necessary to pretreat the sample before the titration, to insure that all the sought substance is reduced to its lower oxidation state and to insure that no reducing agent other than the sought material is in the solution. These conditions may be illustrated by the procedure employed for analysis of iron samples. The iron sample is dissolved and the solution treated with a reducing agent that converts any ferric ion present to ferrous ion. One of the reducing agents often employed is metallic zinc, for which the reaction is

$$2Fe^{+3} + Zn = 2Fe^{+2} + Zn^{+2}$$

After reduction is completed, the metallic zinc is removed from the solution, which is then titrated by a standard solution of $KMnO_4$. The reaction[1] is

$$MnO_4^- + 5Fe^{+2} + 8H^+ = Mn^{+2} + 5Fe^{+3} + 4H_2O$$

2. If the sample is a strong oxidizing agent, it may be analyzed by titrating its solution with a solution of a reducing agent. The reducing agents most widely used in this way are ferrous salt solutions and arsenious acid.

Samples of oxidizing agents that are not water soluble may be analyzed by treating a known weight of the sample with a measured volume of a standard reducing agent and, after reaction is completed, back-titrating the excess of reducing agent in the solution with a

---

[1] The symbol H^+ is used throughout this chapter and the following one to designate the hydronium ion, H_3O^+.

standard solution of an oxidizing agent. Thus, the amount of MnO_2 in a sample is determined by treating a weighed portion with a measured volume of standard $FeSO_4$ solution and back-titrating the excess $FeSO_4$ with standard permanganate solution.

3. An indirect method is frequently used for analyses of oxidizing agents. The sample is treated with a solution of KI, and the liberated iodine is titrated by a solution of $Na_2S_2O_3$ (sodium thiosulfate) which reacts rapidly and quantitatively with I_2

$$I_2 + 2S_2O_3^{-2} = 2I^- + S_4O_6^{-2}$$

So universally applicable are these three general methods that an analytical laboratory frequently keeps a stock of only three standard reagents: (1) $KMnO_4$, a strong oxidizing agent; (2) H_3AsO_3, a strong reducing agent; and (3) $Na_2S_2O_3$, a specific reagent for I_2. The preparation, standardization, and use of these important reagents and of a few others of general interest are treated in some detail in the experimental section of this text.

Indicators for Redox Titrations

Until fairly recently each oxidizing agent had its own specific indicator, a situation tending to restrict the number of oxidizing agents employed in analytical chemistry. In the past twenty-five years, however, a large family of general-purpose redox indicators has been developed. These indicators are organic dyes which can be oxidized or reduced reversibly and which undergo a color change when they pass from one oxidation state to the other. The selection of redox indicators for specific titrations is discussed in the following chapter. Briefly, the general principle is that the indicator must be a substance that is oxidized or reduced at or near the equivalence point of the titration. This means that if we are titrating a reducing agent, the indicator should be a substance that is more difficult to oxidize than the sample material, thus it does not become oxidized and give an end point until all or practically all the sample has first been oxidized.

The two most widely used general-purpose redox indicators are:

1. Diphenylamine sulfonic acid, normally used as the soluble barium or sodium salt.

2. Ferrous complex of orthophenanthroline, sold under the trade name of Ferroin.

The former is the stronger reducing agent and is the one employed with the weaker oxidizing agents, such as dichromate ion. Ferroin is used with the stronger oxidizing agents, such as potassium permanganate and ceric sulfate.

Long before the advent of general-purpose redox indicators, specific indicators had been developed for certain oxidizing agents. These are so useful that they are still preferred for many analyses. The more common specific redox indicators are:

1. Color of the permanganate ion. This ion is so highly colored that a single drop will impart a distinct pink tinge to a liter or more of solution. Also, when reduced the MnO_4^- ion gives (in acid solution, where normally employed) the colorless Mn^{+2} ion. It was this property of self-indication that led to the early prominence of $KMnO_4$ as an oxidation reagent. The colored end point appears with the first drop of MnO_4^- in excess, after all the sample is oxidized.

2. Starch indicator for I_2. A colloidal suspension of starch will, in the presence of iodide ion, give an intense blue color with free I_2. This color is used to detect the end point in iodine titrations.

3. Spot-test indicator. In some titrations spot tests are made as the equivalence point is approached. This is done by withdrawing a drop of solution after each addition of reagent and testing this drop for the constituent that is being oxidized or reduced. The test is made by adding the drop, which is withdrawn on a stirring rod, to a drop of a suitable reagent. An example is the use of potassium ferricyanide as the spot-test indicator in the oxidation of ferrous ion. As long as the titrated solution contains any ferrous ion, the spot test gives the intense blue color characteristic of the reaction of Fe^{+2} with ferricyanide ion. The end point is that point at which the withdrawn drop gives no blue color with the reagent. Titrations with spot-test indicators are tedious and are not used when other indicators are available.

4. Potentiometric end point. As shown in later chapters, the course of most oxidation-reduction titrations can be followed by means of the voltage difference between two electrodes placed in the sample solution. A titration curve can be obtained, similar to the titration curves previously described for acid-base titrations. These can be used to determine the end point.

Oxidation-Reduction Equations

In an ideal educational system every student who has the prerequisites to enter a course in quantitative analysis has already learned how to write and balance oxidation-reduction equations. In practice this ideal is not universally achieved, since the large amount of material to be covered in general chemistry limits the time that may be given to redox equations. Consequently, the writing of such equations is reviewed at this point.

Since oxidation-reduction reactions are attended by the transfer of one or more electrons from the reducing agent to the oxidizing agent, it is convenient to use the electron transfer as the basis for balancing the equations.

The first step in balancing any equation is to know the products formed in the reaction. We do not, however, have to memorize the products for all possible oxidation-reduction reactions. Rather, we learn what products are formed from a selected list of oxidizing agents and from a similar list of reducing agents. Then, for any combination of these agents we know what products to expect. Thus, when MnO_4^- is used as the oxidizing agent in acid solution, we always have Mn^{+2} and H_2O as the product, regardless of the reducing agent used.

A list of the more common oxidizing agents and their products is given in Table 10–1 and a corresponding list of reducing agents in

TABLE 10–1. Common Oxidizing Agents and Their Products

Reactant	Medium	Products
$MnO_4^- + H^+$	acid	$Mn^{+2} + H_2O$
$Cr_2O_7^{-2} + H^+$	acid	$Cr^{+3} + H_2O$
Ce^{+4}	acid	Ce^{+3}
$ClO_3^- + H^+$	acid	$Cl^- + H_2O$
$BrO_3^- + H^+$	acid	$Br^- + H_2O$
$IO_3^- + H$	acid	$I^- + H_2O$
$NO_3^- + H^+$	acid	$NO + H_2O*$
$H_2O_2 + H^+$	acid	H_2O
$MnO_2(s) + H^+$	acid	$Mn^{+2} + H_2O$
$PbO_2(s) + H^+$	acid	$Pb^{+2} + H_2O$
Fe^{+3}	acid	Fe^{+2}

*This is the most common reaction in dilute solution. Other products frequently obtained are NO_2, N_2, and NH_4^+.

Table 10–2. These brief lists should be memorized, for with them the student can readily write most of the redox equations he will encounter. The formulas of these tables are given in ionic form, i.e., only the essential ions are shown for strong electrolytes. Molecular formulas are given for weak electrolytes, and solids are designated by the italic (s). The abbreviation H^+ is used for the hydronium ion.

TABLE 10–2. Common Reducing Agents and Their Products

Reactant	Medium	Products
$HNO_2 + H_2O$	acid	$NO_3^- + H^+$
I^-	acid or neutral	I_2
$C_2O_4^{-2}$	acid	CO_2
H_2S	acid	$S(s) + H^+$
$SO_2 + H_2O$	acid	$SO_4^{-2} + H^+$
$H_3AsO_3 + H_2O$	acid or neutral	$H_3AsO_4 + H^+$
$S_2O_3^{-2}$	acid or neutral	$S_4O_6^{-2}$*
Fe^{+2}	acid	Fe^{+3}
Sn^{+2}	acid	Sn^{+4}
$Zn(s)$	acid	Zn^{+2}

* This reaction is specific for I_2. Other oxidizing agents may oxidize $S_2O_3^{-2}$ to SO_4^{-2}.

Various methods may be used to balance an oxidation-reduction equation for which the reactants and their products are known. The following is simple and easy to remember.

Step 1. Write the oxidizing agent and its products as a half reaction. Using the permanganate ion in acid solution as an example, we have from Table 10–1

$$MnO_4^- + H^+ = Mn^{+2} + H_2O$$

Step 2. Balance the half-reaction equation

$$MnO_4^- + 8H^+ = Mn^{+2} + 4H_2O$$

Step 3. Add the number of electrons needed to make the sum of the charges on the left equal the sum of the charges on the right. In the half reaction above we have on the left $-1 + 8 = +7$ charges. On the right we have $+2$. We therefore add $5e^-$ to the left-hand side. The completed half reaction is now

(1) $$MnO_4^- + 8H^+ + 5e^- = Mn^{+2} + 4H_2O$$

Step 4. Develop the balanced half reaction for the reducing agent in the same way. Using ferrous ion as an example, we have

(2) $$Fe^{+2} = Fe^{+3} + e^-$$

Step 5. Balance the number of electrons used in the two half reactions. In equation (1) we have five electrons and in (2) we have one electron. To balance we must multiply equation (2) by 5. This gives

(3) $$5Fe^{+2} = 5Fe^{+3} + 5e^-$$

Step 6. Add the two half reactions. This gives the balanced final equation

(4) $\quad MnO_4^- + 8H^+ + 5Fe^{+2} = Mn^{+2} + 4H_2O + 5Fe^{+3}$

Note that when the two half reactions are added the electrons cancel. This is a necessary condition. Another condition is that the sum of the ionic charges on the two sides of the equation be equal. Applying this test to equation (4), we have a net of 17^+ charges on each side, which verifies the equation.

In this treatment we have used only the essential ions, which greatly simplifies writing and balancing the equation. Often we need to express a redox equation in molecular form, particularly when it is used as the basis for stoichiometric calculations. After working out the ionic equation, it is a simple matter to develop the molecular equation, merely taking of each species the number of molecules required to provide the number of ions we find in the balanced ionic equation. For example, we use the ionic equation of (4) to write the equation for the reaction of potassium permanganate with ferrous sulfate in sulfuric acid solution.

(5) $\quad KMnO_4 + 4H_2SO_4 + 5FeSO_4$
$$= MnSO_4 + 4H_2O + \tfrac{5}{2}Fe_2(SO_4)_3 + \tfrac{1}{2}K_2SO_4$$

Note that $4H_2SO_4$ is needed to give $8H^+$, that the products show the Mn^{+2} and the Fe^{+3} ions as the sulfates, since the anion of the solution is the SO_4^{-2} ion, and that the K^+ ion added as $KMnO_4$ appears as K_2SO_4. These are the products obtained if the solution is evaporated to dryness after the reaction is completed.

The fractional coefficients of equation (5) are converted to whole numbers if we multiply through by 2. This gives

$2KMnO_4 + 8H_2SO_4 + 10FeSO_4$
$$= 2MnSO_4 + 8H_2O + 5Fe_2(SO_4)_3 + K_2SO_4$$

As a final check of the correctness of the equation, we count the number of sulfate ions on the two sides of the equation. We find eighteen SO_4^{-2} ions on each side.

When one of the reactants is a weak electrolyte, we write it in the molecular form. To illustrate, we write the equation for reduction of $K_2Cr_2O_7$ in H_2SO_4 solution by H_2S. The two half reactions are (Table 10–1 and 10–2)

(6) $\quad Cr_2O_7^{-2} + 14H^+ + 6e^- = 2Cr^{+3} + 7H_2O$

(7) $\quad H_2S = S + 2H^+ + 2e^-$

Multiply equation (7) by 3, to balance the electrons, and add.

$$Cr_2O_7^{-2} + 14H^+ + 3H_2S = 2Cr^{+3} + 7H_2O + 6H^+$$

Since hydronium ion appears on both sides of the equation, we simplify by subtracting $6H^+$ from each side. This gives the equation

(8) $$Cr_2O_7^{-2} + 8H^+ + 3H_2S = 2Cr^{+3} + 7H_2O + 3S$$

This is used to write the molecular equation

(9) $K_2Cr_2O_7 + 4H_2SO_4 + 3H_2S$
$$= Cr_2(SO_4)_3 + 7H_2O + 3S + K_2SO_4$$

Oxidation Numbers. It is frequently convenient to compute the number of electrons involved in redox reactions from the changes in the oxidation numbers of reactants. The oxidation number of an atom is a number representing the electric charge that atom must have if the electrons in a compound are assigned in a definite manner. Thus, if a neutral atom has gained an electron, its oxidation number is -1; or if it has lost an electron, its oxidation number is $+1$. The following general rules are used in assigning oxidation numbers.

1. The oxidation number of an element is 0, regardless of its molecular formula.

2. The oxidation number of a monatomic ion is the charge of the ion. Thus, the oxidation numbers are $+1$ for Na^+ and H^+ ions, $+2$ for Cu^{+2} ion, etc.

3. The oxidation number of H is $+1$ except in hydrides of metals, for which it is -1.

4. The oxidation number of oxygen in compounds is -2 except in peroxides, for which it is -1.

5. The oxidation number of the central atom in a coordination complex is the charge that atom needs for the molecule to be electrically neutral, taking into account the oxidation number assigned to other atoms of the complex. Thus, in H_2SO_4 the oxidation number of S is $+6$, since we have two hydrogen atoms of oxidation number $+1$ and four oxygen atoms of oxidation number -2.

It should be noted that the assignment of oxidation numbers is entirely arbitrary and not necessarily related to true valences of atoms. For example, in Fe_3O_4 the oxidation number of the Fe atoms is $\frac{8}{3}$, since three Fe atoms carry the plus charges to balance the eight negative charges of the four oxygen atoms. If Fe_3O_4 is oxidized, one electron is lost per molecule, to give three Fe^{+3} ions.

Changes in oxidation numbers show the number of electrons gained and lost per molecule, atom, and ion of reactant. Thus, when MnO_4^-

ion is reduced, the product is Mn^{+2}. The oxidation number of Mn changes from $+7$ to $+2$, which requires the gain of five electrons. Note that we arrived at the same conclusion in balancing the half reaction, as described in the preceding section.

Stoichiometry of Redox Reactions

All the calculations of redox reactions may be handled by use of molarities of solutions, as described in Chapter 5, but in titrimetric analyses it is convenient to use normalities, as was done in the calculations for neutralization reactions. The advantage in using normalities of solutions and equivalents or milliequivalents is that the number of equivalents of the oxidizing agent is equal to the number of equivalents of reducing agent.

The equivalent weight in a redox reaction is defined as the weight of reagent required to oxidize a gram-atom of hydrogen or to reduce a gram-atom of hydrogen ion. Since oxidation or reduction of a hydrogen atom or ion requires gain or loss of one electron, we have the simple relation

$$\text{Equivalent weight} = \frac{\text{molecular weight}}{\text{number of electrons gained or lost}}$$

Thus, the equivalent weight of $KMnO_4$ is $\frac{1}{5}$ mole and the equivalent weight of $K_2Cr_2O_7$ is $\frac{1}{6}$ mole. Likewise a milliequivalent of $KMnO_4$ is $\frac{158.04}{5}$ mg and a milliequivalent of $K_2Cr_2O_7$ is $\frac{294.21}{6}$ mg.

Typical Calculations

Problem.   Concentration of Solutions.   What weight of pure potassium dichromate is needed for the preparation of 500 ml 0.1000 N solution?

SOLUTION: When potassium dichromate is used as an oxidizing agent, six electrons are gained per molecule. The equivalent weight is, therefore,

$$\frac{K_2Cr_2O_7}{6} = \frac{294.21}{6} = 49.04 \text{ g}$$

The milliequivalent weight is 49.04 mg. In 500 ml 0.1000 N solution the total number of milliequivalents is

$$500 \times 0.1000 = 50.00 \text{ meq}$$

Since the weight of 1 meq is 49.04 mg, the total weight in milligrams is given by the product,

$$50.00 \times 49.04 = 2452 \text{ mg or } 2.452 \text{ g}$$

Problem. Standardization of Solution. A solution of permanganate is standardized by comparison with sodium oxalate. From the following data calculate the normality of the permanganate.

$$\text{Volume permanganate} = 40.41 \text{ ml}$$
$$\text{Weight sodium oxalate} = 0.2538 \text{ g}$$
$$\text{Purity of oxalate} = 99.60\%$$

SOLUTION: The half reaction for oxidation of oxalate ion is (Table 10–2)

$$C_2O_4^{-2} = 2CO_2 + 2e^-$$

Thus the equivalent weight of $Na_2C_2O_4$ is $\frac{1}{2}$ mole.

$$\text{Weight pure sodium oxalate} = 0.2538 \times 0.996 = 0.2528 \text{ g}$$
$$= 252.8 \text{ mg}$$

Weight 1 meq sodium oxalate = $134.0/2 = 67.0$
Milliequivalents sodium oxalate = $252.8/67.0 = 3.773 = $ milliequivalents $KMnO_4$
Normality permanganate = milliequivalents per milliliter $= 3.773/40.41$
$$= 0.0934$$

Problem. Analysis of Sample. A sample containing iron is dissolved, all the iron is reduced to the ferrous condition, and the resulting solution is titrated with standard permanganate solution. From the following data calculate the percentage of iron, as ferric oxide, in the sample.

$$\text{Weight of sample} = 1.097 \text{ g}$$
$$\text{Volume permanganate solution} = 37.63 \text{ ml}$$
$$\text{Normality permanganate solution} = 0.1050$$

SOLUTION: As in all redox titrations, we have the relation

Milliequivalents Fe = milliequivalents $KMnO_4$ = 37.63 ml $\times$ 0.1050 meq/ml
$$= 3.951 \text{ meq}$$

The weight of 1 meq Fe_2O_3 is $Fe_2O_3/2$ ($159.70/2 = 79.85$ mg), since there are two atoms of Fe per mole of Fe_2O_3, and each changes from Fe^{+2} to Fe^{+3} in the titration

$$Fe^{+2} = Fe^{+3} + e^-$$

Weight Fe_2O_3 = 3.951 meq $\times$ 79.85 mg/meq
$$= 315.4 \text{ mg} = 0.3154 \text{ g}$$

$$\text{Percentage } Fe_2O_3 = \frac{0.3154 \text{ g}}{1.097 \text{ g}} \times 100 = 28.75\%$$

This type of problem is often confusing when encountered for the first time. Although iron as ferric oxide cannot be titrated by an oxidizing agent, since the iron is already in an oxidized condition, we have expressed the equivalent weight in terms of ferric oxide. An alternative method of solving would be to compute the weight of iron titrated by the given volume of permanganate solution, and from the weight of iron to compute the weight of ferric oxide. Such a solution is unnecessarily long. Since each atom of iron present in the form of ferric oxide gives an atom of ferrous ion when the sample is prepared for analysis, it is obvious that a mole of ferric oxide will give two equivalents of ferrous ion. Therefore the most direct method for the solution of the problem is to compute the weight of ferric oxide from the number of milliequivalents, as was done previously.

Problem. Indirect Oxidation Titration. A solution of permanganate is 0.1000 N; 40.00 ml of this solution is added to an acidified solution which contains excess potassium iodide. The iodine liberated is titrated by 35.00 ml sodium thiosulfate solution. What is the normality of the latter?

SOLUTION: The reactions involved are:

$$2MnO_4^- + 16H^+ + 10I^- = 2Mn^{+2} + 5I_2 + 8H_2O$$
$$I_2 + 2S_2O_3^{-2} = S_4O_6^{-2} + 2I^-$$

Milliequivalents $KMnO_4$ = 40.00 × 0.1000 = 4.000
= milliequivalents I_2 liberated
= milliequivalents $Na_2S_2O_3$
Therefore, milliequivalents $Na_2S_2O_3$ = milliequivalents $KMnO_4$
= 4.000
N of $Na_2S_2O_3$ = 4.000/35.00 = 0.1143

Problem. Oxidation Equivalence versus Acidimetric Equivalence. A solution of potassium acid oxalate, KHC_2O_4, is found to be 0.1200 N when standardized by titration with sodium hydroxide. If 42.15 ml of this solution require 38.26 ml potassium permanganate solution for titration, what is the normality of the permanganate?

SOLUTION: In the acidimetric reaction, the equivalent weight of KHC_2O_4 is the molecular weight, since there is only one hydrogen atom per molecule. As a reducing agent, the equivalent weight of KHC_2O_4 is one-half the molecular weight, since the half reaction for oxidation of $C_2O_4^{-2}$ ion is

$$C_2O_4^{-2} = 2CO_2 + 2e^-$$

Accordingly, the normality as a reducing agent is *twice* the normality as an acid.

$$\text{Normality } KMnO_4 = 0.1200 \times 2 \times \frac{42.15}{38.26} = 0.2644$$

Questions

1. Define the following terms: oxidation, reduction, reducing agent, oxidizing agent, valence, oxidation number.

2. Complete and balance the following:

$$MnO_4^- + H_2O_2 + H^+ =$$
$$Cr_2O_7^{-2} + H_2S + H^+ =$$
$$Ce^{+4} + H_2C_2O_4 =$$
$$BrO_3^- + I^- + H^+ =$$
$$MnO_4^- + H_2SO_3 =$$
$$ClO_3^- + Sn^{+2} + H^+ =$$

$$IO_3^- + HNO_2 + H^+ =$$
$$Ag + NO_3^- + H^+ =$$
$$BrO_3^- + H_3AsO_3 + H^+ =$$
$$Cr_2O_7^{-2} + Fe^{+2} + H^+ =$$
$$S_2O_3^{-2} + I_2 =$$
$$H_2O_2 + Sn^{+2} + H^+ =$$
$$MnO_4^- + HNO_2 + H^+ =$$
$$Cr_2O_7^{-2} + I^- + H^+ =$$
$$I_2 + H_2S =$$
$$VO_3^- + Fe^{+2} + H^+ = VO^{+2} +$$

3. Why is nitric acid not a suitable oxidizing agent for volumetric determinations?

Problems

1. Arrange in tabular form (a) the molecular weight, (b) the equivalent weight, and (c) the milligrams per milliliter of 0.1250 N solution for the following substances: $KMnO_4$ (used in acid solution), $K_2Cr_2O_7$, $Na_2S_2O_3 \cdot 5H_2O$, H_2SO_3, SO_2, HNO_2 (as a reducing agent), $Ce(SO_4)_2$, $Na_2C_2O_4$.

2. What is the equivalent weight of each of the following metallic oxides? The reactions by which the determinations are made are given in parentheses in unbalanced form.

$$As_2O_3 \ (H_2AsO_3^- + I_2 + H_2O = H_2AsO_4^- + I^- + H^+)$$
$$Fe_3O_4 \ (Fe^{+2} + Ce^{+4} = Fe^{+3} + Ce^{+3})$$
$$Cu_2O(Cu^{+2} + I^- = CuI + I_2)$$
$$V_2O_5 \ (VO^{+2} + MnO_4^- + H_2O = VO_3^- + Mn^{+2} + H^+)$$
$$Mo_2O_3 \ (Mo^{+3} + MnO_4^- + H_2O = MoO_4^{-2} + Mn^{+2} + H^+)$$

3. (a) How many milligrams of $K_2Cr_2O_7$ are present in 400.0 ml 0.1037 N solution? What is the molarity of the solution?

(b) How many grams of $Na_2S_2O_3 \cdot 5H_2O$ need to be dissolved in 650 ml to prepare a 0.200 N solution?   Ans. (a) 2034 mg; 0.01729 M; (b) 32.27 g.

4. How much of substance B will react with the stated amount of substance A?

A	B
(a) 50.00 ml 0.1200 N H_2S	_____ ml 0.01200 M $K_2Cr_2O_7$
(b) 25.00 ml 0.0400 M $KMnO_4$	_____ mg Fe_3O_4
(c) 30.00 ml 0.01000 M H_2S	_____ ml 0.05000 N $KMnO_4$
(d) 55.85 mg Fe	_____ ml 0.02000 N Ce^{+4}
(e) 25.00 ml 0.1000 N I_2	_____ mg As_2O_3
(f) 35.00 ml 0.1200 N $KMnO_4$	_____ g H_2O_2
(g) 3.000 mM KHC_2O_4	_____ ml 0.2000 N KOH
	_____ ml 0.1000 N $KMnO_4$
(h) 0.2000 g CaC_2O_4	_____ ml 0.1500 N $KMnO_4$

Ans. (a) 83.33 ml; (e) 123.6 mg.

5. What is the Fe_2O_3 titer of 0.1125 N $Ce(HSO_4)_4$ solution?

6. How many milliliters of a solution of 25.00 g $KMnO_4$ per liter will react with 3.400 g $FeSO_4 \cdot 7H_2O$? *Ans.* 15.48 ml.

7. What is the iron titer of a solution made by dissolving 5.017 g $K_2Cr_2O_7$ in 1.750 liters? *Ans.* 3.265 mg.

8. How many grams of As_2O_3 must be dissolved in concentrated NaOH to produce 500.0 ml of a solution which is 0.05000 N in $NaAsO_2$? *Ans.* 1.236 g.

9. From the following data compute the normalities and molarities of the permanganate and ferrous solutions:

$$\begin{aligned} \text{Weight } Na_2C_2O_4 &= 0.2000 \text{ g} \\ \text{Volume } KMnO_4 &= 42.00 \text{ ml} \\ \text{Volume } FeSO_4 &= 5.00 \text{ ml} \\ 30.00 \text{ ml } KMnO_4 &= 35.00 \text{ ml } FeSO_4 \end{aligned}$$

10. What is the percentage of iron in an ore sample which weighs 0.5250 g and which requires 40.00 ml 0.1193 N ceric sulfate solution for titration?

11. How many milliliters of 0.0973 N $K_2Cr_2O_7$ solution are needed to titrate a 0.3563-g sample of an iron ore which is 78.33 per cent Fe_2O_3? *Ans.* 35.92 ml.

12. How many milliliters of 0.1000 M $KMnO_4$ are equivalent in oxidizing power to 50.00 ml 0.1200 M $K_2Cr_2O_7$ solution? Both reagents are to be used in acid solution. *Ans.* 72.00 ml.

13. 40.00 ml 0.1000 N $KMnO_4$ is equivalent to 35.00 ml KHC_2O_4 solution; how many milliliters of 0.1000 N KOH are needed to titrate 50.00 ml of the KHC_2O_4 solution?

14. 20.00 ml $KHC_2O_4 \cdot H_2C_2O_4$ solution reacts with 30.00 ml 0.2000 N NaOH solution. How many milliliters of 0.1000 N ceric sulfate solution will react with 40.00 ml of the oxalate solution? *Ans.* 160.0 ml.

15. What is the molarity of a $KMnO_4$ solution if 25.00 ml of it reacts with 50.00 ml 0.1250 M KHC_2O_4 solution? *Ans.* 0.1000 M.

16. From the following calculate the normality of the permanganate and thiosulfate solutions:

25.00 ml HCl = 0.5465 g AgCl
24.36 ml HCl = 27.22 ml NaOH
24.62 ml NaOH = 30.17 ml KHC_2O_4
41.54 ml KHC_2O_4 = 37.82 ml $KMnO_4$
40.00 ml $KMnO_4$ is added to 100 ml of a molar KI solution, and the liberated I_2 requires 35.17 ml $Na_2S_2O_3$

CHAPTER 11

Equilibria in
Oxidation-Reduction Reactions

The half reaction of a redox reagent

$$\text{Reduced form} = \text{oxidized form} + ne^-$$

is usually reversible, or can be made to go in either direction. In the presence of an electron acceptor the reaction goes to the right, or in the presence of an electron donor it goes to the left. For example, if I^- is added to an equimolar mixture of Fe^{+2} and Fe^{+3}, the reaction is

$$Fe^{+3} + I^- = Fe^{+2} + \tfrac{1}{2}I_2$$

Here the I^- ion donates an electron to the Fe^{+3} ion, thus reducing it. But if Cl_2 is added to a similar mixture, the reaction is

$$Fe^{+2} + \tfrac{1}{2}Cl_2 = Fe^{+2} + Cl^-$$

and the Fe^{+2} ion is now the electron donor.

In any mixture of redox reagents the direction of the reaction is determined by the relative tendencies of the reduced forms present to lose electrons. The substance that has the greatest tendency to lose electrons reduces other substances present that can accept electrons. The tendency of a reducing agent (atom, ion, or molecule) to lose electrons can be measured by electrical means, in terms of an electromotive force, emf. We shall describe in this chapter the measurement of reduction emf's of redox systems and the use of these emf's to compute equilibrium conditions in mixtures of redox reagents.

A list of the measured reduction emf's for common oxidation reduction reagents is given in Table A–5, Appendix. The numerical values are the voltages associated with the system when all reactants are at conditions known as "standard state," which we shall define later. When all reactants are at standard state, each substance will reduce all others whose emf values are less than the emf of that substance.

Electromotive Force of Half-Cells

As previously shown, an oxidation-reduction reaction requires the transfer of electrons from the reducing agent to the oxidizing agent. Ordinarily the two reagents are in contact and the transfer of electrons takes place directly. If, however, we so separate the reagents that the transfer of electrons is through a connecting wire, we can use the reaction to produce an electric current. When this is done, the potential difference between the electrodes in contact respectively with the reducing and the oxidizing agent is determined by the relative tendencies of the two reagents to give up electrons.

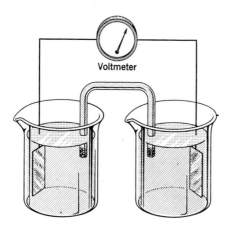

Voltmeter

A simple form of cell for an electrochemical reaction is shown in Fig. 11–1. The reducing agent, metallic zinc in contact with a solution of zinc ions, is in the left-hand beaker.

FIGURE 11–1 Production of current by oxidation-reduction reaction.

The oxidizing agent, cupric ions in contact with a metallic copper electrode, is in the right-hand beaker. As zinc forms zinc ions according to the reaction

$$Zn = Zn^{+2} + 2e^-$$

electrons are released. These electrons flow through the conducting wire to the copper electrode, where they are taken up by cupric ions.

$$Cu^{+2} + 2e^- = Cu$$

For the reaction to proceed we must have (1) a conducting material connecting the two electrodes, and (2) a salt bridge (an inverted U-tube filled with a solution of an electrolyte) connecting the two solutions. The function of the salt bridge is to permit ions to flow from one solution to the other. This is necessary because positive ions are formed in one beaker and removed in the other. As the current flows negative ions migrate to the left-hand beaker and positive ions to the right-hand beaker. The Cu and Zn electrodes are the two poles of the cell. Copper is the positive pole, the one to which the electrons flow. The emf of the cell is 1.10 volts when the concentrations of Cu^{+2} and Zn^{+2} ions are each $1M$.

The electromotive force of a cell depends on several factors.

1. The inherent tendencies of the oxidizing agent to take up electrons and of the reducing agent to give up electrons.

2. The concentrations of the reacting ions in the solutions at the positive and the negative poles. The voltage of the cell in Fig. 11–1 can be changed at will by changing the concentrations of the two solutions. If, for example, the concentration of Zn^{+2} in the left-hand beaker is lowered by precipitating an insoluble Zn salt, the voltage rises because a decrease in the concentration of Zn^{+2} ions at the electrode surface increases the tendency of the Zn atoms to go into solution. Conversely, a reduction in the concentration of Cu^{+2} in the right-hand beaker decreases the emf, by decreasing the tendency of cupric ions in the solution to take up electrons. If cyanide ion is added to the solution in the right-hand beaker, the concentration of cupric ion becomes so small that the polarity of the cell is reversed, copper now forming cupric ion and zinc ion taking up electrons to form Zn.

3. The absolute temperature. We shall discuss only cells operating at room temperature, near $25°C$.

4. The junction between two solutions due to differences in the rates at which various ions migrate across such a junction. This effect, small in magnitude, is neglected in the treatment of this text.

The potential difference between the two electrodes of a cell such as we have described is measured in volts, by use of a voltmeter or a potentiometer. Since the quantity we measure is the difference in potential between the two electrodes, it is convenient to ascribe to each electrode a half-cell emf which is related to the over-all emf by the equation

$$E_{l-r} = E_l - E_r$$

where E_l and E_r are the two half-cell voltages and E_{l-r} is the difference between the two. We find in Table A–5 that the half-cell emf's are 0.76 volt for the cell whose reaction is $Zn = Zn^{+2} + 2e^-$ and -0.34 volt for the cell whose reaction is $Cu = Cu^{+2} + 2e^-$. The difference between the two is 1.10 volts. It should be noted that the absolute values of the two half-cell emf's are not factors in the measured value of E_{l-r} but only the difference between the two half-cell values. The emf values of Table A–5 are on a relative basis only, with an arbitrarily chosen zero point. The situation is akin to the measurement of height of an object. When we say that a desk is 3 ft high, we mean with reference to the floor of the room. This may be at sea level or a mile above sea level—in either event the top of the desk is 3 ft higher than the floor level. The arbitrarily chosen zero in the table of half-cell voltages is the voltage of a cell composed of hydrogen gas at 1 atm in equilibrium with hydronium ion at unit activity (a term which we shall define later). All other values of Table A–5 are in relation to this standard.

A thoughtful student may at this point wonder what is meant by a half-cell emf, since a half-cell such as a metal in contact with a solution of its ions has no voltage that can be measured. We can only measure the voltage between the electrodes of two half-cells. However, if one of the two is a hydrogen half-cell, arbitrarily assigned an emf value of zero, then the measured emf of the whole cell must by definition be that of the other half-cell. To measure the half-cell emf of the zinc–zinc ion system, we connect this to a hydrogen half-cell and measure the potential difference between the two electrodes. If all reactants are at standard state, we find a measured value of 0.76 volt, the value given for the zinc–zinc ion system in Table A–5.

In practice we do not actually use a standard hydrogen reference cell for the measurement, because it is difficult to maintain exact concentrations of hydrogen gas and hydrogen ion in the system. Rather we use standard reference half-cells whose emf's have been accurately determined relative to that of a standard hydrogen electrode. The reference cells most widely used contain a metal in contact with a saturated solution of its slightly soluble salt in an electrolyte solution whose anion is the same as that of the metal salt. In such a system the concentration of the metal ion in solution can readily be maintained at a constant value. The most widely used reference half-cells and their accepted voltages at 25°C are:

$$\text{Ag, AgCl, HCl } (1\ M) \qquad E = -0.222 \text{ volt}$$
$$\text{Hg, Hg}_2\text{Cl}_2,\ \text{KCl (saturated)} \quad E = -0.242 \text{ volt}$$
$$\text{Hg, Hg}_2\text{Cl}_2,\ \text{KCl } (1\ M) \qquad E = -0.280 \text{ volt}$$
$$\text{Hg, Hg}_2\text{Cl}_2,\ \text{KCl } (0.1\ M) \qquad E = -0.334 \text{ volt}$$

The latter three are known respectively as the saturated calomel, the normal calomel, and the decinormal calomel electrodes.

The emf values of Table A–5 are designated as $E°$, read as "E standard." They are the voltages when the temperature is 25°C and all reactants are at concentrations of unit activity. We use these numbers to compute the emf values at other conditions, by application of the Nernst equation as described later. Strictly speaking, the activities or effective concentrations should be used in all such calculations. It is, however, beyond the scope of an elementary textbook to deal with the activities of the reactants, and it is necessary to make some approximations that are in many instances not greatly in error but may in others be several times the true activity. With these reservations we shall describe activities as follows.

1. The mole fraction of solids and liquids is taken as the activity. Thus, in a solution containing 23.0 g sodium in 200.6 g mercury, a gram-atom of each, the activity is 0.5 for each.

2. The partial pressure of gases, in atmospheres, is taken as the activity.

3. The molarity of ions or molecules in aqueous solution is taken as the activity. (The activities of many ions are often less than one-tenth their molarities, and this approximation may at times introduce large errors into the computations. Fortunately, however, the errors often cancel one another, and a measured cell voltage for 1 M solutions may agree quite closely with the standard state values.)

The Use of Standard Half-Cell Voltages

The half-cell values of Table A–5 provide the data needed for the equilibrium calculations of redox reactions, in much the same way as the tables of equilibrium constants and solubility products were used in the calculations of preceding chapters. We shall in this section show various useful applications of the half-cell reactions and the emf values associated with these.

A half-cell is described by the notation

$$\text{Electrode, solution } (xM)$$

in which the comma indicates a phase boundary. If more than two phases are present, a comma is used to indicate each phase boundary. Thus, the notation for the silver–silver chloride half-cell is

$$\text{Ag, AgCl, HCl } (0.1\ M)$$

and that for a Zn-Zn^{+2} half-cell is

$$Zn, Zn^{+2} \ (xM)$$

When the oxidation-reduction reaction of a half-cell is between two species of ions in solution, an inert electrode, usually a platinum foil, is put into the solution to collect or give out electrons. Thus, for the half-cell involving the equilibrium between ferrous and ferric ions we have the notation

$$Pt, Fe^{+2} \ (xM) + Fe^{+3} \ (yM)$$

When two half-cells are combined to form a whole cell, the liquid junction between the two solutions is indicated by a pair of vertical bars. The cell of Fig. 11–1 is designated as

$$Zn, Zn^{+2} \ (1\,M) \ | \ | \ Cu^{+2} \ (1\,M), \ Cu$$

The voltage of a half-cell is designated by the symbol E° when all reactants are at standard state and by the symbol E when not at standard state. The voltage of a whole cell composed of two half-cells is given by the relation used above

$$E_{l-r} = E_l - E_r$$

where E_l is the voltage of the left-hand half-cell and E_r is that of the right-hand half-cell.

Conventions. In everyday experience we think of the voltage of a cell as positive only, for we must connect the positive terminal of a cell to the voltmeter terminal marked by a + sign to obtain a voltmeter reading. But if we used a voltmeter which gave readings to both sides of the zero, we could by reversing the connections obtain a minus voltage of the same magnitude as the plus voltage. It is necessary, therefore, to designate just how the measurement is made in describing the voltage of a cell; in other words, we must adopt certain conventions and then adhere closely to them. There is no particular merit to any given set of conventions, provided the set is self-consistent, and the student may in other courses encounter conventions opposite to those we use here. We shall follow those used by Latimer,[1] which are found in most American texts on physical chemistry.

1. Each half-cell reaction is written with the electron on the right-hand side.

$$Reduced \ form = oxidized \ form + n \ electrons$$

[1] W. M. Latimer, *The Oxidation States of the Elements and Their Potentials in Aqueous Solutions,* 2nd edition, Prentice-Hall, 1952.

2. A plus value for $E°$ shows that in the reduced form the reactant is a better reducing agent than H_2 at standard state. Each substance in Table A–5 will, at standard state, reduce all substances below it.

3. In combining two half-cells:

(a) Write each half-cell reaction as in Table A–5.

(b) Subtract the half-cell reaction of the right-hand electrode from that of the left-hand electrode and subtract the corresponding $E°$ values. Before subtracting the half-cell reactions, we must balance them, so that the same number of electrons appears in each, just as we have done in the preceding chapter.

For the cell of Fig. 11–1 we have

$$Zn = Zn^{+2} + 2e^- \qquad E° = \quad 0.76 \text{ volt}$$

$$Cu = Cu^{+2} + 2e^- \qquad E° = -0.34 \text{ volt}$$

$$\overline{Zn + Cu^{+2} = Zn^{+2} + Cu \qquad E°_{l-r} = 1.10 \text{ volts}}$$

4. When two half-cells are combined as shown above, the following conventions are applied.

(a) The sign of the cell voltage indicates the polarity of the right-hand electrode. In the cell above, copper is the positive electrode and zinc the negative. Electrons flow from the zinc to the copper electrode.

(b) A plus value of the cell voltage indicates that the cell reaction goes from left to right as written. Conversely, a minus value indicates that the reaction goes from right to left. We note that in the reaction above, zinc reduces cupric ion. Had we reversed the two half-cells, we would have obtained a value of -1.10 volts and the equation would have been reversed. It is, therefore, immaterial which half-cell we select as the right- or left-hand part of the assembly.

(c) A value of zero for the cell voltage indicates that the reaction is at equilibrium. No current flows, and the concentrations of reactants in the solutions remain unchanged.

Effect of Concentration on Cell Voltage. The voltage of a half-cell at conditions other than standard state may be computed from the $E°$ value by use of the Nernst equation

$$E = E° - \frac{RT}{nF} \ln Q$$

where R is the gas law constant, T is the absolute temperature, n is the number of electrons involved in the cell reaction, F is the faraday

(96,500 coulombs), and Q is the quotient of the activity products of substances on the right-hand side of the cell reaction equation divided by the activity products of the substances on the left-hand side, each raised to a numerical power equal to the number of moles involved in the reaction.

It is convenient to combine the constants, to convert natural logs to base 10, and at 25°C to use the equation in the form

$$E = E° - \frac{0.059}{n} \log_{10} Q$$

Problem. Compute the voltage of the half-cell

$$Pt, H_2(1 \text{ atm}), H_2O$$

SOLUTION: Since $[H^+]$ in pure water at 25°C is $10^{-7} M$ the half-cell reaction is

$$H_2 = 2H^+(10^{-7} M) + 2e^-$$

Application of the Nernst equation gives

$$E = E° - \frac{0.059}{2} \log \frac{(a_{H^+})^2}{(a_{H_2})}$$

Substituting values of the terms

$$E° = 0.00$$

$$a_{H^+} = 10^{-7} \quad \text{(the concentration of hydronium ion}^2 \text{ in water)}$$

$$a_{H_2} = 1 \quad \text{(the pressure is 1 atm)}$$

This gives

$$E = 0.00 - \frac{0.059}{2} \log (10^{-7})^2$$

$$= 0.00 - 0.059 \log 10^{-7}$$

$$= +0.41 \text{ volt}$$

The emf when two half-cells are combined is given by the equation

$$E_{l-r} = E°_l - E°_r - \frac{0.059}{n} \log Q$$

where Q is the quotient of the activity products on the right-hand side of the equation for the chemical reaction divided by the activity products of the substances on the left-hand side, each raised to the proper power.

Problem. Iron filings are added to a solution containing Fe^{+2} and Cd^{+2}, each at a concentration of 0.1 M. Use the Nernst equation to determine whether Fe will reduce Cd^{+2}.

---

[2] The symbol H^+ is used for the hydronium ion in Table A–5 and throughout this chapter.

SOLUTION: Select from Table A–5 the half-cells that will give the desired reaction. Combine these to determine the emf of the cell.

$$\text{Fe} = \text{Fe}^{+2} + 2e^- \qquad E° = 0.44 \text{ volt}$$

$$\text{Cd} = \text{Cd}^{+2} + 2e^- \qquad E° = 0.40 \text{ volt}$$

$$\overline{\text{Fe} + \text{Cd}^{+2} = \text{Fe}^{+2} + \text{Cd}} \qquad E°_{l-r} = 0.04 \text{ volt}$$

Apply the Nernst equation

$$E_{l-r} = 0.04 - \frac{0.059}{2} \log \frac{(a_{\text{Fe}^{+2}})(a_{\text{Cd}})}{(a_{\text{Fe}})(a_{\text{Cd}^{+2}})}$$

The activities of the pure metals are 1; those of the ions are taken as the molar concentrations. Substituting these values

$$E_{l-r} = 0.04 - \frac{0.059}{2} \log \frac{0.1}{0.1}$$

$$= 0.04$$

The positive potential shows that the reaction goes to the right, or that Fe will reduce Cd^{+2}.

Problem. A solution contains Fe^{+3}, Fe^{+2}, and I^- ions, each at a concentration of 0.1 M, and is saturated with I_2. Does Fe^{+3} oxidize I^- or does I_2 oxidize Fe^{+2}?

SOLUTION: The half-cell reactions are (Table A–5)

$$2\text{I}^- = \text{I}_2 + 2e^- \qquad E° = -0.54 \text{ volt}$$

$$2\text{Fe}^{+2} = 2\text{Fe}^{+3} + 2e^- \qquad E° = -0.77 \text{ volt}$$

$$\overline{2\text{I}^- + 2\text{Fe}^{+3} = \text{I}_2 + 2\text{Fe}^{+2}} \qquad E°_{l-r} = 0.23 \text{ volt}$$

Applying the Nernst equation

$$E_{l-r} = 0.23 - \frac{0.059}{2} \log \frac{(a_{\text{I}_2})(a_{\text{Fe}^{+2}})^2}{(a_{\text{I}^-})^2(a_{\text{Fe}^{+3}})^2}$$

The activity of I_2 is 1, since it is present as pure solid in equilibrium with the saturated solution. Substitute the molar concentrations for the activities of the ions. This gives

$$E_{l-r} = 0.23 - \frac{0.059}{2} \log \frac{(1)(0.1)^2}{(0.1)^2(0.1)^2}$$

$$= 0.23 - 0.059 \log \frac{1}{(0.1)(0.1)}$$

$$= 0.23 - 0.059 \log 100$$

$$= 0.23 - 0.12 = 0.11$$

The sign of the emf shows that the reaction goes to the right, or that Fe^{+3} oxidizes I^- under these conditions.

Equilibrium Constant for Redox Reactions. If a cell is set up to give an oxidation-reduction reaction, the reaction reaches a condition of equilibrium when the emf of the cell becomes zero. When this occurs, the activity quotient Q is equal to the equilibrium constant (K_e) for the reaction. The Nernst equation is used to evaluate the constant from the half-cell emf values. At equilibrium we have

$$0 = E^\circ_l - E^\circ_r - \frac{0.059}{n} \log K_e$$

Solving,

$$\log K_e = \frac{nE^\circ_{l-r}}{0.059}$$

Problem. What is the equilibrium constant for the reaction

$$Zn + Cu^{+2} = Zn^{+2} + Cu$$

SOLUTION: The half-cells giving the desired reaction are

$$Zn = Zn^{+2} + 2e^- \qquad E^\circ = 0.76 \text{ volt}$$

$$Cu = Cu^{+2} + 2e^- \qquad E^\circ = -0.34 \text{ volt}$$

$$\overline{Zn + Cu^{+2} = Zn^{+2} + Cu \qquad E^\circ_{l-r} = 1.10 \text{ volt}}$$

$$\log K_e = \frac{2 \times 1.10}{0.059} = 37.3$$

$$K_e = 10^{37.3} = \frac{[Zn^{+2}]}{[Cu^{+2}]} = 2 \times 10^{37}$$

Note that the activities of the pure metals are not included in the equilibrium constant, since each is 1.

Problem. Metallic zinc is added to a 0.1 M solution of $CuSO_4$. What is the final concentration of Cu^{+2} remaining in solution?

SOLUTION: Let $x = [Cu^{+2}]$. Then $0.1 - x$ is $[Zn^{+2}]$, since each mole of Cu^{+2} that is reduced is replaced in solution by a mole of Zn^{+2}. Substitute these concentrations into the equilibrium constant equation of the preceding section. This gives

$$\frac{(0.1 - x)}{x} = 2 \times 10^{37}$$

Neglecting x in the numerator, since it is much less than 0.1, and solving, we have

$$x = [Cu^{+2}] = \frac{0.1}{2 \times 10^{37}} = 0.5 \times 10^{-38} M$$

When we consider a redox reaction as the basis for a titrimetric analysis, we may use the equilibrium constant to compute the degree of completeness of the reaction at the end point, i.e., to evaluate the suitability of the method for a quantitative determination. A computation of this type is illustrated in the following problem.

Problem. A sample containing 5mM Fe^{+2} is titrated with 0.02 M $KMnO_4$ in an acid solution whose hydronium ion concentration is 1 M. The end point is observed when there is one drop, 0.05 ml, of permanganate in excess. The final volume of the solution is 200 ml. What fraction of the ferrous ion remains unoxidized at the end point?

SOLUTION: The reaction is

$$MnO_4^- + 8H^+ + 5Fe^{+2} = Mn^{+2} + 4H_2O + 5Fe^{+3}$$

The half reactions are (Table A–5)

$$5Fe^{+2} = 5Fe^{+3} + 5e^- \qquad\qquad E^\circ = -0.77 \text{ volt}$$

$$4H_2O + Mn^{+2} = MnO_4^- + 8H^+ + 5e^- \qquad E^\circ = -1.52 \text{ volt}$$

$$\overline{MnO_4^- + 8H^+ + 5Fe^{+2} = 5Fe^{+3} + 4H_2O + Mn^{+2} \qquad E^\circ{}_{l-r} = 0.75 \text{ volt}}$$

Omitting the substances that are at standard state and replacing activities of ions by their molar concentrations, the equilibrium constant equation is

$$K_e = \frac{[Fe^{+3}]^5[Mn^{+2}]}{[MnO_4^-][H^+]^8[Fe^{+2}]^5}$$

The value of K_e is computed from the half-cell emf's.

$$\log K_e = \frac{0.75 \times 5}{0.059}$$

$$= 63.6$$

$$K_e = 10^{63.6}$$

Let x equal the millimoles of Fe^{+2} remaining unoxidized at the end point and $5 - x$ the millimoles of Fe^{+3} at the end point, since we started with 5mM Fe^{+2}. The concentrations to use in the computation are

$$[Fe^{+2}] = \frac{x}{200} = 5 \times 10^{-3}x$$

$$[Fe^{+3}] = \frac{5 - x}{200}$$

Neglecting x, which is much smaller than 5, we have

$$[Fe^{+3}] = \frac{5}{200} = 2.5 \times 10^{-2}$$

$$[Mn^{+2}] = \frac{1}{200} = 5 \times 10^{-3}$$

Note that we have one Mn^{+2} ion produced for five Fe^{+3} ions.

$$[MnO_4^-] = \frac{0.05 \text{ ml} \times 0.02 \text{ mM/ml}}{200 \text{ ml}}$$

$$= 5 \times 10^{-6} M$$

$$[H^+] = 1 M$$

$$\frac{(2.5 \times 10^{-2})^5(5 \times 10^{-3})}{(5 \times 10^{-6})(1)^8(5 \times 10^{-3}x)^5} = 10^{63.6}$$

$$\frac{(2.5)^5}{(5x)^5} = \frac{10^{63.6} \times 10^{-21}}{10^{-13}} = 10^{55.6}$$

$$\frac{2.5}{5x} = \sqrt[5]{10^{55.6}} = 10^{11.1} = 1.3 \times 10^{11}$$

$$x = \frac{0.5}{1.3 \times 10^{11}} = 0.4 \times 10^{-11} \qquad \text{or } 4 \times 10^{-12} \, \text{mM Fe}^{+2}$$

The fraction of the ferrous ions remaining unoxidized is

$$\frac{4 \times 10^{-12} \, \text{mM}}{5 \, \text{mM}} = 0.8 \times 10^{-12} \qquad \text{or } 0.8 \times 10^{-10}\%$$

Thus, the reaction is essentially complete at the end point.

The completeness of a redox reaction at the equivalence point is computed in the same manner as in the preceding problem.

Problem.   A solution containing 5 mM Fe^{+2} is titrated by a standard solution of ceric sulfate, $Ce(SO_4)_2$. The final volume is 200 ml. What weight of ferrous ion remains unoxidized at the equivalence point?

SOLUTION: The half reactions are

$$Fe^{+2} = Fe^{+3} + e^- \qquad\qquad E^\circ = -0.77 \text{ volt}$$

$$\underline{Ce^{+3} = Ce^{+4} + e^- \qquad\qquad E^\circ = -1.61 \text{ volts}}$$

$$Ce^{+4} + Fe^{+2} = Ce^{+3} + Fe^{+3} \qquad E^\circ_{l-r} = 0.84 \text{ volt}$$

$$\log K_e = \frac{0.84}{0.059} = 14.24$$

$$K_e = 10^{14.24}$$

Let x equal the millimoles of Fe^{+2} remaining unoxidized at the equivalence point. Then $5 - x$ equals the millimoles of Fe^{+3} formed. Since each mole of Fe^{+3} formed causes reduction of a mole of Ce^{+4}, x equals the millimoles of Ce^{+4} remaining in the solution and $5 - x$ equals the millimoles of Ce^{+3} formed. The concentrations in 200 ml of solution are

$$[Fe^{+3}] = [Ce^{+3}] = \frac{5 - x}{200} \sim \frac{5}{200}$$

$$[Fe^{+2}] = [Ce^{+4}] = \frac{x}{200}$$

Substituting these concentrations gives

$$\frac{\left(\dfrac{5}{200}\right)^2}{\left(\dfrac{x}{200}\right)^2} = 10^{14.24}$$

$$\frac{5}{x} = \sqrt{10^{14.24}} = 10^{7.12} = 1.3 \times 10^7 = \frac{[Fe^{+3}]}{[Fe^{+2}]}$$

$$x = \frac{5}{1.3 \times 10^7} = 4 \times 10^{-8}\,\text{mM}$$

The weight of ferrous ion remaining unoxidized at the equivalence point is 4×10^{-8} mM $\times$ 55.8 mg per millimole, or 224×10^{-8} mg, a negligible quantity.

Oxidation-Reduction Indicators

Redox indicators are dyes that undergo a reversible color change upon oxidation or reduction. The first to be so used was diphenylamine, as indicator for titration of Fe^{+2} by $Cr_2O_7^{-2}$ in acid solution.[3] The indicator remains in its reduced form until essentially all the ferrous ion is oxidized. Then, as more dichromate ion is added, the indicator is oxidized and shows a color change that serves as the end point for the titration.

The general indicator reaction is

$$\text{Ind} = \text{Ind}^{n+} + ne^-$$

where n is usually 1 or 2. Each redox indicator has its characteristic half-cell emf, just as any other oxidation–reduction couple. This emf is used to select the indicator best suited to a given titration, just as an acid-base indicator is selected to show its color change at the pH of the equivalence point.

A partial list of redox indicators and their transition emf values is given in Table 11–1. The emf values are not necessarily the E°

TABLE II–I. Transition Half-Cell Voltages for Selected Redox Indicators

Indicator	Color		Emf at Transition Point, volts
	Reduced	Oxidized	
Diphenylamine	colorless	violet	-0.76
Diphenylamine sulfonic acid, as barium salt	colorless	violet	-0.84
Ferroin (ferrous complex of orthophenanthroline)	red	faint blue	-1.14
Nitroferroin (ferrous complex of nitro-orthophenanthroline)	red	faint blue	-1.25

values, for E° is the emf when the concentrations of the oxidized and reduced forms are the same. Many of the indicators show a color change when the amount of the oxidized form is materially less than half the total present.

[3] J. Knop, *J. Am. Chem. Soc.*, **46**, 263 (1924).

Selection of Indicator for Titration. If a titration curve has been experimentally determined, as described in Chapter 22, the indicator is chosen whose emf at the transition point most nearly coincides with the emf of the equivalence point in the titration. If such data are not available, we may compute the desired indicator transition emf by considering the equilibrium between the indicator and the reducing agent that is titrated. This method is illustrated in the following problem.

Problem. A sample containing Fe^{+2} is titrated by an oxidizing agent. What must be the transition emf of an indicator which will show a color change when 99.9 per cent of the ferrous ion is oxidized?

SOLUTION: The emf of the $Fe^{+2} - Fe^{+3}$ system when 99.9 per cent is oxidized is given by the equation

$$E = E° - \frac{0.059}{1} \log \frac{Fe^{+3}}{Fe^{+2}}$$

where $E°$ is -0.77 for the reaction

$$Fe^{+2} = Fe^{+3} + e^-$$

Substituting, we have

$$E = -0.77 - 0.059 \log \frac{99.9}{0.1}$$

$$= -0.77 - 0.059 \log 999$$

$$= -0.95 \text{ volt}$$

The indicators of Table 11–1 whose transition emf's more closely agree with this value are diphenylamine sulfonic acid ($E' = -0.84$) and Ferroin ($E' = -1.14$).

It will be noted that in the preceding calculation no reference is made to the half-cell emf of the oxidizing agent used. This agent must be one that will oxidize the reductant almost completely at the equivalence point; within this limitation the specific reagent used does not affect the computation of the E^t value for the indicator.

Measurement of Concentrations by Cell Voltage

We have seen that the emf of a cell is related to the concentrations of the ions participating in the cell reaction. This relation enables us to use the measured emf of a cell to compute concentrations so small that direct measurement by other means is not possible. The following problems illustrate use of emf measurements for evaluation of dissociation constants, solubility product constants, and complex-ion constants.

Problem.   The emf of the cell

$$\text{Pt, H}_2(1 \text{ atm), HA (0.1 } M) \parallel \text{KCl(1 } M), \text{Hg}_2\text{Cl}_2, \text{Hg}$$

is $+0.58$ volt. Compute the concentration of H^+ ion in the HA solution and from this the dissociation constant for HA.

SOLUTION: For this cell we have the relation

$$E_{l-r} = E_{\text{hydrogen}} - E_{\text{calomel}}$$

since E_{calomel} is -0.28 volt (page 216)

$$0.58 = E_{\text{hydrogen}} - (-0.28)$$

$$E_{\text{hydrogen}} = 0.30$$

For the hydrogen half-cell we have

$$\text{H}_2 = 2\text{H}^+ + 2e^- \qquad E° = 0.00$$

Substituting 0.30 volt for E_{hydrogen} and applying the Nernst equation

$$0.30 = 0.00 - \frac{0.059}{2} \log [\text{H}^+]^2$$

$$= 0.00 - 0.059 \log [\text{H}^+]$$

Solving,

$$[\text{H}^+] = 10^{-5.08} = 8.3 \times 10^{-6}$$

Since $[\text{H}^+] = [\text{A}^-]$ we have

$$K_i = \frac{[\text{H}^+][\text{A}^-]}{[\text{HA}]} = \frac{(8.3 \times 10^{-6})^2}{0.1} = 6.9 \times 10^{-10}$$

Problem.   Compute the K_{sp} of AgI from the half-cell emf's (Table A–5)

$$\text{Ag} = \text{Ag}^+ + e^- \qquad E° = -0.80 \text{ volt}$$

$$\text{I}^- + \text{Ag} = \text{AgI} + e^- \qquad E° = 0.15 \text{ volt}$$

SOLUTION: Combine the half-cell reactions in the usual manner. This gives

$$\text{AgI} = \text{Ag}^+ + \text{I}^- \qquad E°_{l-r} = -0.95 \text{ volts}$$

From this,

$$\log K_e = \frac{0.95}{0.059} = -16.1$$

Since $K_e = [\text{Ag}^+][\text{I}^-] = K_{\text{sp}}$ for AgI, we have

$$K_{\text{sp}} = 10^{-16.1} = 8 \times 10^{-17}$$

In this problem we have used the established half-cell values to compute E_{l-r}, from which we calculate K_{sp}. If the half-cell values were not available, we could obtain the same result by setting up the cell

$$\text{Ag, Ag}^+ (1\,M) \parallel \text{I}^- (1\,M), \text{AgI, Ag}$$

and experimentally determining the emf. A cell of this type is known as a concentration cell. We have the same electrode substance in the two half-cells, but with different concentrations of the ions in the two halves. Essentially the cell is

$$\text{Ag, Ag}^+ \ (M_1) \ || \ \text{Ag}^+ \ (M_2), \text{Ag}$$

Treating the reaction in the usual manner gives

$$E_{l-r} = 0 - 0.059 \log \frac{M_1}{M_2}$$

Knowing the measured value of E_{l-r} and one of the molarities, we can readily compute the other.

Problem. K_{sp} for AgBr is 8×10^{-13}. What is the emf of the following cell,

$$\text{Ag, AgNO}_3(1 \ M) \ || \ \text{KBr}(1 \ M) + \text{AgBr (sat.), Ag}$$

SOLUTION: In 1 molar KBr the concentration of Br^- is $1 \ M$. Substituting this value into the K_{sp} expression

$$[\text{Ag}^+] = \frac{8 \times 10^{-13}}{1} = 8 \times 10^{-13}$$

Using the formula above, for a concentration cell,

$$E_{l-r} = 0 - 0.059 \log \frac{1}{8 \times 10^{-13}} = -0.71 \text{ volt}$$

Problem. Use the following half-cell values from Table A–5 to compute the dissociation constant for the $\text{Ag(NH}_3)_2^+$ complex.

$$\text{Ag} = \text{Ag}^+ + e^- \qquad\qquad E^\circ = -0.80 \text{ volt}$$

$$2\text{NH}_3(\text{aq}) + \text{Ag} = \text{Ag(NH}_3)_2^+ + e^- \qquad E^\circ = -0.37 \text{ volt}$$

SOLUTION: Combining the half-cells in the usual manner gives

$$\text{Ag(NH}_3)_2^+ = \text{Ag}^+ + 2\text{NH}_3$$

$$\log K_e = \frac{-0.43}{0.059} = -7.28$$

$$K_e = \frac{[\text{Ag}^+][\text{NH}_3]^2}{[\text{Ag(NH}_3)_2^+]} = 10^{-7.28} = 5.3 \times 10^{-8}$$

Questions

1. What is meant by the statement that a potential difference exists between two electrodes?

2. Describe cells whereby the following chemical reactions could be made to produce electric energy:

(a) $Zn + 2H^+ = Zn^{+2} + H_2$

(b) $Pb + 2Ag^+ = Pb^{+2} + 2Ag$

(c) $Cl_2 + 2I^- = 2Cl^- + I_2$

3. If the normal calomel electrode were taken as the reference standard for single-electrode voltages, with assigned value of zero, how would the values of the standard electrode potentials be changed?

4. Why are calomel and silver–silver chloride electrodes better experimental standard reference electrodes than the hydrogen electrode?

5. Describe an experimental procedure for determining the standard electrode emf's of the following half-cells. Specify the concentrations to be employed.

(a) Cu, Cu^{+2}

(b) Pt, $Fe^{+3} + Fe^{+2}$

(c) Pt, Cl_2, Cl^-

(d) Pt, Br_2, Br^-

(e) Pt, I_2, I^-

6. What effect will addition of hydrogen sulfide to the left-hand portion of the cell,

$$Zn, Zn^{+2} \parallel Cu^{+2}, Cu$$

have on the emf? What effect will the addition of hydrogen sulfide to the right-hand half-cell have on the emf? The addition of potassium cyanide to the right-hand half-cell will cause the polarity of the cell to reverse. Explain.

7. In practice the copper-zinc cell is often used without a mechanical salt bridge. Consult a physics textbook for the design employed. Explain the operation of this cell if both the copper and zinc solutions are in the same container.

8. Formulate cells which may be used to evaluate the following equilibrium constants:

(a) The solubility product of silver iodide.

(b) The dissociation constant of the argentocyanide complex ion.

(c) The ion product constant for water.

(d) The hydrolysis constant for sodium cyanide.

(e) The ionization constant of a weak acid.

(f) The constant for the reaction of permanganate with ferrous ion.

Problems

1. In which direction will the following reactions proceed if all substances are at standard state:

(a) $Sn^{+4} + Cd = Sn^{+2} + Cd^{+2}$

(b) $Ce^{+4} + Br^- = Ce^{+3} + \frac{1}{2}Br_2$

(c) $2Fe^{+3} + Cd = 2Fe^{+2} + Cd^{+2}$

(d) $Sn^{+4} + 2Ce^{+3} = Sn^{+2} + 2Ce^{+4}$

(e) $S^{-2} + 2Cr^{+3} = S + 2Cr^{+2}$

(f) $2MnO_4^- + 5O_2 + 6H^+ = 5O_3 + 3H_2O + 2Mn^{+2}$

(g) $5Cl_2 + I_2 + 6H_2O = 2IO_3^- + 10Cl^- + 12H^+$

Ans. (a) right; (b) right.

2. Compute the voltage of each of the following cells. Designate the polarity of the electrodes.

(a) Cu, Cu^{+2} (0.001 M) || Zn^{+2} (0.1 M), Zn

(b) Ag, $Ag^+(10^{-10} M)$ || KCl (1 M), Hg_2Cl_2, Hg

(c) Pt, H_2 (0.90 atm), H^+ (0.01 M) || KCl (0.10 M), Hg_2Cl_2, Hg

(d) Pt, H_2 (1 atm), HCl (0.005 M), Cl_2 (1 atm), Pt

(e) Pt, H_2 (0.5 atm), HBr (0.1 M), Br_2 (l), Pt

(f) Pt, Ce^{+4} (0.1 M) + Ce^{+3} (0.2 M) || Fe^{+3} (0.01 M) + Fe^{+2} (0.05 M), Pt   *Ans.* (a) -1.04 volts, Cu positive.

3. What is the voltage of the cell:

$$Ag, Ag^+ (1 M) || H^+ (1 M), H_2 (1 atm), Pt$$

What is the spontaneous cell reaction? Which is the $+$ pole of the cell?

4. For the cell,

$$Pt, Fe^{+2} (10^{-2} M) + Fe^{+3} (10^{-5} M) || I^- (10^{-3} M), I_2, Pt$$

(a) What is the numerical magnitude of the cell voltage?

(b) Which is the negative electrode?

(c) If Fe^{+2}, Fe^{+3}, and I^- are mixed together in solution at the concentrations above and the solution is kept in contact with solid I_2, will Fe^{+3} oxidize I^- or will Fe^{+2} reduce I_2?   *Ans.* (a) 0.12 volt.

5. (a) Compute the equilibrium constant for the reaction,

$$Fe + Cu^{+2} = Fe^{+2} + Cu$$

(b) Compute the equilibrium constant for the reaction,

$$Fe + Cd^{+2} = Fe^{+2} + Cd$$

(c) An acidified solution contains cupric and cadmium ions at a concentration of 0.05 M. Iron is added and the solution shaken until equilibrium is reached. Compute the final concentration of the cupric and cadmium ions, assuming a final concentration of 0.3 M for the ferrous ion (a considerable quantity of ferrous ion is formed by the reaction of iron with hydrogen ion).

(d) Compute the equilibrium constant for the reaction,

$$Cd + Cu^{+2} = Cd^{+2} + Cu$$

(e) What is the ratio of Cu^{+2} to Cd^{+2} at equilibrium?

Ans. (a) 2.5×10^{26}.

6. A solution containing ferrous and hydrogen ions at concentrations of 0.1 M each is brought into contact with a strip of pure iron. What pressure of hydrogen gas must be applied to prevent liberation of hydrogen by the action of iron on hydrogen ion?

7. A 0.1 M solution of copper sulfate is treated with zinc dust. What is

the final concentration of cupric ion remaining in solution? (*Note:* The final concentration of zinc ion may be assumed to be $0.1 M$. Explain.)

Ans. $5 \times 10^{-39} M$.

8. If a $0.1 M$ solution of ferrous sulfate of pH 7 is treated with pure iron, should any hydrogen be evolved? At what concentration of hydrogen ion is hydrogen gas liberated at a pressure of 1 atm?

9. Metallic silver is shaken with a $0.1 M$ Fe^{+3} solution. Find the ratio $[Fe^{+3}]/[Fe^{+2}]$ at equilibrium.

10. Find the equilibrium constant and the ratio $[Sn^{+4}]/[Sn^{+2}]$ at equilibrium for the reaction:

$$2Ce^{+4} + Sn^{+2} = 2Ce^{+3} + Sn^{+4}$$

if 1 mM Sn^{+2} and 2 mM Ce^{+4} are mixed in a volume of 100 ml.

11. Compute the equilibrium constant for the reaction,

$$Fe^{+3} + I^- = Fe^{+2} + \frac{1}{2}I_2 \qquad\qquad Ans.\ 8 \times 10^3.$$

12. Ferric nitrate is added to a solution of potassium iodide until the final concentration of ferric ion is $10^{-5} M$. Compute the equilibrium concentration of iodide ion, assuming the solution to be saturated with iodine, and the final concentration of Fe^{+2} to be $0.1 M$. Would you conclude that iodide ion may be determined by titration with ferric ion?

13. A solution of potassium iodide is treated with excess ferric nitrate so that the final concentration of Fe^{+3} is $0.1 M$. Assuming the final concentration of Fe^{+2} to be $0.1 M$ and the solution to be saturated with iodine, compute the equilibrium concentration of iodide ion.

14. Find the concentrations of Fe^{+3}, Fe^{+2}, and I^- present when equilibrium is reached in the mixture described in the cell of problem 4. Assume that solid I_2 is always in contact with the solution.

15. 5 mM $FeSO_4$ is dissolved in 100 ml of water and titrated with $0.1000 N$ $KMnO_4$ until two drops (0.10 ml) have been added in excess. Assume the $[H^+] = 1$, and calculate the weight of ferrous sulfate which has not reacted with the $KMnO_4$.

16. Compute the equilibrium constant for the reaction,

$$MnO_4^- + H^+ + Sn^{+2} = Mn^{+2} + Sn^{+4} + H_2O$$

17. If 5 mM Sn^{+2} is titrated with permanganate in acid solution, what is the concentration of Sn^{+2} at the equivalence point? Assume the following conditions:

Final volume of solution: 100 ml
$\quad [Sn^{+4}] = 0.05$, since practically all the stannous ion is oxidized
$\quad [Mn^{+2}] = 0.02$, since 2 moles Mn^{+2} are formed for 5 moles Sn^{+4}
$\quad [Sn^{+2}] = x$
$[MnO_4^-] = 0.4x$ (see equation)
$\quad [H_2O] = 1$
$\quad [H^+] = 1$ \hfill *Ans.* $3.6 \times 10^{-35} M$.

18. If the titration of problem 17 is made to an end point, with 0.05 ml 0.02 M permanganate present in excess, what is the final concentration of Sn^{+2}? Assume all concentrations except that of permanganate to be the same as in problem 17.

19. What should be the transition emf of a redox indicator suitable for use in the titration of problem 17?

20. A solution of ferrous sulfate contains 10^{-5} per cent of the iron in the ferric condition. What is the half-cell emf of this solution?

Ans. −0.36 volt.

21. What is the half-cell emf of a ferric salt solution in which 10^{-5} per cent of the iron is in the ferrous condition?

22. The solubility product of AgCl is 1×10^{-10}. Compute the emf of the cell

$$\text{Ag, AgCl, KCl } (0.05\ M), \text{Hg}_2\text{Cl}_2, \text{Hg} \qquad \textit{Ans. } 0.05 \text{ volt.}$$

23. (a) From the standard silver–silver ion emf compute the voltage of the half-cell,

$$\text{Ag, AgCl, KCl } (1\ M)$$

(b) Compute the voltage of the half-cell,

$$\text{Ag, AgCl, KCl } (0.01\ M)$$

24. From the two half-cell voltages

$$2\text{Hg} = \text{Hg}_2{}^{+2} + 2e^- \qquad E° = -0.79 \text{ volt}$$

$$2\text{Hg} + 2\text{Cl}^- = \text{Hg}_2\text{Cl}_2 + 2e^- \qquad E° = -0.27 \text{ volt}$$

compute the solubility product for mercurous chloride, Hg_2Cl_2.

Ans. 2.2×10^{-18}.

25. The emf of the cell,

$$\text{Ag, AgX, KX } (0.05\ M) \,||\, \text{KCl } (1\ M), \text{Hg}_2\text{Cl}_2, \text{Hg}$$

is $+0.30$ volt. What is the solubility product for AgX?

26. $E°$ for $\text{Pb} + \text{SO}_4{}^{-2} = \text{PbSO}_4 + 2e^-$ is 0.36 volt. From this and Table A–5, compute the K_{sp} for PbSO_4.

27. From a table of standard electrode voltages and the K_{sp} for Cu(OH)_2 develop a value of $E°$ for

$$\text{Cu} + 2\text{OH}^- = \text{Cu(OH)}_2 + 2e \qquad \textit{Ans. } 0.22 \text{ volt.}$$

28. From a table of standard electrode voltages compute the following:
   (a) K_{inst} for $\text{HgI}_4{}^{-2}$.
   (b) K_{sp} for AgI.
   (c) K_{sp} for Mg(OH)_2. \qquad *Ans.* (a) 6.3×10^{-31}.

29. The ionization constant of the acid HA is 5×10^{-6}. Compute the emf of the cell,

$$\text{Hg, Hg}_2\text{Cl}_2, \text{KCl } (1\ M) \,||\, \text{HA } (0.1\ M), \text{H}_2 \text{ (1 atm), Pt}$$

Ans. −0.47 volt.

30. What is the emf of the cell,

Normal calomel electrode $||$ NaA (0.05 M), H_2 (1 atm), Pt

when NaA is a salt of the acid HA of problem 29?

31. Using the data of Table A–4 and A–5, Appendix, compute the emf of the cells:

(a) Zn, $ZnCl_2$ (0.01 M) + NH_3 (0.1 M) $||$ normal hydrogen electrode

(b) Ag, AgCl, KCl (0.01M)$||$ KCN (1 M) + $AgNO_3$ (0.1 M), Ag

32. The emf of the cell,

Normal calomel electrode $||$ R_2NH_2Cl (0.1 M), H_2 (1 atm), Pt

is -0.52 volt. What is the ionization constant of the base R_2NH?

33. Compute the emf of the half-cells:

(a) Hg + Na (1 g Na per 200.6 g Hg), NaCl (0.1 M)

(b) Hg + Na (1 mole per cent Na), NaCl (0.1 M)   *Ans.* (a) 2.69 volts.

PART THREE—TITRIMETRIC

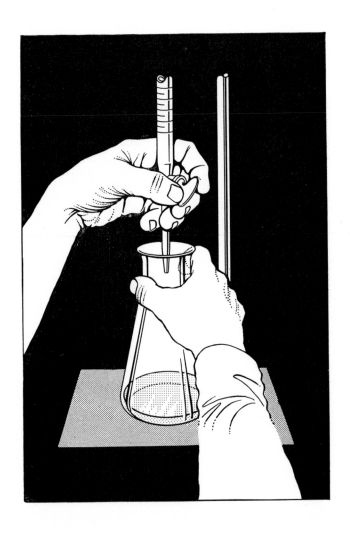

METHODS OF ANALYSIS

CHAPTER 12

General Methods
of Volumetric Analysis

As stated in the introduction, one of the widely used methods of analysis is to titrate a solution of the sample with a reagent that reacts with some ion of the sample. The amount of the sought substance is determined indirectly, by computations based on the volume (or weight) of reagent used and its concentration. Determinations based in this way on the amount of solution used are known as volumetric or titrimetric analyses.

A great advantage of titration analyses is speed. Once the proper reagent solutions are prepared, the titration of an individual sample can be done in a matter of minutes. A skilled technician can perform hundreds of determinations in a day's time. Consequently more analyses are done by titrations than by any other method.

Titration analyses utilize various types of chemical reactions. Neutralization of hydrogen or hydroxide ion is used in analysis of acids or bases, oxidation of ferrous ion in determination of iron, precipitation of AgCl in determinations of silver or chloride ion, formation of a chelate complex in the determination of the calcium content of hard water, etc. In fact, almost any type of chemical reaction can be adapted for use in a volumetric analysis. The reaction, however, must satisfy the following conditions:

1. There must be a single, definite reaction which can be expressed by a chemical equation.

2. The reaction must be rapid and must go essentially to completion in the course of the titration.

3. An indicator must be available to locate the end point of the titration. This end point should occur at or near the equivalence point.

Not all neutralization, oxidation-reduction, or precipitation reactions meet these criteria, but a sufficient number do so to provide the analyst with thousands of tested volumetric methods. We shall not in this course attempt a thorough survey of all the known methods. Rather, we shall study a few typical methods intensively. An understanding of these will give the student sufficient background to enable him to use any volumetric method that specific problems may require.

Standard Solutions

The starting point for every volumetric analysis is the standard solution used to titrate the sample. This is a solution whose concentration is accurately known. The process of determining the concentration of the standard solution is known as *standardization*. Standardization may be carried out in one of three ways.

1. Direct preparation of the standard solution by dissolving a weighed amount of a pure dry chemical and diluting the solution to an exactly known volume. For example, a 1 molar silver nitrate solution may be prepared by dissolving 169.89 g of dry $AgNO_3$ crystals and diluting the solution in a volumetric flask to exactly 1 liter.

2. Titration of a weighed portion of a pure, dry chemical by the solution to be standardized. The weighed material used for this purpose is known as a *primary standard*. If a solution of $AgNO_3$ is standardized by determining the volume of the solution required to react with a known weight of pure dry NaCl, the latter serves as a primary standard.

3. Titration of a measured volume of a solution that has itself been standardized previously. The solution titrated is known as a *secondary standard*. A standard solution of NaCl can be used as a secondary standard for an $AgNO_3$ solution. The concentration of the $AgNO_3$ solution cannot in this titration be known to any greater accuracy than that of the NaCl solution.

Since it is much easier to compare two solutions by a direct titration than to dry and weigh out portions of a primary standard reagent, a well-equipped analytical laboratory will maintain a supply of reference standard solutions for use in standardization of other solutions as they may be required. The solutions used as permanent reference standards must be stable on storage, i.e., their concentrations must not change appreciably with time. The solution should not react with glass or with constituents of the atmosphere and it should not be affected by light. Many of the standard solutions that can be used safely within a few days of their preparation are not sufficiently stable for use as permanent reference standards.

Analysis of Samples

All titrimetric analyses require the following steps.

1. Preparation of the sample solution. The solution may be prepared either by weighing out and dissolving individual portions of the sample or by measuring definite volumes of a previously prepared solution. In either case the analysis itself is a titration of a prepared solution by a standard reagent solution.

In some analyses the sample needs no prior treatment other than preparation of a solution. In others, the titration is preceded by separations of specific substances from other materials in the sample. Such separations may be done in various ways, such as precipitation, volatilization, extraction, electrodeposition, etc.

In the preliminary treatment prior to titration it is often necessary to oxidize or reduce the sample material to insure that it is in the proper state for the titration. For example, in an analysis of iron by titration with an oxidizing agent, the iron is reduced to Fe^{+2} before the titration.

2. Titration of the sample solution. The solution is titrated by a standard solution of the reagent, according to the procedures given in Chapter 4. These discussions should be reviewed before the student performs his first titrations, and care should be taken to use proper techniques for each of the operations.

The titration may be made either to (a) a dead-stop end point or (b) past the end point, followed by back titration with sample solution or another standard solution. The choice of methods is governed by conditions. When a fixed amount of sample has been weighed out or measured from a pipet, a dead-stop end point is often desirable. There is the disadvantage, however, that if the end point is overrun the determination is spoiled. This possibility can be avoided by back titrations, which can be employed when the titration is for aliquot portions that can be measured from a buret, or when a second standard solution is available that can be used to titrate the excess reagent. When back titration is used, the total volume of solution must be corrected for the amount used in excess of the end point, to give the net amount that has reacted with the sample.

3. Computations. The volume and concentration of standard solution used in the titration are the data employed to compute the results of an analysis. The final result can be expressed in a variety of ways, such as (a) the molarity or normality of the sample solution, (b) the weight of a given substance present in the titrated solution, and (c) the percentage of a given constituent of the sample. All these methods will be illustrated in the computations of the analyses to be performed in this course.

Sources of Error in Titrimetric Operations

There are many potential sources of error in the exacting techniques of titrimetric analyses. These should always be kept in mind so that suitable precautions can be taken to avoid them.

1. Loss of sample or solution by spilling in weighing, spilling solution in transfers, leaking burets, faulty pipetting, etc.

2. Contamination or dilution of solutions by failure to properly rinse burets, pipets, and bottles.

3. Faulty mixing of solutions after they have been diluted.

4. Impurities in the primary standard on which the concentration values are based.

5. Errors in weighing.

6. Errors in reading burets.

7. Use of wrong indicator for titrations.

8. Poor drainage from buret or pipet because apparatus has not been cleaned.

CHAPTER 13

Neutralization Methods

The neutralization of hydronium or hydroxide ion to form water is widely used as the basis for volumetric determinations of acids, bases, and salts of weak acids. The reaction is characterized by a rapid change in pH near the equivalence point, a change that is readily detected by an acid-base indicator or that can be followed electrically by use of a pH meter.

Standard Solutions

The standard solutions used for neutralization titrations are usually solutions of *strong* acids or bases. The most frequently used acids are hydrochloric and sulfuric, either of which is suitable for use as a permanent reference standard, provided the solution is protected from loss by evaporation. Sodium, potassium, and barium hydroxides are the most frequently used bases. Bases, however, are not as good as acids for permanent standards because they absorb carbon dioxide whenever they come in contact with air. Solutions of bases also react with glass on long storage. Base solutions can, however, be stored in Pyrex or plastic bottles (such as polyethylene) if the bottle is equipped with a siphon and the inlet tube is equipped with a tube of soda lime to remove CO_2 from the entering air.

For many purposes it is desirable to remove the carbonate when preparing standard solutions of bases. The presence of carbonate ion interferes with some titrations by so buffering the solution that the end point is not sharp. The carbonate can be sufficiently removed either by precipitation as $BaCO_3$ or by initially making a saturated solution of sodium hydroxide, in which sodium carbonate has a fairly low solubility, and filtering out the carbonate before diluting to the desired strength. A 50 per cent reagent NaOH solution of low carbonate content has recently been put on the market. This can be used, without further removal of carbonate, for preparation of standard solutions.

Standardization

Many substances are available for primary standards in acidimetry-alkalimetry; only a few of those more widely used are discussed here, inasmuch as these are sufficient to take care of almost any need. Since both an acid and a base solution are generally prepared, it makes no difference which is compared with the primary standard when the solutions are intended for general laboratory use. The normality of the solution not compared with the primary standard can be easily determined from the ratio of the strengths of the acid and base solutions.

When a solution is prepared for the analysis of a particular substance, the standard is chosen that best corresponds with the conditions to prevail in the analyses. For example, if a sodium hydroxide solution is to be used chiefly for the analysis of samples of acetic acid, it is preferable to standardize by potassium acid phthalate. Similarly, an acid solution to be used for the analysis of carbonate samples is preferably standardized by sodium carbonate.

Potassium Acid Phthalate ($KHC_8H_4O_4$, mol. wt. 204.2). This is probably the best generally available primary standard. Assayed samples of purity 99.95 per cent or better can be obtained from the Bureau of Standards or from any chemical supply house. Advantages are a high equivalent weight, stability on drying, slight affinity for water vapor, and ready availability. Standard solutions of potassium acid phthalate may be kept for long periods of time without danger of decomposition. Phenolphthalein indicator is employed with this standard, and only carbonate-free strong bases should be used. In neutralization reactions potassium acid phthalate behaves as a monoprotic organic acid, comparable to acetic acid in strength.

Sodium Carbonate $(Na_2CO_3$, mol. wt. 106.0). Sodium carbonate is the best standard for acids that are to be used for titrations of samples containing carbonate. High-grade carbonate, of assay value 99.95 per cent, is available, or, if preferred, sodium bicarbonate may be purified by recrystallization and converted to carbonate by heating at 260–270°C until constant weight is attained.

Gravimetric Standardization. If no primary standard reagent is available, acid-base solutions may be standardized by determining the chloride ion of a hydrochloric acid solution gravimetrically, as described in Chapter 19. This HCl solution is then used to standardize base solutions.

Other Standards. Other standards are used for acid-base solutions, including the following:

1. *Tris(hydroxymethyl)aminomethane* $(CH_2OH)_3CNH_2$. This is a recently developed[1] primary standard for acid solutions. It is stable, non-hygroscopic, and has a high equivalent weight (121.4).

2. *Calcite.* Clear crystals of calcite are nearly pure calcium carbonate and as such they may be used to standardize acids. A weighed portion is treated with excess acid until solution is complete. The remaining acid is then determined by titration with a base solution.

3. *Sulfamic Acid.* $(HNH_2SO_3$, mol. wt. 97.10). This is a strong acid, available in primary standard grade. Practically all its salts are soluble.

Analyses

The analyses that can be made by neutralization methods may be grouped in two general classes: (1) determination of the acid or base (or anhydrides of these) in a sample by titration with a standard base or acid, and (2) determination of salts of weak acids (or bases) by titration with a standard solution of a strong acid (or base). Not all acids and bases can be satisfactorily titrated. It is shown in Chapter 7 that the titration of very weak acids or bases is not feasible because a sharp end point cannot be obtained. Usually when an acid is too weak to be titrated with standard base, a salt of the acid can be titrated with standard acid. For example, carbonic acid cannot be titrated by sodium hydroxide, but sodium carbonate can be titrated

[1] J. H. Fossum, P. C. Markunas and J. A. Riddick, *Anal. Chem.*, **23,** 491 (1951).

by hydrochloric acid, according to the equation:

$$CO_3^{-2} + 2H_3O^+ = CO_2 + 3H_2O$$

Acid-Base Indicators

In all neutralization titrations, results are dependent on the selection of the proper indicator. A comprehensive discussion of the properties of indicators and the selection of an indicator for a titration is given in Chapter 7. Since this material may not be studied before starting acid-base analyses, we shall consider briefly the properties of indicators at this point.

The indicators used in neutralization titrations are highly colored organic compounds which have the property of changing in color when the hydrogen ion concentration of the solution is changed over a certain range. The most widely used are phenolphthalein, methyl orange, methyl red, and bromcresol green.

The hydrogen ion concentrations at the point of color change (end point) are widely different for various indicators. Phenolphthalein is colorless in solutions in which the concentration of hydrogen ion is greater than 10^{-8} molar and pink in solutions in which the concentration of hydrogen ion is less than 10^{-10} molar. If the hydrogen ion concentration changes from 10^{-10} to 10^{-8} molar, the phenolphthalein indicator will change from pink to colorless. If the solution assumes a hydrogen ion concentration intermediate between these two values, the color is also intermediate between deep pink and colorless. It is recalled that in pure water the concentration of hydrogen and of hydroxyl ion is 10^{-7} molar. Thus phenolphthalein shows a color change on the basic side of neutrality.

Methyl orange, methyl red, and bromcresol green give color changes on the acid side of neutrality, in the region of hydrogen ion concentration 10^{-3} to 10^{-5} molar. Intermediate between phenolphthalein and these indicators are many others, with color changes at various hydrogen ion concentrations. A list of the more common ones is given in Table 7–3, page 160.

The proper indicator for a titration is one that will exhibit a color change at the hydrogen ion concentration found at the equivalence point of the titration. In the titration of strong acids by strong bases, the salt formed does not hydrolyze. Therefore the hydrogen ion concentration at the equivalence point is that of pure water, or 10^{-7} molar. When a weak acid, such as acetic acid, is titrated by sodium hydroxide, the salt formed will hydrolyze to give a basic reaction and the hydrogen ion concentration will be approximately 10^{-9} molar. Therefore phenolphthalein is a suitable indicator. In the titration of a weak base, such as ammonia, by a strong acid, the salt formed will hydrolyze to give an acid reaction. Methyl red is the most suitable indicator. In the titration of a salt of a weak acid by a strong acid, for example sodium carbonate by hydrochloric acid, there is formed at the equivalence point a solution of the weak acid. The solution is, therefore, slightly acidic, and methyl orange and methyl red are suitable indicators.

From the foregoing it might seem that neither phenolphthalein nor methyl orange would be a suitable indicator for the titration of hydrochloric acid by sodium hydroxide, since neither of the indicators shows a color change at the neutral point. It happens, however, that, at the equivalence point in this titration, the addition of a very small volume of reagent will greatly change the hydrogen ion concentration. If an indicator blank is used, as described in the following paragraphs, either indicator is satisfactory.

Indicator Blank. If the indicator color change occurs exactly at the hydrogen ion concentration found at the equivalence point in an analysis, it is not necessary to use an indicator blank. Since it is not always possible to select the indicator so that the end point exactly coincides with the equivalence point, it is advisable either to standardize the solution with an analyzed sample of the constituent which is to be determined by this solution or to use an indicator blank in both the standardization and the analysis. The first process is illustrated in the determination of carbonate by titration with hydrochloric acid. If the hydrochloric acid is standardized by pure sodium carbonate, the same indicator error will occur in both standardization and analysis, provided the concentration of carbonic acid is the same at the end point in the two processes. Therefore exact results will be obtained in the analysis.

Use of an indicator blank may be illustrated by results found for the titration of hydrochloric acid by sodium hydroxide. Duplicate titrations were made, one with methyl orange indicator and the other with phenolphthalein indicator.

1. When methyl orange indicator was used, 25.00 ml of acid was required to titrate 24.80 ml of base. A blank was determined by adding methyl orange to a solution of NaCl and titrating to the end point. This required 0.20 ml of acid. Subtracting the value of the blank from the total volume of HCl gives a net volume of 24.80 ml.

2. When phenolphthalein indicator was used, 25.02 ml NaOH was

required to titrate 25.00 ml HCl. The indicator blank was found to be 0.02 ml NaOH. Subtracting this from the total volume of NaOH gives a net volume of 25.00 ml to titrate 25.00 ml of acid.

If the blank correction had not been made, it would appear in the methyl orange titration that 1.000 ml of acid is equivalent to 0.992 ml of base and in the phenolphthalein titration that 1.000 ml of acid is equivalent to 1.001 ml of base. When the blank correction is used, we find in both titrations that 1.000 ml of acid equals 1.000 ml of base.

Preparation of Acid and Base Solutions

Directions are given for preparation of 1 liter of each solution. This will suffice for each student's needs in comparing his solutions, standardizing them, and analyzing three or four portions each of an acid and a base sample. Directions are for approximately 0.1 N solutions.

Procedure. Measure out 9 ml (note 1) concentrated HCl solution, with a small graduated cylinder. Pour into a clean 1-liter glass-stoppered bottle. Fill the bottle almost to the neck with distilled water. Stopper and mix thoroughly by repeatedly inverting and shaking.

Prepare carbonate-free base solution by one of the following methods. The instructor will designate the one to use.

1. Obtain from the stock bottle approximately 100 ml 1 N stock solution. Pour into a clean 1-liter bottle, add boiled distilled water almost to the neck, stopper, and mix by inversion and shaking. The stock solution is prepared[2] by dissolving about 45 g (note 2) solid NaOH for each liter of solution wanted. After the NaOH is all dissolved, add a solution containing about 5 g barium chloride for each liter of final solution, mix well, and allow to stand overnight in a stoppered bottle. The precipitated $BaCO_3$ will settle to the bottom. Siphon off the clear liquid into the stock bottle. Fit the bottle with a two-hole stopper carrying a siphon delivery tube in one hole and a soda lime tube in the other (to remove CO_2 from incoming air).

[2] The authors recommend preparation of the stock solution from 50 per cent reagent NaOH solution (low in carbonate) if this is available. When this solution is used, no $BaCl_2$ treatment is needed. Use boiled water.

Attach a short length of rubber tubing at the end of the siphon tube. This is closed with a pinch clamp except when the clamp is opened for delivery. The stock solution may be preserved indefinitely in this way.

2. Prepare an asbestos mat in a Gooch crucible,[3] as directed in Chapter 4. Wash the mat with a little water but do not dry. Empty the suction flask, suspend a clean dry test tube beneath the filter, as illustrated in Fig. 4–10, and replace the filter.

Weigh on a trip scale about 10 g of reagent-grade NaOH pellets. Put an equal weight of water (10 ml) in a Pyrex test tube and add the NaOH pellets. Close the tube tightly with a rubber stopper, and mix (note 3) the solution by repeatedly inverting the tube until practically all the NaOH has dissolved. Cool the tube in tap water and filter, with suction, through the Gooch crucible. Stop the suction as soon as most of the liquid has run through the filter. Remove and stopper the test tube containing the filtrate. This should be clear. Fill a clean 1-liter bottle almost to the neck with distilled water which has been boiled to remove CO_2. Pour about half the NaOH solution into the bottle, stopper, and mix thoroughly by inverting and shaking.

Reserve the remainder of the NaOH solution in a stoppered test tube until a comparison of the acid and base strengths has been made. Should it be found that the base solution is appreciably weaker than the acid solution, add the estimated amount of the 1:1 solution to make the two approximately the same strength. After this is done, discard the remainder of the 1:1 NaOH solution.

Notes. 1. The concentrated HCl solution, of density 1.18 g per milliliter, is about 12 N. If the acid is measured from a graduated cylinder, the retention in the cylinder will make the net amount delivered just about the proper amount for a 0.1 N solution.

2. If the NaOH were pure, the weight required for a liter of 1 N solution would be 40 g. A slight excess is taken, to compensate for the water and Na_2CO_3 content of the pellets.

3. The tube becomes quite warm as the NaOH dissolves. Hold in a towel and keep the stopper firmly in place while shaking and inverting the tube. The concentrated caustic solution is dangerous. Wear goggles to protect the eyes should any spray or splash from the tube.

[3] If available, a porcelain filtering crucible may be used instead of the Gooch. The filtering crucible should be washed with water and finally with dilute HCl immediately after the filtration is completed, since the highly concentrated NaOH solution will ruin the filter if allowed to stand in it.

Comparison of Acid and Base Solutions

The student should be able to do the comparison, the standardization, and the analysis of an unknown in a single afternoon if samples are prepared well in advance. It is wasteful of time and solutions to fill up burets and make the comparison one day and then to wait until another period to standardize. Before starting the comparison, the student should prepare dried portions of his primary standard material and his first unknown. It is his responsibility to look ahead and plan his work in advance so that waste of time is avoided.

Procedure. Reread the general instructions for the operations of volumetric analysis, Chapter 4. Drain the water from both burets (note 1) and fill them, respectively, with the 0.1 N acid and base solutions, *with proper rinsing*. Read and record the positions of both menisci.

Run into a clean (but not necessarily dry) wide-mouth 250-ml conical flask between 35 and 45 ml of the acid (note 2). Do not read the buret accurately at this time. Touch the wall of the flask to the tip to remove pendent drops. Add one or two drops of phenolphthalein indicator, and titrate with base until a faint pink color is obtained. Do not take inordinate care to stop the addition of base just at the end point. Rather, titrate rapidly and, if the end point is overrun, back-titrate by adding acid until the solution is just colorless; then add base dropwise until a single drop imparts a pink color which pervades the entire solution when the liquid is well mixed by swirling. The color should persist at least 15 seconds. Read both burets and record the data in the notebook as shown in Table 13–1. No indicator blank is needed for this titration.

Compute the volume of acid equivalent to 1 ml of base:

$$1 \text{ ml base} = \frac{\text{volume acid (with buret corrections)}}{\text{volume base (with buret corrections)}}$$

If the ratio is less than 0.8, the base solution should be made stronger by addition of more of the 1:1 solution, and the comparison repeated. If the ratio is greater than 1.2, the base solution has been made too strong. In this case, the acid solution should be made stronger by addition of the computed amount of 12 N HCl.

Obtain three comparison values. The range (Chapter 6) should not exceed 4 parts per 1000 and usually it is not this large. When a satisfactory comparison is obtained, proceed with standardization,

TABLE 13-1. Comparison of Strength of Acid and Base

Notebook Record

HCl		NaOH		Ml HCl* Ml NaOH
37.92 (−0.06)†	37.86	37.07 (+0.05)	37.12	
1.10 (0.00)	1.10	0.83 (0.00)	0.83	
				1.013
	36.76		36.29	
41.10 (−0.08)	41.02	40.45 (+0.06)	40.51	
0.26 (0.00)	0.26	0.15 (0.00)	0.15	
				1.010
	40.76		40.36	
41.15 (−0.08)	41.07	40.01 (+0.06)	40.07	
0.73 (0.00)	0.73	0.17 (0.00)	0.17	
				1.011
	40.34		39.90	
41.25 (−0.08)	41.17	40.41 (+0.06)	40.47	
0.56 (0.00)	0.56	0.26 (0.00)	0.26	
				1.010
	40.61		40.21	
			Average	1.011

$$\text{Range} = 0.003 = \frac{0.003}{1.011} = \frac{3}{1000}$$

* Note that 1.000 ml base = 1.011 ml acid.
† Buret correction.

following the method specified by the instructor. If this cannot be done on the same day, empty the burets and discard the solutions taken from them. Fill the burets with distilled water, after a preliminary rinsing, and allow to stand.

Notes. 1. An alternative method, preferred by many teachers, is to measure out a 25-ml portion of the acid by a calibrated pipet and to use only one buret in the determination. If instructed to follow this method, fill the buret with the base solution and titrate to a dead-stop end point. If the end point is overrun in the first titration, discard that sample but use the data to compute the approximate amount of NaOH to add in succeeding samples. It is advisable to pipet out five samples and to have one or two in reserve should a titration be spoiled. Pipet all the desired samples at the same time, for otherwise it is necessary to rinse the pipet before taking each sample.

The loss of a determination by overrunning the end point can be avoided by removing a small portion of solution before the end point is reached, adding reagent until a color change occurs in the remainder of the sample, and then returning the small portion that was removed. After pipetting out the 25-ml portion, titrate until near the end point. Then pour about 1 ml from the flask into a clean beaker. Titrate to a pink color. Wash the portion from the beaker back into the flask, and add base a drop at a time until the final end point is reached.

2. There is less absorption of CO_2 during the titration if the acid is titrated by a base, for if the base is titrated by an acid, the swirling solution has an opportunity to absorb an appreciable amount of CO_2 in the early stages of the titration.

Standardization

Potassium Acid Phthalate

This primary standard is recommended as the best of the generally available ones, provided the base is free of carbonate. It will not give good results for a base solution containing an appreciable amount of carbonate. Phenolphthalein indicator is used.

Procedure. Weigh out 4 g potassium acid phthalate (KHP) on a trip scale. Put in a clean weighing bottle and dry in an oven at 110°C for an hour or more. Stopper the bottle and keep in a desiccator.

Weigh out three portions of 0.7–0.9 g KHP (to the nearest half milligram) and dissolve each in about 50 ml of water in a 250-ml widemouth conical flask. Add a drop of phenolphthalein indicator and titrate with the base solution to a faint pink color. (As the end point is approached withdraw a small portion of solution, as described in note 1, page 247, to prevent overrunning the end point.)

Compute the normality of the NaOH solution as follows. Apply buret corrections if needed and compute the volume of NaOH required to titrate the weighed portion of KHP. From the weight and purity of KHP compute the number of milliequivalents in each sample. Divide the number of milliequivalents by the net volume of NaOH to obtain the normality of the NaOH (milliequivalents per milliliter). The relative range in the three titrations should not exceed 4–5 parts per 1000. Record the average of the three determinations on the label of the NaOH bottle.

From the NaOH normality and the previously determined acid-base ratio, compute the normality of the HCl. Record the value on the bottle label.

Notebook Record. Show all original data in the notebook, as taken. This includes weights of the weighing bottle before and after samples are withdrawn, buret readings, etc. On a separate page give a summary of the standardization, as shown in Table 13–2.

TABLE 13–2. Standardization of Base and Acid Solutions with Potassium Acid Phthalate

Notebook Record

Purity of KHP, 99.95%
1.000 ml base = 1.011 ml acid

Sample No.	I	II	III
Weight KHP	0.7835	0.8460	0.8035
Corrected weight (for purity)	0.7831	0.8456	0.8031
Corrected volume of base	38.85	42.00	39.96
Milliequivalents KHP	3.835	4.141	3.933
Normality of base	0.0987	0.0986	0.0984
Average normality of base	0.0986		

$$\text{Range} = 0.0003 \text{ or } \frac{3}{986}$$

Normality of acid = 0.0976

Errors. Review the list of general titration errors in Chapter 12. Specific errors in this titration are (1) impurities in the standard and (2) presence of sufficient carbonate in the base solution to cause an uncertain end point.

It should be noted that the NaOH solution always contains a small amount of carbonate. As discussed in Chapter 7, when a carbonate solution is titrated to a phenolphthalein end point, the carbonate ion is but half neutralized, to form $HCO_3{}^-$. Consequently the standardization value obtained by KHP for the NaOH is not the total basic strength of the solution. But when the solution is used in analyses, with phenolphthalein indicator, the conditions are identical to those used for standardization and the normality value we have obtained is the correct one. If, however, this same NaOH solution were titrated by HCl to a methyl orange end point, which occurs when the carbonate ion has been completely neutralized, the basic strength of the solution would be greater than the value we have obtained in the KHP titration.

Sodium Carbonate

When an acid solution is to be standardized directly, without comparison with a base solution, sodium carbonate is the best of the generally available primary standards.

Procedure. Weigh out three samples of 0.2–0.25 g each, to the nearest 0.1 mg. Put each in a 250-ml wide-mouth conical flask and dissolve in about 50 ml of distilled water. Titrate each sample with the HCl solution, as follows.

Add three drops of methyl red–bromcresol green mixed indicator and titrate to a faint pink color. Place the flask on a wire gauze and heat just to gentle boiling. Cool under running tap water or in a beaker of ice water to room temperature. If the end point was not overrun, the indicator turns back to its original green color (note 1). Continue the titration, a drop at a time, to a sharp color change.

Apply buret corrections (if needed) and determine the volume of HCl used for each sample. From the weight and purity of the Na_2CO_3 compute the number of milliequivalents of pure Na_2CO_3 (the equivalent weight is 53.00 g). Divide the milliequivalents of Na_2CO_3 by the volume of HCl to obtain the normality. Summarize the data in the notebook, using the general form of Table 13–2. The average for the three determinations is recorded on the label of the HCl bottle. The relative range in the three determinations should not exceed 4–5 parts per 1000.

Compute the normality of the NaOH solution from the acid normality and the previously determined acid-base ratio. Record the value on the label of the bottle. Record all data for the determination and computation of the normalities in tabular form in the notebook.

Notes. 1. If the end point was overrun before boiling the solution, the color remains pink after cooling. Either discard the sample or, if preferred, run in a previously compared NaOH solution from a buret until the color changes to green. Record the volume of NaOH used. Continue the titration with HCl to the end point. Use the previously determined acid-base ratio to compute the volume of HCl equivalent to the NaOH that is added. Subtract this from the total volume of HCl to obtain the net volume that has reacted with the Na_2CO_3.

Errors. The specific errors of this procedure are (1) loss of acid on boiling the solution if there is much excess HCl present at this time, (2) failure to cool the solution to room temperature before completing the titration, and (3) impurities in the sodium carbonate.

Analyses

The analyses of this section illustrate the more common types of acid-base determinations: (1) weak acids, (2) salts of weak acids, and (3) ammonia or ammonium salts. The determination of strong acids and bases has been illustrated in the comparison of the acid and base solutions.

Determination of Weak Acids

The procedures of this section are applicable to the analysis of any sample of strong acid or any weak acid whose ionization constant is not less than about 10^{-5}. For very weak acids, it is better to titrate a salt of the acid with a strong acid, as in the determination of carbonates.

Potassium Acid Phthalate.   The sample for analysis may be issued either as a weighed portion of pure KHP or as an analyzed sample, which is a mixture of KHP with an inert diluent such as NaCl. The authors recommend the first method because it is difficult to prepare absolutely uniform mixtures of KHP with a diluent. The student is issued a beaker containing a 4–5-g sample, weighed to the nearest milligram. A small amount of some inert substance is added to the weighed sample before it is issued. The entire sample is dissolved and diluted to volume. Aliquot portions are taken for the separate analyses. If an analyzed mixture is issued, the student weighs out individual portions for each determination.

Procedure.   Obtain an unknown sample, as directed in the laboratory instructions. Record the sample number in the notebook. Prepare portions for analysis by one of the following methods, as directed.

1. *Weighed Unknown.* Transfer the entire sample to a 250-ml beaker, taking care to avoid any loss of powder. Add 50–100 ml of distilled water and warm to bring all the sample into solution. Transfer the solution quantitatively to a 250-ml volumetric flask, as directed in Chapter 4. Dilute to volume and pipet out four 50-ml aliquot portions for analysis.

2. *Analyzed Mixture.* Dry at 110°. Weigh out four portions, of the size directed by the instructor (the size of sample to take is determined by the approximate composition of the sample), and transfer each to a wide-mouth conical flask.

Titration. Titrate each portion of sample with standard NaOH solution, using phenolphthalein indicator. If the end point is overrun, either discard that portion or use the standard HCl solution to back-titrate. Compute for each portion the milliequivalents of NaOH used. If HCl was used for back titration, compute the milliequivalents of acid and subtract from the milliequivalents of base to obtain the net milliequivalents of base used for the sample.

If the sample was a weighed portion, compute the average number of milliequivalents of NaOH for each aliquot portion. Multiplication of this average value by 5 gives the total milliequivalents of KHP in the sample. Multiply this by 204.2 to obtain the weight of KHP received. Report the analysis on a 3x5-in. card, as in Table 13–3.

TABLE 13–3. Sample Report Card for Weight of KHP Sample Received

Name_____Date_____Desk No._____

    Sample No._____   Normality of NaOH <u>0.1005</u>

Net Vol. NaOH	Meq NaOH
36.38 ml	3.656
36.27	3.645
36.31	3.649
	Average 3.650

Weight KHP $= 3.650 \times 5 \times 204.2 = 3727$ mg

Range $= 0.011 = \dfrac{11}{3650} = \dfrac{3}{1000}$

If the sample was an analyzed mixture, use the net milliequivalents of NaOH for each portion to compute the weight of KHP in that portion and the percentage of KHP in the portion. Report the average percentage of KHP on a 3x5-in. card, as in Table 13–4.

TABLE 13–4. Sample Report Card for Solid Acid Unknown

Date_____Name_____Desk No._____

    Sample No._____   Normality of base <u>0.1032</u>

Wt. Sample	Vol. Base	% KHP
3.175 g	38.73	25.70
3.562	43.36	25.65
3.485	42.34	25.60
		Average 25.65

Range $= 0.10 = \dfrac{10}{2565} = \dfrac{4}{1000}$

Errors. If the analysis results show good precision but fail to agree with the instructor's record, the most likely sources of error are (1)

incorrect standardization of the NaOH solution and (2) a mistake in the calculations. Experience has shown that beginning students are prone to make mistakes not in the arithmetical operations but in such fundamental things as inverting the acid-base ratio, etc. If the reported result is not accepted, check the computations carefully before repeating the analysis.

Analyzed Solution. The unknown sample is an analyzed solution of an acid. It is dispensed in a stoppered bottle. Measured portions are taken by a buret or pipet for titration with the standard base. The normality of the solution is reported.

Procedure. Obtain the sample in a stoppered bottle. Record the number. Pipet out four portions into wide-mouth conical flasks. Use a 25-ml pipet which has been calibrated. Alternatively, the portions for titration may be taken from a buret.

Titrate each portion with the standard NaOH solution, as in the standardization with KHP. Apply buret corrections if needed. From the volume and normality of the NaOH and the volume of the acid unknown sample, compute the normality of the acid. The report should show the value obtained for each sample, the average value, and the relative range in parts per 1000.

Errors. The most probable error is in standardization of the NaOH. If the pipet is not properly calibrated, a small error of the same size may be introduced for each portion of sample.

Titration of Carbonates by Acids

As discussed in Chapter **7**, carbonic acid cannot be titrated directly by a strong base because the pH change at the equivalence point is not sufficiently great to give a sharp end point. Carbonates may, however, be titrated by strong acids. As shown in Fig. 7–5, there are two inflection points in the titration curve. The first, near pH **8**, corresponds to completion of the reaction

$$CO_3^{-2} + H_3O^+ = HCO_3^- + H_2O$$

and the second, at pH near 4, to the reaction

$$HCO_3^- + H_3O^+ = 2H_2O + CO_2$$

The second break is sharper than the first one.

Reference to the table of indicators on page 160 shows that phenolphthalein gives a color change at the pH of the first break and methyl orange or methyl red at that of the second. We do not, however, obtain a sharp end point at the first break, for the solution is buffered by the mixture of carbonate and bicarbonate ions and the titration curve slopes instead of giving a sharp break. There is also some buffer effect as the second break is approached, but not nearly so much as for the first.

The sharpness of the end point at the second break can be increased by using the procedure given for standardization of an acid by Na_2CO_3. Acid is added to near the equivalence point, and the solution is boiled to remove CO_2 before the final end point is obtained.

Determination of the Basic Strength of Soda Ash. The analysis of soda ash, crude sodium carbonate, has for many years been a standard student exercise in acidimetry. Titration with HCl to the proper end point gives the total basic strength, regardless of whether the sample contains NaOH, $NaHCO_3$, or Na_2CO_3. The results are generally reported as the percentage of Na_2CO_3 in the sample, but they might just as well be reported as Na_2O or as NaOH; the report represents merely the method of computation, and it does not imply that the basic material of the sample is in the form reported.

The following procedure may be used for the analysis of carbonate samples and for the analysis of base solutions which contain carbonates.

Procedure. Obtain the sample in a weighing bottle, and dry in the oven. Preserve in a desiccator until needed. Weigh out three samples of 0.4–1.5 g as directed by the instructor. Transfer each to a 250-ml wide-mouth conical flask and dissolve in about 50 ml of distilled water.

Titrate each portion with standard HCl solution. Add three drops of methyl red–bromcresol green mixed indicator and titrate to a faint pink color. Place the flask on a wire gauze and heat to gentle boiling. Cool under running tap water or in a beaker of ice water to room temperature. If the end point was not overrun, the indicator turns back to the original green color. Add acid, a drop at a time, until a sharp color change is obtained.

Apply buret corrections as needed. From the volume and normality of the HCl, compute the milliequivalents of acid used for each sample. Multiply by the equivalent weight of Na_2CO_3 (53.0) to obtain the weight of Na_2CO_3 in milligrams. Divide by the weight of sample and multiply by 100 to obtain the percentage. Report the percentage of Na_2CO_3 for each portion, the average value, the range and the relative range in parts per 1000. The latter should not exceed 3–5 parts per 1000.

Titration with Two Indicators

The existence of two breaks in the titration curves of polyprotic acids or their salts permits simultaneous determination of two constituents of simple mixtures, by a single titration employing two separate indicators to give two separate end points. To illustrate the principle, we shall discuss the determination of carbonate and hydroxide in the same sample, an analysis of practical value for crude alkali samples.

When a solution containing both NaOH and Na_2CO_3 is titrated by HCl to a phenolphthalein end point, the amount of acid used is that required to react with all the OH^- ion plus that required to convert CO_3^{-2} to HCO_3^-. The volume of acid required for this end point is noted, then methyl orange or another indicator turning at pH 4–5 is added, and the titration is continued to the end point of the second indicator. Since phenolphthalein is colorless at pH below 8, its presence does not affect the second end point. The volume of acid used to titrate from the first end point to the second is the amount required to convert HCO_3^- to CO_2.

Problem. It is found that 15.00 ml 0.1000 N HCl titrates a sample containing both NaOH and Na_2CO_3 to the phenolphthalein end point. An additional 10.00 ml of the acid is required to continue the titration to the methyl orange end point. Compute the weights of NaOH and Na_2CO_3 in the sample.

SOLUTION: The total volume of HCl used is 15.00 + 10.00 or 25.00 ml. Total milliequivalents of HCl = 25.00 × 0.1000 = 2.500. Since 10.00 ml or 1.000 meq HCl was needed to convert HCO_3^- to CO_2 and H_2O (the amount used to titrate from the first end point to the second), an equal amount of HCl was used in converting CO_3^{-2} to HCO_3^-. Thus the total HCl used by carbonate is 2.000 meq. The weight of Na_2CO_3 is 2.000 meq × 53.0 mg per milliequivalent or 106.0 mg.

Since 2.000 meq HCl was used to titrate the carbonate of the sample, the amount of HCl used for the NaOH is 2.500 minus 2.000 or 0.500 meq. The weight of NaOH is 0.500 meq × 40.0 mg per milliequivalent, or 20.0 mg.

The two-indicator method described in this problem is not precise in actual titrations because of the fading of the phenolphthalein end point, caused by the buffered solution. More accurate results can be obtained by a modification of the procedure. Separate aliquot portions of sample are taken. The total basic strength of one portion is determined as in the standardization of HCl with Na_2CO_3. The other portion is treated with $BaCl_2$ solution, which precipitates the carbonate as $BaCO_3$. The NaOH, remaining in solution, is then titrated to the phenolphthalein end point. It is not necessary to filter out the $BaCO_3$ before this titration, for the phenolphthalein end point occurs when the solution is still slightly basic and the hydrogen ion concentration is not high enough to give any reaction with the precipitate.

A mixture of $NaHCO_3$ and Na_2CO_3 can be analyzed by a similar procedure. One portion is completely titrated with HCl to give the total basic strength. The other portion is treated with a known volume of standard NaOH, in excess of the amount needed to convert all HCO_3^- to CO_3^{-2}. The carbonate is then precipitated by $BaCl_2$, and the amount of NaOH remaining in solution is determined by HCl titration to a phenolphthalein end point. The difference between the milliequivalents of NaOH added and the milliequivalents titrated by the HCl gives the milliequivalents of NaOH required to react with the HCO_3^- originally present in the sample.

Procedure.   Determination of Carbonate and Hydroxide in the Same Sample.   Obtain the sample, which is an analyzed solution, in a stoppered bottle or flask. Transfer to a 250-ml volumetric flask, dilute to volume, and take six 25-ml aliquot portions with a pipet that has been calibrated relative to the flask. Determine the total basic strength of three portions by titration with HCl, according to the procedure of page 250 for standardization of acid with sodium carbonate.

Prepare a solution of barium chloride containing about 5 g of crystals per 50 ml of water. Add a drop of phenolphthalein indicator, and make the solution just neutral by adding base or acid as needed.

To each of the remaining three aliquot portions of the sample, add barium chloride solution, drop by drop from a pipet, until no further precipitation occurs. It may be necessary to allow the precipitate to settle after each addition of barium chloride, to determine when precipitation is complete. No harm is done if a slight excess of barium chloride is added.

To each portion add phenolphthalein indicator, and titrate with standard HCl. If the end point is overrun, back-titrate with carbonate-free base.

From the net amount of acid used for the samples treated with barium chloride, compute the milliequivalents of NaOH. The difference between the total basic strength as previously determined and the milliequivalents of NaOH gives the milliequivalents of carbonate. Report the total milliequivalents of NaOH and Na_2CO_3 in the sample.

Kjeldahl-Gunning Method for the Determination of Nitrogen in Organic Compounds

Nitrogen in the form of amines or amides may be quantitatively determined by being converted into an ammonium salt and analyzed by the usual volumetric procedures. The process of conversion into ammonium salts is accomplished by digesting the organic compound with concentrated sulfuric acid. Carbon and hydrogen are oxidized to carbon dioxide and water, respectively, while nitrogen is converted into ammonium sulfate. After digestion is complete, as shown by the disappearance of carbon, the solution is made alkaline with sodium hydroxide, and the liberated ammonia is distilled into a measured volume of standard acid solution. The amount of ammonia is determined by titrating the remaining acid with a standard base solution. This method for the determination of nitrogen is known as the Kjeldahl process. Nitrogen present as azo compounds or as nitro derivatives is not detected in analyses performed by the method as outlined, but, if the sample is first reduced, so as to convert the nitrogen into an amine, the Kjeldahl method may be used. Nitrates may be included in a regular Kjeldahl analysis, if means are provided for preventing the escape of nitric acid and reducing the nitrogen to ammonia. Treatment of the sample with a salicylic acid–sulfuric acid mixture serves to prevent escape of nitric acid, because salicylic acid reacts with the nitric acid liberated, giving nitrated derivatives. After fixation of the nitrogen in this manner, it may be reduced by treating the sulfuric acid solution with sodium thiosulfate, prior to digestion. The thiosulfate serves to reduce the nitrated salicylic acid to corresponding amine derivatives.

Many modifications of the Kjeldahl method have been proposed, and many catalysts have been used to accelerate the digestion process. In the Gunning modification, which is widely used, no catalyst is employed, but the sulfuric acid solution is nearly saturated with sodium or potassium sulfate. The salt elevates the boiling point, and thus, by permitting higher digestion temperatures to be employed, accelerates the process.

Directions are given in the following procedures for the analysis of dried-blood samples. They do not contain nitrates. If commercial fertilizers or other samples which may contain nitrates are to be analyzed, modifications of the procedure are necessary. See standard reference books, such as *Official and Tentative Methods of Analysis* of the Association of Official Agricultural Chemists, for the test for nitrates and the necessary modifications. Wheat flour can also be analyzed by the following procedure.

Procedure. Digestion of Sample. Obtain the sample in a weighing bottle, and dry at 110°C.

Weigh out three portions of 0.3–2.5 g (note 1), as directed. Wrap each sample in a filter paper to prevent it from sticking to the neck of the flask; drop into a Kjeldahl digestion flask. Add 25 ml concentrated H_2SO_4, 15 g powdered anhydrous K_2SO_4, and a catalyst, as directed by the instructor (note 2). Clamp the flask in a slanting position in a hood, and heat cautiously over an open flame until the acid is boiling. Take care not to put the hands beneath the flask during the heating; breakage of the flask could cause a serious burn from the boiling acid. Adjust the flame to keep the solution at a gentle boil, and continue heating until the solution is colorless or a bright straw yellow (note 3). Remove the flame, and allow the solution to cool, shaking occasionally if it begins to solidify (note 4).

Notes. 1. Samples may vary from 2–15 per cent nitrogen. For the determination we need about 4 meq or 4×14 mg nitrogen.

2. A catalyst is not essential but will materially shorten the digestion time. A drop of mercury or a small crystal of copper sulfate may be used. Still better is a "Hengar granule" prepared for this purpose. It contains an inert solid impregnated with selenium. Add only a single granule to each flask.

3. Blood samples can be digested in half an hour or less if a catalyst is used.

4. If much of the acid has been boiled away, the solution may solidify on cooling. It should not be allowed to solidify into a solid lump which is difficult to redissolve.

Distillation and Titration. After the sample in the digestion flask is thoroughly cooled, add 250 ml of water, with frequent shaking. Heat, if necessary, until solution is complete, but disregard any small precipitate which may persist. Cool the sample after dilution. Prepare, for each sample analyzed, a solution of 45 g sodium hydroxide (note 1) in 75–100 ml of water, and cool thoroughly. Set up a distillation apparatus according to Fig. 13–1. The bulb surmounting the digestion flask is known as a *Kjeldahl trap.* Its purpose is to prevent alkaline solution from being carried over as a spray. The bottom end of the condenser is connected, by a rubber stopper, to a bent adapter which dips into the receiving vessel, an Erlenmeyer flask with a mouth wide enough to permit the tip of the adapter to reach nearly to the bottom of the flask. Clamp the digestion flask, condenser, and receiving flask firmly into position. Remove the receiver, and accurately measure into it 50 ml standard 0.2 N sulfuric or hydrochloric acid.

Replace in position, taking care that the tip of the adapter reaches below the surface of the liquid. Disconnect the digestion flask, and carefully pour down the side, so as to form two liquid layers, the prepared solution of sodium hydroxide (note 2). Drop into the flask several pieces of granulated zinc (note 3) and a strip of litmus paper.

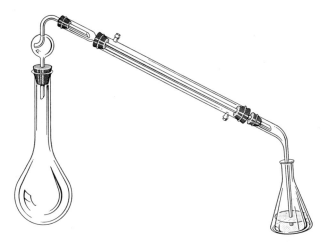

FIGURE 13–1.   Distillation of ammonia in Kjeldahl determination of nitrogen.

Mix the solutions by gently swirling the flask, and immediately connect the flask to the condenser. Observe, from the litmus paper, whether the solution is basic after mixing (note 4). If not, more sodium hydroxide must be added.

Bring the solution to a boil, and distil about one-half of the liquid. Constant supervision must be maintained in order to prevent temperature fluctuations which might cause acid from the receiving flask to be drawn back into the alkaline solution in the digestion flask. When distillation is complete, lower the receiver (note 5), remove the flame, disconnect the Kjeldahl trap, and rinse the condenser with a little water. Remove and rinse the adapter. Titrate the remaining acid in the receiving flask with standard 0.1 N NaOH, using methyl red indicator. From this titration and the amount of acid originally present, compute the milliequivalents of acid neutralized by the ammonia which was liberated from the sample. Run a blank (note 6) determination by digesting and analyzing a 1-g sample of some non-nitrogenous organic substance such as sugar. Subtract the milliequivalents of acid neutralized in the blank from the amount found in the analysis. Multiply the net milliequivalents of acid by the atomic weight of nitrogen in order to obtain the weight of nitrogen in milligrams. Compute the percentage of nitrogen in the sample. Duplicate results should agree within 0.3 per cent (note 7).

Notes. 1. If it is available, use "sodium hydroxide flake, 76 per cent for nitrogen determination." This is considerably cheaper than the reagent-grade pellets and is just as satisfactory.

2. All manipulations, after the alkali solution is mixed, should be performed rapidly in order to prevent loss of ammonia. For this reason also the solutions must be cold before mixing.

3. Granulated zinc aids in preventing bumping during distillation.

4. The test is very important for beginning students. Often too much acid is used, and an attempt is made to distil ammonia from a solution which is still acidic.

5. This is to prevent the solution from sucking back into the Kjeldahl flask as it cools

6. A blank is essential, since contamination of reagents or absorption of ammonia during the analysis may readily lead to results that are far too high.

7. The precision to be desired is not 3 parts per 1000, but rather an agreement of 0.3 per cent in the percentage of nitrogen in the sample. Thus, 7.60 and 7.90 per cent represent acceptable results, although the agreement among these results is only 30 parts per 790, or 38 parts per 1000.

Errors. The more frequent errors, in addition to those common to all acidimetric determinations, are due to the following mistakes:

1. Spray is carried over into the receiver from the alkaline solution in the digestion flask. If a very high result is obtained, it is almost certain to be due to this error.

2. Ammonia may be absorbed during digestion of the sample or by the standard sulfuric acid during the distillation.

3. If the initial distillation is made too rapidly, ammonia may escape through the acid solution without being absorbed. See H. S. Miller, *Ind. Eng. Chem., Anal. Ed.,* **8,** 50 (1936), for a discussion of this error. Steam distillation minimizes this source of error.

Boric Acid Modification. This method, suggested by Winkler in 1913, has in recent years received very favorable attention. It appears to be fully as accurate as the older method of absorption of ammonia in a measured volume of sulfuric acid, and it saves a great deal of time. The liberated ammonia is absorbed in a 4 per cent boric acid solution, the volume and concentration of which need not be accurately determined. The amount of ammonium borate formed is determined by titration with standard hydrochloric acid. The indicator used, bromcresol green, gives an end point at a hydrogen ion concentration corresponding to a solution of ammonium chloride. Boric acid itself is so weak that it has no appreciable influence on the hydrogen ion concentration. The reactions are:

$$NH_3 + HBO_2 = NH_4{}^+ + BO_2{}^-$$

$$H_3O^+ + BO_2{}^- = HBO_2 + H_2O$$

Procedure. Carry out digestion, distillation, and so on, just as in the preceding directions, but replace the sulfuric acid absorbent with approximately 50 ml 4 per cent boric acid solution. When the distillation is complete, remove the absorption flask, as previously directed, and titrate with standard 0.1 N hydrochloric acid, using bromcresol green indicator.

Determination of Ammonia in Soluble Salts

Ammonia, in the form of a soluble salt, may be determined by a modification of the Kjeldahl procedure. No digestion is needed. The solution is placed in a Kjeldahl flask and treated with excess sodium hydroxide solution, and the liberated ammonia is distilled into a measured volume of standard sulfuric acid. The excess acid is then determined by titration with a standard base solution. The boric acid method may advantageously be used if desired (see previous paragraph). This exercise is often substituted for a Kjeldahl determination in elementary courses because no digestion is needed; the sample can be a measured solution of ammonium sulfate of known concentration.

Procedure. Dilute the sample, which contains 20–25 meq of ammonium sulfate, to a volume of 250 ml, and measure out three 50-ml aliquot portions with a pipet. Add to each portion 150 ml of water, and place each portion in a 500-ml Kjeldahl flask. Set up a distillation apparatus as described in the preceding section, with a measured 50-ml portion of 0.2 N sulfuric acid (or boric acid) in the receiving vessel. Pour into the Kjeldahl flask a solution of 10 g sodium hydroxide in 50 ml of water, mix by shaking, and immediately connect the flask to the distilling apparatus. Slowly distil about half of the solution, and complete the analysis as described in the preceding exercise. Compute the total number of milliequivalents of ammonium salt received.

Questions

1. Does the volume of water in which the KHP is dissolved for titration influence the normality determined for the NaOH?

2. HCl is standardized by Na_2CO_3, using methyl orange indicator, and titrated cold. Is an indicator blank required? Why?

3. An organic acid has a molecular weight of approximately 200. It is water-soluble. Give detailed directions for the determination of its equivalent weight. How is it necessary to modify the procedure if the acid is not readily soluble in water? If the equivalent weight is found to be 98.7, what can be said about the molecular weight and the structure of the acid? Give illustrative calculations for this determination, assuming that the equivalent weight is 98.7.

4. Give a detailed procedure for the determination of the percentage of ammonia in crude ammonium sulfate. Illustrate your procedure by a typical computation.

5. Although in the standard Kjeldahl procedure the titration which follows the distillation is that of a strong acid versus a strong base, a serious error will result if phenolphthalein is selected as the indicator. Explain.

6. Concentrated sulfuric acid, which contains about 95 per cent acid by weight, is to be analyzed by use of a $0.1\,N$ base solution. Give detailed directions, including approximate weights and volumes, for this determination. How would the directions be modified if the standard base solution were $1\,N$?

7. A sample may be one of the following: NaOH, Na_2CO_3, $NaHCO_3$, NaOH and Na_2CO_3, or $NaHCO_3$ and Na_2CO_3. If A represents the milliliters of standard acid to reach the phenolphthalein end point and B the volume of acid to go from this end point to the methyl red end point, state the composition of the sample for each of the following cases:

    (a) $B = 0$
    (b) $A = 0$
    (c) $A = B$
    (d) $A > B$

8. The following normalities were found for a solution which contains 5.000 meq of H_2SO_4 in 20.00 ml:

$$0.2452, \; 0.2450, \; 0.2454$$

Which of the following errors are likely to be involved in this analysis?

    (a) Failure to apply an indicator blank.
    (b) Erratic end points.
    (c) Spilling a few drops of solution in one titration.
    (d) Use of an uncalibrated volumetric flask.
    (e) Parallax.

Problems

1. A solution of sodium hydroxide is standardized by potassium acid phthalate of assay value 99.95 per cent. From the following data compute the normality of the base:

Weight potassium acid phthalate	0.8573 g
Buret reading before titration	0.00
Buret reading after titration	45.53
Buret correction	−0.03

Ans. 0.0922 *N*.

2. The base solution of problem 1 was compared with an acid (HCl) solution. From the following data compute the normality of the acid.

Readings:	Acid Buret	Base Buret
After	45.63	43.25
Before	0.17	0.00
Correction:	0.06	−0.03

Phenolphthalein indicator was used. The blank required 0.02 ml of the base solution.

3. Acid and base solutions were standardized by use of sodium carbonate. From the following data compute the normality of the two solutions:
Comparison of acid and base, with methyl orange indicator.

Readings:	Acid Buret	Base Buret
After	38.85	42.46
Before	0.14	0.12
Correction:	0.04	0.03

The indicator blank requires 0.23 ml of the acid solution.

$$\text{Weight sodium carbonate} = 0.2547 \text{ g}$$
$$\text{Purity sodium carbonate} = 99.9 \text{ per cent}$$

Back titration is used; the buret readings are as follows in the standardization, which employs methyl orange indicator.

Readings:	Acid Buret	Base Buret
After	44.35	0.57
Before	0.04	0.06
Correction:	0.06	0.00

The solution is boiled to expel carbon dioxide, and the indicator blank correction is applied.   *Ans.* 0.1099 *N* acid; 0.0999 *N* base.

4. A sodium hydroxide solution is standardized with benzoic acid. A 1.8810-g sample of the acid is neutralized by 32.25 ml of the base solution. What is the normality?

5. A weighed crystal of calcite, $CaCO_3$, is dissolved in excess HCl and the solution boiled to remove carbon dioxide. The unneutralized acid is then titrated by a base solution which has previously been compared with the acid solution. From the following data compute the normality of the acid and base solutions:

Weight calcite	2.100 g
Volume HCl added	47.00 ml
Volume NaOH used in back titration	15.75 ml
Comparison of acid and base	40.15 ml acid = 30.00 ml base

6. A 0.8350-g sample of KHP was incorrectly recorded in the notebook as 0.8530 g. 37.16 ml of NaOH was required for titration. By how many parts per thousand does the apparent value of the normality differ from the true value? *Ans.* 22/1000.

7. 37.93 ml 0.1157 N HCl is required to titrate the Na_2CO_3 in a sample. The end point is overrun by three drops (0.12 ml). What error is made in terms of milligrams of Na_2CO_3 reported?

8. Would a 0.4000-g sample of Na_2CO_3 be of satisfactory size for the standardization of half normal HCl? Explain.

9. What range of sample weights would you suggest for the standardization of 0.5 N NaOH by use of KHP? To what precision should these samples be weighed?

10. A sample contains approximately 4 per cent Na_2CO_3, and the remainder is inert material. How large a sample should be taken for analysis if the titration is to be carried out with 0.1 N HCl? To what precision should the sample be weighed?

11. What is the succinic acid $(H_2C_4H_4O_4)$ titer of 0.02500 M $Ba(OH)_2$ on the basis of complete neutralization? *Ans.* 2.952 mg.

12. What weight of a barium hydroxide sample shall be taken for analysis so that each milliliter of 0.1000 N HCl represents 0.5000 per cent $Ba(OH)_2$ in the sample? *Ans.* 1.714 g.

13. What is the normality of a solution of H_2SO_4 such that each milliliter is equivalent to 2.000 per cent Na_2O in a 0.2500-g sample of soda ash? *Ans.* 0.1613 N.

14. 8.993 g of constant-boiling HCl prepared at 755 mm (180.1 g contains 1 mole HCl) is diluted to 500 ml. 21.65 ml of this acid is equivalent to 38.73 ml NaOH. What are the normalities of the acid and base solutions?

15. A 16.24-ml sample of vinegar of density 1.060 requires 48.24 ml 0.3564 N base for titration. What is the percentage by weight of acetic acid in the vinegar?

16. From the following data calculate the percentage purity of a sample of $KHSO_4$:

Weight of sample   1.2118 g
Volume NaOH used   26.28 ml
Volume HCl used   1.53 ml
1.000 ml HCl = 1.206 ml NaOH
1.000 ml HCl = 0.02198 g Na_2CO_3 (99.5% pure)

Ans. 93.9%.

17. A solution of HCl is of such strength that 50.00 ml react with 0.2600 g calcite. This acid is used to determine the percentage purity of a sample of $Ba(OH)_2$ from the following data.

Weight of sample   0.4213 g
Volume HCl used   34.51 ml
Volume NaOH used   2.44 ml
1.100 ml HCl = 1.220 ml NaOH   *Ans.* 68.28%.

18. Obtain the percentage of CaO in a sample of limestone from the following data:

Weight of sample	1.5000 g
Volume HCl used	40.00 ml
Volume NaOH used	3.00 ml
1.000 ml HCl = 1.500 ml NaOH	
Normality NaOH	0.3333

Ans. 35.52%.

19. Express the results of problem 18 as percentage of $CaCO_3$ in the limestone.

20. The analysis of a mixture of sodium chloride and partially hydrated sodium carbonate produced the following data. Calculate the percentages of NaCl, Na_2CO_3, and H_2O in the sample.

Weight of sample as received	1.2000 g
Weight after drying at 110°C	1.1000 g
Volume of acid	43.65 ml
Volume of base	3.65 ml
Normality of acid	0.1000
Normality of base	0.1250

21. From the following data decide whether the sample is $H_2C_2O_4 \cdot 2H_2O$, KHP, or citric acid, $H_3(C_6H_5O_7)$:

Weight of sample	0.2245 g
Volume NaOH used	30.00 ml
31.75 ml NaOH = 35.11 ml 0.1056 N HCl	

22. 9.05 g NH_3 solution is treated with 50.00 ml 0.4982 N H_2SO_4; 21.64 ml 0.1016 N NaOH is required for back titration. What is the percentage of NH_3 in the solution?   *Ans. 4.274%.*

23. 45.50 ml 0.2020 N NaOH is required to neutralize a 0.5000-g sample of a crystalline organic acid. What is the equivalent weight of the acid? If the acid is dibasic and completely neutralized in the titration, how many grams of it are required to prepare 300 ml 0.400 M solution?

24. The "saponification number" of a fat or oil is defined as the number of milligrams of KOH required to saponify 1 g of fat or oil. To a sample of butter weighing 2.010 g is added 25.00 ml 0.4900 N KOH solution. After saponification is complete, 8.13 ml 0.5000 N HCl is found to be required to neutralize the excess alkali. What is the "saponification number" of butter?
Ans. 228.6.

25. 50.00 ml HCl solution yielded a precipitate of AgCl which weighed 1.1200 g.

(a) How many milliliters of the HCl solution are needed to prepare 100.0 ml 0.1000 N solution?

(b) To what volume should 100.0 ml HCl solution be diluted to prepare a 0.1500 N solution?   *Ans. (a)* 63.98 ml; *(b)* 104.2 ml.

26. A solution of H_2SO_4 which is approximately 52 per cent by weight is in equilibrium with air of average moisture content and is recommended as a

stock solution to be prepared, standardized, and preserved for the preparation of standard solutions of desired concentrations.

(a) By reference to a handbook, what is the approximate specific gravity to which concentrated sulfuric acid should be diluted to prepare the acid?

(b) How many grams of water should be added to 100 g of 95 per cent H_2SO_4 to prepare the solution?

(c) If 1.000 g of an approximate 52 per cent solution gave a precipitate of $BaSO_4$ which weighed 1.245 g, what weight of solution should be taken to prepare 500.0 ml of exactly 0.1000 N solution?

27. How many milliliters of 0.5193 N NaOH should be added to exactly 1 liter of 0.0975 N NaOH to make a 0.1000 N solution?   Ans. 5.96 ml.

28. To what volume must 50.00 ml concentrated nitric acid, density 1.4450 g per milliliter containing 78.00 per cent HNO_3 by weight, be diluted to contain 10.00 g HNO_3 per 100.0 ml (10.00 weight-volume per cent)?

29. Compute the percentage by weight in the diluted nitric acid solution of problem 28. Use a handbook for the necessary density table, interpolating if necessary.   Ans. 9.51%.

30. How much water must be added to 50.00 ml of the concentrated nitric acid solution of problem 28 to prepare a solution which is exactly 10 per cent by weight?

31. 1.000 g of a crude ammonium salt is treated with concentrated KOH, and the liberated NH_3 is distilled and collected in 50.00 ml 0.5000 N HCl. 1.50 ml 0.5010 N NaOH is required for back titration. Compute the percentage of NH_3 in the sample.   Ans. 41.29%.

32. A 0.5000-g sample of urea, $CO(NH_2)_2$, is treated by the Kjeldahl procedure and the NH_3 distilled into 50.00 ml 0.4861 N H_2SO_4. How much 0.4967 N KOH will be required for back titration?

33. What weight of sample should be taken for a Kjeldahl analysis by the boric acid modification so that the volume of 0.5051 N H_2SO_4 used will be equal to the percentage of nitrogen in the sample?   Ans. 0.7076 g.

34. A fertilizer sample is known to contain about 5 per cent nitrogen. A Kjeldahl analysis is to be made, using boric acid as the absorbent for ammonia. If the available acid solution is 0.1 N, what weight of sample will be taken for the analysis? Would it be desirable to weigh out a large sample for the digestion and to use a ⅕-aliquot portion for the distillation? Explain your answer.

35. Nitrates may be determined by a procedure similar to that of the Kjeldahl method. The sample, dissolved in sodium hydroxide solution, is treated with Davorda's alloy (50 per cent Cu, 5 per cent Zn, 45 per cent Al), which reduces the nitrogen to ammonia. The ammonia is distilled from the sodium hydroxide solution and absorbed in an acid solution, just as in the Kjeldahl analysis. From the following data compute the percentage of potassium nitrate in a sample:

Weight of sample	2.123 g
Volume of acid	50.00 ml
Normality of acid	0.5000
Base for back titration	12.01 ml, 0.4620 N

36. A sample of sodium hydroxide contains 95.0 per cent NaOH and 3.00 per cent sodium carbonate. If 4.25 g of this material is dissolved and diluted to exactly 1 liter, (a) what is the normality of the solution when it is employed in titrations with phenolphthalein indicator; (b) what is the normality when methyl orange indicator is used?   Ans. (a) 0.1021 N; (b) 0.1033 N.

37. A sample contains sodium hydroxide and sodium carbonate: 20.00 ml 0.1000 N acid is needed to titrate it to the phenolphthalein end point and an additional 5.00 ml of acid is needed to complete the titration from the phenolphthalein to the methyl orange end point. How many grams of each constituent are present in the sample?

38. A sample which weighed 1.1179 g contains Na_2CO_3, NaOH, and NaCl. It is dissolved and titrated with 0.3000 N HCl. 38.16 ml of acid is required to reach the phenolphthalein end point, and an additional 24.08 ml of acid is needed to reach the methyl red end point. Calculate the percentages of each constituent in the sample.

39. A solution of crude caustic soda contains 5.000 g in 500.0 ml; 50 ml of this solution is titrated by 26.42 ml 0.3760 N acid to the methyl orange end point. Another 50.00-ml portion, after addition of barium chloride, required 22.48 ml of acid for titration to the phenolphthalein end point. Calculate the percentages of sodium hydroxide and sodium carbonate in the sample.
Ans. 67.62% NaOH; 15.70% Na_2CO_3.

40. From the following data find the percentages of Na_2CO_3 and $NaHCO_3$ in a mixture of these chemicals and inert constituents:
(a) A 0.5000-g sample required 45.50 ml 0.1050 N HCl to reach the methyl red end point.
(b) 25.00 ml 0.1000 N KOH is added to 0.2500 g of the sample; $BaCl_2$ is added to precipitate the $BaCO_3$, and 12.00 ml 0.1050 N HCl is required to reach the phenolphthalein end point.

41. To 50.00 ml H_2CO_3 solution is added 25.00 ml 0.0500 N Ba(OH)$_2$. 13.25 ml 0.0510 N HCl is required to titrate to the phenolphthalein end point. What is the molarity of the H_2CO_3 solution?   Ans. 0.00574 M.

42. Titration of Na_3PO_4 with HCl to the phenolphthalein end point produces Na_2HPO_4; subsequent titration to the methyl orange end point produces NaH_2PO_4. Find the milligrams of Na_3PO_4 and Na_2HPO_4 present in a sample which required 25.00 ml 0.1200 N H_2SO_4 to reach the phenolphthalein end point and an additional 35.00 ml of the acid to attain the methyl orange end point.

43. A base solution was prepared in the usual manner for removal of carbonate. It was found, however, that some carbonate was present. The following titration data were obtained:
49.00 ml of an acid solution is needed to titrate 50.00 ml of the base solution to the phenolphthalein end point.
49.50 ml of the acid solution is needed to titrate 50.00 ml of the base solution to the methyl orange end point.
The volumes of acid have been corrected for the indicator blanks.
47.50 ml of the acid solution is needed to titrate a solution of 0.2125 g pure sodium carbonate in 50 ml of water to the methyl orange end point. No indicator blank is taken; the titration is made in the cold.

(a) Compute the normality of the base solution when it is used with methyl orange indicator.

(b) Compute the normality of the base solution when it is used with phenolphthalein indicator.

(c) The base solution is separately standardized by potassium acid phthalate of assay value 95.5 per cent. What weight of phthalate will react with 50.00 ml of the base?   *Ans.* (a) 0.0836 N.

44. An acid is 0.1000 N at 20°C. What is its normality at 30°C if it is assumed that the density of the solution is the same as that of water?

45. A mixture contains only $CaCO_3$ and $MgCO_3$. 0.5000 g of the mixture is dissolved in 50.00 ml 0.2500 N H_2SO_4, and 15.00 ml 0.0960 N KOH is needed for back titration to a methyl orange end point. Find the composition of the mixture.   *Ans.* 42.68% $CaCO_3$.

46. Oleum or fuming sulfuric acid is a mixture of H_2SO_4 and SO_3. If 1.5000 g oleum require 32.63 ml 0.998 N NaOH for titration, what is the composition of the sample?

CHAPTER 14

Oxidation by Potassium Permanganate

Potassium permanganate has been the work horse of redox analysis for more than a century. It is a strong oxidizing agent which quantitatively oxidizes most of the common reducing agents when added in equivalent amount. The intense purple color of the permanganate ion serves as a self-indicator; one drop in excess will impart a distinct color to a large volume of solution. These two properties enable us to use $KMnO_4$ titrations for most reducing agents that are not themselves colored.

There are, however, some undesirable qualities of permanganate. Care must be taken to prepare a standard solution of this reagent that is free of reducing agents, for any reduction occurring after the solution is prepared causes MnO_2 to precipitate, and the solid MnO_2 particles in the solution catalyze further decomposition. Light also causes decomposition of $KMnO_4$ solutions; they should be kept in dark bottles if intended for permanent standards. Even with precautions $KMnO_4$ solutions may show some decomposition with time, and it is advisable to restandardize the solution at frequent intervals. Because of these undesirable qualities there is a growing tendency to replace $KMnO_4$ as a standard solution by $Ce(SO_4)_2$, an equally strong oxidizing agent that is quite stable on storage. The increased cost of the ceric sulfate is more than offset by the time saved in the restandardizations required for permanganate solutions.

Standards. The most widely used primary standards for permanganate solutions are arsenious oxide, sodium oxalate (which forms oxalic acid when the solution is acidified), and iron wire. All three are available in primary standard grade, with the assay value printed on the package label. None of them should be used as the standard if the assay value is not given. A solution of oxalic acid, previously standardized as an acid, can be used as a secondary standard for permanganate.

Of the three primary standards listed, the authors prefer arsenious oxide. Student results with it have been consistently better than with the other two. However, if the $KMnO_4$ solution is to be used only for iron analyses, it may well be standardized with iron wire, since then the same errors will appear both in the standardization and in the analysis, so that they cancel one another. In skilled hands $Na_2C_2O_4$ is a reliable standard, but if directions are not rigorously followed there may be undesirable side reactions that lead to loss of intermediate products and attendant error. If a permanganate solution is to be used solely for titration of oxalic acid, as in the volumetric determination of calcium, then $Na_2C_2O_4$ is recommended as the standard so that errors of standardization will cancel in the analysis.

Analyses. Most of the common reducing agents can be titrated directly by permanganate in acid solution. A partial list of oxidations for which $KMnO_4$ has been used follows.

Reduced Form		Oxidized Form
Fe^{+2}	=	Fe^{+3}
H_3SbO_3	=	H_3SbO_4
H_3AsO_3	=	H_3AsO_4
Sn^{+2}	=	Sn^{+4}
Ti^{+3}	=	Ti^{+4}
Mo^{+3}	=	MoO_3 (or MoO_4^{-2})
U^{+4}	=	UO_2^{+2}
H_2SO_3	=	$SO_4^{-2} + 2H^+$
VO^{+2}	=	VO_3^-
H_2S	=	$S + 2H^+$
H_2O_2	=	$H_2O + O_2$
$H_2C_2O_4$	=	$H_2O + CO_2$
HNO_2	=	$NO_3^- + H^+$

Ordinarily the titration is done in either HCl or H_2SO_4 solution.

It will be noted that in many of the determinations listed the permanganate ion is used to oxidize a substance from the -ous to the -ic state. In such titrations it is essential to reduce all the reductant to its lower oxidation state prior to the titration and to remove any excess of the reducing agent before the titration is begun. The following three methods are those most widely used for this reduction.

1. Reduction by metals, such as zinc, cadmium, aluminum, or silver. The metal may be used in the form of a spiral coil, which is immersed in the sample solution until reduction is complete, or it may be in the form of granules packed in a column through which the sample solution is passed. The Jones reductor, described later, is an example.

2. Reduction by stannous chloride, followed by removal of excess stannous ion by reaction with mercuric chloride:

$$Sn^{+2} + 2HgCl_2 = Sn^{+4} + Hg_2Cl_2 + 2Cl^-$$

The mercurous chloride formed in this reaction is insoluble and is not reoxidized during the titration, since it is present as solid. This method is used chiefly for reduction of ferric ion to ferrous ion, as in the Zimmermann-Reinhardt procedure.

3. Reduction by gases, such as SO_2 and H_2S. The excess of reducing gas is removed from the solution by boiling. In general, this method is not widely used because reaction proceeds slowly, and it is difficult to determine when reduction is complete.

In analyses where the reductant is volatile, such as SO_2, H_2S, HNO_2, etc., the procedure is to add the sample (from which the gas is released upon acidification) to a measured volume of the standard $KMnO_4$ solution. As gas is released, it is oxidized by the permanganate and after reaction is complete the excess permanganate that has not reacted is determined by titration with a standard reducing agent.

Permanganate titrations can also be used for reactions in neutral or basic solutions. If Ba^{+2} is present, reduction proceeds only to the hexavalent state and $BaMnO_4$ precipitates, thus preventing further reduction. If Ba^{+2} is not present, the permanganate is reduced to MnO_2, which precipitates. Among the analyses carried out in neutral and basic solutions are those of cyanides, alcohols, aldehydes, glycol, and sugars.

Preparation and Standardization of Solutions

Permanganate solutions are ordinarily used at a concentration of about $0.1 N$ ($0.02 M$), but solutions as dilute as $0.01 N$ are employed on occasion. When the very dilute solutions are used, it is advisable to employ an indicator such as orthophenanthroline since the permanganate color no longer provides a sensitive end point.

Procedure.   Preparation of 0.1 N Permanganate Solution.   Weigh on a watch glass approximately 3.2 g (note 1) reagent-grade potassium permanganate crystals.   Dissolve in 1 liter of distilled water, and either heat in a covered beaker for 2 hours (note 2) on a steam bath or boil the solution for 10–15 minutes.   Allow the covered beaker to stand overnight in the desk, filter (note 3) through a filtering crucible (note 4), and store in a clean glass-stoppered bottle.   Keep in the dark except when in use.   Subsequent decomposition of the solution can be recognized by the appearance of a brown coating on the walls and bottom of the bottle.

Notes.   1. The equivalent weight is 158.04/5, or 31.61 g.

2. Distilled water may contain organic matter which will reduce permanganate. The solution is heated in order to hasten the oxidation of this material and coagulate the precipitate of manganese dioxide which forms as a reduction product.   In neutral solution the reaction is

$$MnO_4^- + \text{reducing agent} = MnO_2 + \text{oxidation product}$$

3. All particles of manganese dioxide should be removed since these particles catalyze further decomposition of the solution.

4. Filter paper should not be used, since particles of the paper suspended in the solution would lead to further reduction.

Standardization

Arsenious Oxide.   A weighed sample of arsenious oxide is dissolved in sodium hydroxide solution.   This solution is acidified with hydrochloric acid, and the acid solution of arsenious acid is titrated with permanganate.   Permanganate oxidizes arsenious acid slowly if no catalyst is used, but addition of a trace of ICl, iodide, or iodate ion causes the reaction to proceed rapidly, and each drop of permanganate is quickly decolorized.   Apparently I_2 or ICl oxidizes the arsenite ion rapidly and the resulting I^- ion is reoxidized by MnO_4^-, so that in effect I_2 takes electrons from $H_2AsO_3^-$ and donates them to MnO_4^- more rapidly than the process occurs in the absence of iodine.   The titration of arsenious acid can be made in the presence of moderate amounts of HCl without oxidation of chloride ion.   The reactions of the determination are:

$$As_2O_3 + 2OH^- + H_2O = 2H_2AsO_3^-$$
$$H_2AsO_3^- + H^+ = H_3AsO_3$$
$$2MnO_4^- + 5H_3AsO_3 + 6H^+ = 2Mn^{+2} + 5H_3AsO_4 + 3H_2O$$

Procedure. Dissolve 20 g reagent-grade NaOH pellets in 80 ml of water. Weigh out to the closest 0.1 mg three 0.2-g portions (note 1) of dried arsenious oxide of known purity into 250-ml wide-mouth conical flasks. Add 10 ml of the sodium hydroxide solution, and allow to stand 8–10 minutes with occasional swirling, or until the sample is all dissolved. Dilute to 100 ml, and add 10 ml concentrated hydrochloric acid and 1 drop 0.002 M potassium iodide or iodate solution (note 2). Titrate with permanganate solution until a faint pink color persists for 30 seconds or more (note 3). Run a blank determination (note 4). From the net volume of permanganate, corrected for buret calibrations and the blank, and from the weight and purity of the standard, compute the normality of the permanganate. The relative range for the three results should not exceed 3 parts per 1000.

Notes. 1. The equivalent weight is $As_2O_3/4$ or 49.45 g.

2. Prepare the solution by dissolving about 0.3 g potassium iodide or iodate in 1 liter distilled water. The exact amount of catalyst used is not important provided care is taken not to add enough to react with a detectable volume of the permanganate. It is immaterial whether iodide or iodate is used. If iodate is used, it is immediately reduced to iodide by the arsenious acid.

3. A more sensitive end point can be obtained by use of ferrous orthophenanthroline indicator, but for ordinary work this is not necessary. One drop of 0.025 M indicator solution should be used. It is added when it appears that the end point is near, as shown by the slowness with which each drop of permanganate is decolorized. The indicator gives a color change from pink to a very faint blue.

4. A blank is necessary to insure that the NaOH is free of reducing agents. Carry out the determination just as in the standardization, except that no As_2O_3 is present. The first drop of permanganate should impart a detectable color.

Errors. Guard against the usual volumetric errors (page 238). Insufficient acidity will cause error, but, if instructions are followed, the solution should contain excess acid at a concentration of about 0.5 N.

Iron Wire. An analyzed grade wire, labeled "for standardization," should be used.

Procedure. Obtain a spool of analyzed wire from the instructor. It should be bright and free of rust. Weigh out to the closest 0.1 mg three portions of about 0.2 g (note 1), and place in 100-ml beakers (note 2). Add 10 ml concentrated HCl and 10 ml of water, and heat on the steam bath, with the beaker covered, until all the iron is dissolved. A slight residue of carbon may be disregarded. When the sample is all dissolved, rinse off the cover and remove it. Follow the procedure of page 276 for reduction of the sample by $SnCl_2$ and titration with the permanganate solution.

From the net volume of permanganate, corrected for buret calibrations and the end point blank, compute the normality of the solution. The relative range of the three results should not exceed 4–5 parts per 1000.

Notes.   1. Since the valence change of iron is 1, from $+2$ to $+3$, the equivalent weight is 55.85 g.

2. The sample is originally placed in a beaker, instead of in the final titration flask, because it should be evaporated to small volume before reduction with stannous chloride. Evaporations cannot be done efficiently from flasks.

Errors.   Aside from purity of standard and the usual volumetric errors, the errors of this method are those of the Zimmermann–Reinhardt procedure. Read the discussion on page 277.

Sodium Oxalate.   The over-all reaction of permanganate with sodium oxalate in acid solution is given by the equations:

$$C_2O_4^{-2} + 2H^+ = H_2C_2O_4$$
$$2MnO_4^- + 5H_2C_2O_4 + 6H^+ = 2Mn^{+2} + 10CO_2 + 8H_2O$$

The mechanism of the reaction is complicated and not at all like what the equations would indicate. The reaction is catalyzed by manganous salts; the first portions of permanganate added react slowly, but, after some manganous ion is formed, succeeding portions of permanganate are decolorized almost instantaneously. If manganous ion is added before the titration, the reaction is rapid from the start. The reaction is slow at room temperature, and it is necessary to carry out the last portions of the titration, at least, at elevated temperatures so that the reaction is sufficiently fast to enable the analyst to determine when the end point is reached.

Procedure.   Weigh out to the closest 0.1 mg three 0.25-g samples (note 1) of dry $Na_2C_2O_4$, and transfer to wide-mouth 250-ml conical flasks. Prepare approximately 1.5 N sulfuric acid solution by pouring 13 ml concentrated H_2SO_4 into 300 ml of water. Dissolve the sodium oxalate sample in 75 ml 1.5 N sulfuric acid solution.

Heat the solution (note 2) to 80–90°C, and titrate slowly with permanganate, with constant swirling. The end point is a permanent faint pink color. The temperature should not fall below 60° during the titration. Obtain a blank by heating 75 ml 1.5 N sulfuric acid to 80° and titrating with permanganate. From the corrected volume of permanganate and the weight and purity of the standard, compute the normality of the permanganate. The relative range for three portions should not exceed 4–5 parts per 1000.

Notes.   1. The equivalent weight is 134.00/2 or 67.00 g.

2. Place a thermometer in the flask for accurate measurement of the temperature. Remove and rinse off the thermometer with a small stream of water before the titration. Recheck the temperature just before the end point is reached; if necessary, reheat.

Errors. In addition to the usual volumetric errors there is danger of error from insufficient acidity of the solution. If the permanganate is added rapidly, the hydrogen ions may be exhausted locally. When this happens, a brown color is obtained resulting from precipitation of colloidal manganese dioxide. A determination should be discarded if this occurs. Care must be taken to keep the temperature above 60°.

Analyses with Permanganate

Iron

Titration of ferrous ion by permanganate is the most widely used method for analysis of iron samples, both raw ores and iron products. The sample is brought into solution by suitable methods, and all the iron is reduced to the ferrous state by one of the methods already listed. We shall describe two procedures, reduction by stannous chloride in hydrochloric acid solution (Zimmermann-Reinhardt method) and reduction by metallic zinc when the solution is prepared with sulfuric acid (Jones reductor).

Analysis of Soluble Samples by Zimmermann-Reinhardt Process. The basis for the method is reduction of iron by stannous chloride and the use of preventive solution to inhibit oxidation of chloride ion by permanganate. Ordinarily, in solutions of about $1 N$ HCl, permanganate will not oxidize chloride ion when present in the small excess required to give an end point. However, in the presence of ferric ion, there is an appreciable oxidation of chloride ion before the stoichiometric point for the iron titration is reached. Consequently, too much permanganate may be consumed in the titration, and the analysis may give high results. Further, the reaction with chloride ion is slow, and the apparent end point may "fade" so that it is difficult to say when it is reached. By using preventive solution and proper technique, these effects may be avoided and accurate titrations obtained. The preventive solution contains manganous ion, which inhibits oxidation of chloride ion, and phosphoric acid, which forms a complex ion with ferric ion. This complex formation serves a dual purpose; oxidation of ferrous ion is aided by removal of ferric ion, and the end point is made sharper when the yellow ferric ion is converted to nearly colorless ferric phosphate complex ion.

The student sample may be either an analyzed soluble iron salt or a solution containing a known weight of iron. The instructor will designate the procedure to be used for preparing the sample.

Soluble Salt. Obtain the sample in a stoppered weighing bottle. Do not dry. Weigh out three portions of proper size to contain about 0.2 g Fe (as directed by instructor). Put each portion in a 100-ml beaker, add 10 ml concentrated HCl and 10 ml of water. Cover the beaker and heat on a steam bath until the sample is completely dissolved.

Weighed Iron Sample. Obtain the sample in a 250-ml volumetric flask which has been calibrated relative to a 50-ml pipet. The sample is a weighed portion of pure iron wire, dissolved in 30–50 ml HCl. Remove the stopper from the flask and place on the steam bath until all the iron is dissolved (disregard a small residue of carbon particles that may be found). After solution is complete, cool to room temperature, dilute to volume, mix thoroughly, and pipet out four 50-ml portions into 100-ml beakers.

Procedure.   Reduction and Titration.   Place the beaker containing the sample on a steam bath and evaporate to a volume of about 10 ml (note 1). The solution will probably show the yellow color of ferric ion at this stage. While still hot, add stannous chloride solution (note 2) a drop at a time until the color changes to light green. The first drop added may be sufficient to reduce all Fe^{+3} present. Add two drops of stannous chloride solution in excess after the color change is observed. Cool the solution to room temperature, and pour in rapidly 20 ml mercuric chloride solution (notes 3, 4). Allow to stand 2 minutes (note 5), then wash the contents of the beaker quantitatively into a 500-ml conical flask; add water to bring the volume to about 300 ml. Add 20–25 ml preventive solution (note 6), and titrate immediately with permanganate. Run the solution in slowly, with constant swirling. The end point is a faint pink color which pervades the entire solution when it is thoroughly mixed.

Determine a blank by adding two drops of stannous chloride solution to 10 ml 1 : 1 hydrochloric acid in a 500-ml conical flask; then proceed with addition of mercuric chloride, water, and preventive solution, just as in the titration of a sample. Note the volume of permanganate needed to give the same color as the end point reached in the titrations.

Apply buret corrections as needed. Subtract the blank from the total volume used, to obtain the net volume for each portion of sample.

If the analysis is made on a soluble salt, report the percentage of iron in the sample. This is obtained by computing the weight of iron from the volume and normality of the solution used for titration. (The equivalent weight of Fe is 55.85.)

If a weighed iron sample was used for the analysis, report the total weight of iron received, which is five times the weight of the single aliquot portion.

Notes. 1. Estimate the volume by comparison with 10 ml of water in a similar beaker. Reduction by stannous chloride must be carried out in a relatively concentrated solution in order that the point at which reaction is complete can be determined by the change in color.

2. If the reagent is not provided, prepare the necessary amount by dissolving 3 g stannous chloride in 10 ml concentrated HCl and diluting to 20 ml with water. The solution should be reasonably fresh; on long standing it is oxidized by air to $SnCl_4$.

3. A saturated solution of $HgCl_2$ in water.

4. If the reagent is not added all at once, there may be a local excess of stannous ion which will cause reduction of $HgCl_2$ to mercury rather than to Hg_2Cl_2.

5. The precipitate should be small in amount and of a pure white color. A grayish precipitate indicates reduction to mercury. This reacts slowly with permanganate and will not give true titration values. If no precipitate appears, not enough stannous chloride has been added. If a small white precipitate is not obtained, the sample is spoiled and should be discarded.

6. If the preventive solution is not provided, prepare as follows: Dissolve 20 g $MnSO_4 \cdot 4H_2O$ in 100 ml of water, and add a cooled mixture of 40 ml sulfuric acid, 120 ml of water, and 40 ml sirupy phosphoric acid.

Errors. In addition to the usual volumetric errors, the following may cause trouble in this determination:

1. Incomplete reduction of ferric ion.
2. Use of too much excess stannous chloride.
3. Inability to recognize the end point. Some practice is required before we can be certain of this end point. Even with preventive solution there is some fading of color, and it requires skill to titrate at just the right speed and to stop just at the end point.

Iron Ore

The most difficult part of an ore analysis is bringing the sample into solution. In general the samples used for student unknowns are decomposed by hydrochloric acid, but some naturally occurring samples dissolve with great difficulty.

The sample is decomposed by heating with concentrated hydrochloric acid, with or without a little stannous chloride. When decomposition is complete, there will be a colorless residue of silica. If HCl treatment does not dissolve all except the silica, the sample is treated with a mixture of nitric and hydrochloric acids and then evaporated to dryness. Should the mixed acid treatment still not effect complete solution, it is necessary to filter out the silica, fuse it with sodium carbonate, and then treat with hydrochloric acid. The fusion converts the insoluble silicate into sodium silicate and permits the iron to be brought into solution. Student samples will not require this treatment unless specific instructions are given for it.

Procedure.   Dissolution of Sample.   Obtain the sample in a weighing bottle. Dry at 110°C. Weigh out accurately three samples of 0.3–0.6 g and transfer to 100-ml beakers. Add 10–15 ml HCl, cover the beaker with a watch glass, and place on the steam bath until the sample is disintegrated, as shown by a white residue. This will usually require 30–60 minutes. When disintegration is complete, rinse off the watch glass with a little water and evaporate the solution to a volume of about 10 ml on the steam bath.

Procedure.   Titration.   Follow the procedure of page 276 for reduction with stannous chloride, addition of preventive solution, titration, and determination of a blank.

Apply buret corrections if needed and subtract the blank from the total volume of permanganate used. Compute for each portion the percentage of iron in the sample.

Errors.   Read the discussion of page 277. In addition to these errors we may have additional ones from incomplete dissolution of the sample and the presence of reducing agents other than iron in the sample.

Iron by Jones Reductor Method

Despite the fact that it is not widely used in commercial iron analyses, the Jones reductor method is often given as a student exercise because it illustrates the methods and techniques employed in any reduction by metals. Such procedures are of value in many special analyses.

The sample may be either a solution containing a known weight of iron dissolved in dilute sulfuric acid or a soluble salt. After preparation of the solution, the procedure is the same in either case.

For samples which are soluble in dilute sulfuric acid, the Jones reductor method has one great advantage over the Zimmermann–Reinhardt process, namely, the absence of chloride ion and the side reactions caused by chloride ion.

Procedure. Arrange a reductor (note 1) as shown in Fig. 14–1. If it is not charged when obtained from the storeroom, prepare for use as follows. Place a perforated porcelain disk or a plug of glass wool in the bottom of the empty tube, and make a tight filter by pouring a suspension of acid-washed asbestos fiber through the tube until a retentive filter mat is formed, as in the preparation of a Gooch crucible. The mat should be porous enough to permit water to flow through at a rate of 75–100 ml per minute without the use of suction. Amalgamate (note 2) a quantity of 10–20 mesh analyzed-grade zinc by stirring it for 5–10 minutes in a solution of $0.1\,M$ mercuric chloride in $1\,M$ hydrochloric acid; wash the zinc thoroughly by decantation, and pour it into the reductor column. The zinc column should be about 1.5–2 cm in diameter and 35–50 cm in length. Wash the column several times with water, applying gentle suction, and then wash with $1\,N$ sulfuric acid (note 3), taking care that the liquid level never falls below the top of the zinc column (note 4). Continue the washing until a 250-ml portion of acid will not require more than two drops of permanganate solution to impart a distinct pink tinge. The reductor is now ready for use.

If the reductor is already filled with zinc when obtained, it is only necessary to rinse with acid until a satisfactory blank is obtained, before proceeding with the reduction of the samples for analysis. Pour a 250-ml portion of $1\,N$ sulfuric acid through the column of zinc at a rate of 50–75 ml per minute. Pour the acid in at the top at such a rate that the liquid level stands just above the zinc surface, but take care never to allow the level to fall below the surface so that air can come into contact with the zinc. After the 250-ml portion of acid has run through, rinse with 100 ml of water. Add the water in several portions, letting the level fall to a point

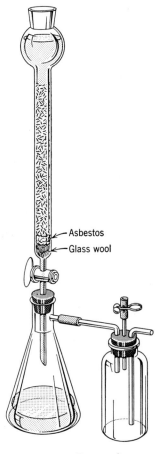

Asbestos

Glass wool

FIGURE 14–1.   Jones reductor.

just above the zinc surface for each portion. Remove the filter flask, and add permanganate from a buret until a barely perceptible pink color is imparted to the solution. If the volume of permanganate needed is greater than two drops repeat the acid treatment until a satisfactory value is obtained. Then proceed immediately with the reduction of the analysis samples.

Prepare sample solutions for analysis by one of the following methods.

1. *Analyzed Salt.* Obtain the sample in a stoppered weighing bottle. Do not dry. Weigh out three portions of proper size, as instructed, placing each in a 250-ml beaker. Add to each 100 ml approximately $1 N$ sulfuric acid, prepared by dilution of the concentrated $18 M$ desk solution.

2. *Weighed Iron Sample.* Obtain the sample in a 250-ml volumetric flask that has been calibrated relative to a 50-ml pipet. The sample is a weighed portion of pure iron wire, dissolved in 50 ml $1 M$ H_2SO_4. Dilute to volume, mix well, and pipet out four 50-ml portions, into 250-ml beakers.

Pour the sample solution through the reductor column at a rate of about 50 ml per minute. Use a stirring rod at the beaker lip to avoid any loss of solution. Use suction, and control the rate of flow by the stopcock at the bottom of the column. Allow the sample to run through until the liquid level is just above the zinc surface, then close the stopcock, add about 50 ml $1 N$ sulfuric acid to the beaker, pour this into the reductor, and allow the acid to run through until the level is just above the zinc surface. In this manner use four 50-ml wash portions. Finally, rinse with 100 ml of distilled water (note 5). Remove the suction flask, and titrate the solution immediately in the suction flask. Permanganate may be added rapidly until the end point is approached, but final addition should be made dropwise so that the end point is not overrun. Subtract from the total volume of permanganate the vloume used in the blank (note 6).

If the sample is an analyzed salt, report the percentage of iron present for each portion, computed from the net volume of permanganate, its normality, and the equivalent weight of iron. If the sample is a solution containing a weighed amount of iron, report the total weight of iron received, five times the weight in each aliquot portion.

Notes. 1. Zinc, cadmium, or aluminum may be used as reducing agents, either in the form of a column or as a spiral which may be introduced into the solution and withdrawn after reduction. Use of the column is recommended because of the certainty of complete reduction. The reaction is

$$Zn + 2Fe^{+3} = Zn^{+2} + 2Fe^{+2}$$

2. Amalgamated zinc does not react so rapidly with $1 N$ sulfuric acid as the pure metal, but it will reduce ferric ion rapidly and quantitatively. The inhibiting effect of amalgamation is probably due to the high overvoltage of hydrogen on a mercury surface. Zinc displaces mercuric ion according to the reaction,

$$Zn + Hg^{+2} = Zn^{+2} + Hg$$

The liberated mercury amalgamates the zinc surface.

3. A large quantity of $1 N$ sulfuric acid should be prepared. Slowly stir 30 ml concentrated sulfuric acid into a liter of water, and set aside to cool.

4. It is very important to exclude air from a prepared reductor, since hydrogen peroxide is formed by the reaction of oxygen with the liberated hydrogen. Hydrogen peroxide reduces permanganate according to the equation,

$$5H_2O_2 + 2MnO_4^- + 6H^+ = 2Mn^{+2} + 8H_2O + 5O_2$$

thereby leading to high results.

5. An acid solution should not be left in the reductor because it will dissolve the zinc.

6. A blank is necessary because of reducing constituents, particularly iron, that may be present in the zinc. It is important that the quantity of solution and the rate of flow for the blank be the same as those employed in analyses. If a satisfactory blank is not obtained, impure zinc is indicated, and the reductor must be refilled with zinc of a good grade.

Errors. In addition to the usual volumetric errors this determination is subject to several specific ones:

1. Incomplete reduction, because of too rapid passage of solution over the zinc, or use of too short a column of zinc.

2. Reoxidation of iron by air. This is not appreciable if the titration is made immediately after reduction.

3. Hydrogen peroxide, formed by air in the reductor.

4. The presence of interfering substances such at titanium, vanadium, chromium, molybdenum, uranium, arsenic, nitrates, and certain nitrogenous compounds. Of the metals likely to occur in iron ores—titanium, vanadium, and chromium—only vanadium will interfere in the stannous chloride reduction, whereas all interfere in reduction by zinc. Nitrates are reduced by zinc to hydroxylamine which reacts with permanganate. If nitrates are present, the solution of the sample should be evaporated until fumes of sulfur trioxide are evolved, then cooled, diluted, and passed through the reductor. Organic compounds are oxidized by this procedure.

5. Loss of sample by incomplete washing from the reductor.

6. Air in the reductor.

Oxalates

Oxalates can be dissolved in dilute sulfuric acid and the oxalic acid determined by titration with permanganate as described previously in sodium oxalate standardization.

Procedure. **Soluble Oxalate.** Obtain the analyzed sample in a weighing bottle. Dry in an oven at 110°C. Weigh out three portions of proper size (ask for instructions) and transfer to 250-ml conical flasks. Dissolve each in 75 ml $1.5\,N$ sulfuric acid, prepared by pouring 13 ml concentrated acid into 300 ml of water.

Before starting the titration, read the discussion of the permanganate–oxalic acid reaction on page 274 and the notes following the procedure of page 274. Heat the solution to 80–90°C and titrate slowly with the permanganate solution, with vigorous and constant swirling. Add only one drop at a time as the end point is approached. Do not allow the temperature to fall below 60°. The end point is a pink color that persists for at least 15 seconds. Determine a blank by titrating 75 ml $1.5\,N$ sulfuric acid heated to above 60°. Subtract the blank from the total volume of permanganate used. From the net volume of permanganate and its normality compute the milliequivalents of $Na_2C_2O_4$ in each sample and the weight of the sample. Report the percentage of $Na_2C_2O_4$.

Volumetric Determination of Calcium as Oxalate. In the absence of cations other than Mg and the alkali metals, calcium can be quantitatively precipitated as $CaC_2O_4 \cdot H_2O$ and determined by dissolving the precipitate in H_2SO_4 and titrating the oxalic acid with permanganate, as in the standardization by sodium oxalate. The equivalent weight of CaO in this method is $\frac{1}{2}$ mole, since each mole CaO gives a mole of $H_2C_2O_4$.

Procedure. The sample is an analyzed mixture containing $CaCO_3$. Obtain the sample in a weighing bottle and dry at 110°C. Weigh out three portions of the specified size and transfer to 600-ml beakers. Cover with a watch glass. Add 10 ml $6\,N$ HCl slowly from a pipet, keeping the cover in place to prevent loss by spattering as CO_2 is evolved. Gently boil the solution to remove CO_2. Rinse the cover glass and the sides of the beaker. Dilute to about 200 ml. Heat the solution to 60–80° and add four or five drops of methyl red indicator. Dissolve 1–1.5 g ammonium oxalate in 50 ml of water and add to the sample. Bring the solution to a boil and add $6\,N$ ammonia solution slowly from a pipet until the indicator shows a color change. Use mechanical stirring if available. As the acid is neutralized the concentration of oxalate ion increases and $CaC_2O_4 \cdot H_2O$ precipitates. To get large and well-formed crystals, it is essential that the neutralization be done a drop at a time, with the solution well stirred. Avoid a large excess of ammonia after the indicator shows a color change. Allow the solution to stand at least half an hour but not longer than an hour before filtration.

Filter through a glass or porcelain filtering crucible, with suction. Use a policeman to transfer the bulk of precipitate to the filter. Chill a flask of distilled water in a beaker of ice and use this for washing, to minimize loss of precipitate by solubility in the water. Wash with 10–20-ml portions of the cold water, adding each to the precipitation beaker and pouring from this into the filtering crucible. Drain each portion of wash before adding the next. After the fourth washing, test each portion with silver nitrate. Continue washing until only a faint cloudiness is obtained when the wash liquid is acidified with HNO_3 and $AgNO_3$ is added. Discard the washings and filtrate.

Remove the crucible from the holder and rinse off the outside with a little water from the wash bottle (to remove any adhering ammonium oxalate solution. Place the crucible in the beaker used for precipitation. At this point the precipitate may be stored for titration later. Add 150 ml of water and 50 ml $6 N$ H_2SO_4 to the precipitate in the beaker, and heat to 80–90°C to dissolve the precipitate. Without removing the filtering crucible, titrate with permanganate, as directed in the procedure for standardization with sodium oxalate, page 274.

Report the percentage of CaO in the sample.

Arsenic and Antimony

Trivalent arsenic and antimony may be determined by titration with permanganate, as in the standardization by arsenious oxide.

Procedure. Soluble Arsenic. Obtain the sample in a weighing bottle and dry at 110°C. Weigh out three portions, using amounts specified by the instructor. Follow the procedure of page 273 for dissolving the samples in NaOH, neutralizing, and titrating with the permanganate solution. Report the percentage of As_2O_3 in the sample.

Antimony Ore. Antimony trisulfide ore, stibnite, provides an excellent student exercise in the use of permanganate oxidimetry. The ore should be free of reducing constituents other than antimony.

The finely ground ore is treated with concentrated sulfuric acid and heated until sulfur and hydrogen sulfide are completely removed. A little potassium sulfate is sometimes added, to facilitate the decomposition. Antimony is usually not oxidized by the hot-acid treatment but is brought into solution in the trivalent form. After the sample is completely dissolved, the solution is diluted, hydrochloric acid is added, and the antimony is determined by titration with standard permanganate solution at a temperature not above 10°C.

If it is doubtful that all the antimony is in the trivalent condition at the start of the titration, the solution should be saturated with sulfur dioxide while in a 1 : 1 sulfuric acid medium and the excess sulfur dioxide expelled by boiling before hydrochloric acid is added. This precaution is not generally necessary in student analyses unless the sulfuric acid treatment for solution of the sample has been unduly prolonged.

If an ore of unknown content, which might contain arsenic and other reducing constituents, is to be analyzed, it is necessary to make preliminary separations before the antimony determination can be carried out. Reference books in mineral analysis should be consulted if this is necessary.

Procedure. Weigh out three 0.7- to 2-g samples of the dried ore (note 1), and transfer to 250-ml beakers. Add 15 ml concentrated H_2SO_4 and a small crystal of potassium sulfate. Place the beaker on a wire gauze, and boil the liquid over an open flame, taking care to avoid spattering (*Hood*). When the solution becomes clear (note 2), cool and cautiously add 20 ml of water. Boil until fumes of sulfur trioxide are evolved. Cool and cautiously add 50 ml of water; then quantitatively transfer the solution from the beaker to a 500-ml conical flask. Add 20 ml hydrochloric acid, and dilute to a total volume of about 200 ml. Cool in ice water to a temperature below 10°. Titrate while cold with standard permanganate solution. Report the percentage of antimony in the ore. The equivalent weight is half the atomic weight.

Notes. 1. The usual samples run from 10 to 30 per cent Sb. The instructor will state what size sample to employ.

2. If the ore is not well disintegrated in the acid treatment, it should be more finely ground before the samples are taken. Usually no further grinding is necessary for student samples.

Determination of Oxidizing Agents

In addition to its use for the analysis of substances that are reducing agents, a standard solution of permanganate can also be used as the reagent for analysis of some oxidizing materials, by the employment of an indirect method. A weighed portion of sample is treated with an excess known amount of a reducing agent, and the excess remaining after reaction is determined by titration with permanganate. Although this procedure may at first sight seem roundabout and involved, it is often preferable to direct titration of the oxidizing agent with a standard reducing agent. This is true because of (1) the difficulty of preserving standard solutions of strong reducing agents, and (2) the fact that the materials determined by this procedure are in solid form and cannot be directly titrated since reaction does not occur instantaneously. In the indirect method time is allowed for complete reaction of the sample with the reducing agent; then the titration of the excess reducing agent is a normal oxidimetric procedure.

The chief applications of this method are in the analysis of manganese dioxide or higher oxides of lead. Several reducing agents can be used, including ferrous sulfate, oxalic acid, and arsenious oxide. The last is preferred since it is not oxidized by air as is ferrous sulfate and since the permanganate titration of it is simpler than that of oxalic acid.

Determination of Available Oxygen in Manganese Ore. The oxidizing power of a manganese ore which contains MnO_2 (pyrolusite) can be determined by the indirect method previously outlined. The result of the analysis is expressed as the percentage of available oxygen in the sample. This is computed by multiplying the milliequivalents of oxidizing agent in the sample by the milliequivalent weight of oxygen, 8.00 mg.

Procedure. Prepare an approximately $0.1 N$ solution of arsenious acid by dissolving 5 g As_2O_3 in 5 g NaOH dissolved in 15 ml of water. When the As_2O_3 has dissolved, dilute to 50 ml, neutralize with $6 N$ HCl to the methyl orange end point, and dilute to 1 liter.

Weigh out three 0.25- to 1-g samples (note 1) of the finely ground ore and transfer to 1-liter conical flasks. Pipet into each flask exactly 50 ml of the arsenious acid solution. Add 10 ml concentrated H_2SO_4, and boil gently (note 2) until the sample is decomposed, as shown by the absence of colored particles in the siliceous residue. Cool the solution, wash down the sides of the flask with a little water (note 3), and add 100 ml of water, 5 ml hydrochloric acid, and a drop of $0.002 M$ potassium iodide solution. Titrate with permanganate as in the procedure of page 273.

Determine the permanganate value of the arsenious acid solution by measuring out three 50-ml portions, adding the same reagents as previously, and titrating with permanganate. The difference in the volume of permanganate required for the solutions without sample and those with sample represents the milliequivalents of oxidizing agent in the sample. From this value compute the percentage of available oxygen in the sample.

Notes. 1. The percentage of available oxygen may range from 2 to 10. The instructor will specify the size of sample to use.

2. Arsenic trichloride is volatile and a little may be lost during the heating. Use of the large flask minimizes the possibility of loss, since the walls act as a condenser.

3. If the residue is large in amount, it is advisable to filter before the titration. Use a filtering crucible with suction.

Procedure. Soluble Chromate. Prepare a solution of 40 g ferrous ammonium sulfate in 950 ml of water and add 50 ml concentrated H_2SO_4. Obtain the sample in a stoppered weighing bottle. Dry and weigh out three 0.5–1-g portions (note 1). Transfer to 250-ml conical flasks, and add from a pipet exactly 50 ml of the ferrous solution. Add about 4 ml concentrated H_2SO_4, and swirl until the sample is dissolved. Titrate immediately (note 2) with standard permanganate (note 3).

Determine the permanganate value of the ferrous solution by measuring out three 50-ml portions and carrying through the procedure above, but without adding the sample. The difference in the permanganate values with and without sample represents the oxidizing equivalents in the sample. Report the percentage of Cr in the sample. The equivalent weight is 52.01/3, or 17.34 g, since chromium undergoes a valence change of three in this reaction.

Notes. 1. Consult instructor for size of sample.

2. If the sample is allowed to stand, atmospheric oxidation of ferrous ion may occur.

3. A better end point is obtained by use of orthophenanthroline indicator, since the green color of chromic ion obscures the permanganate end point somewhat.

Questions

1. Write the balanced ionic equations for the titration reaction with MnO_4^- in acid solution for each of the following:

$$H_3SbO_3, \quad Ti^{+3}, \quad HNO_2, \quad H_2S$$

2. Which of the following could be determined by titration with $KMnO_4$

without treatment of the unknown before titration?

$$Fe^{+2}, Sn, Cl^-, H_2C_2O_4$$

3. Why must acid be present during permanganimetric titrations? What concentration of acid is usually employed?

4. Discuss the suitability of the various standards for permanganate.

5. What would be the effect on the normality of $KMnO_4$ if in the standardization with As_2O_3 the KI or KIO_3 were not added?

6. Why is reduction of iron by stannous chloride carried out in hot solutions? Why is mercuric chloride not added while the solution is hot? Why is the solution diluted and cooled before titration? What error is introduced if too much stannous chloride is used? How is this condition recognized?

7. Explain the purpose of each of the components of preventive solution.

8. Draw a diagram of the Jones reductor, and explain its operation.

9. List the advantages and disadvantages of several different methods for reducing iron to the ferrous condition before titration with permanganate.

10. Predict the effects on the result of an analysis of an iron sample by the Zimmerman–Reinhardt procedure of each of the following:
(a) Failure to add mercuric chloride.
(b) Titration of a sample in which a gray precipitate formed after the addition of mercuric chloride.
(c) Carrying out the reduction with stannous chloride at room temperature.
(d) Failure to add preventive solution.

11. Predict the effect of the result of the analysis of an iron ore sample by the use of a Jones reductor of each of the following:
(a) Passing the solution through the reductor too rapidly.
(b) Admitting air to the reductor.
(c) The presence of vanadium in the ore.

12. Which of the following constituents, if present in large amounts, will interfere in permanganimetric determinations of iron: silica, aluminum, magnesium, manganese, calcium, alkalies, chromium, sulfur, titanium, vanadium?

13. How may both the ferrous iron and the total iron be determined in the analysis of a sample of magnetite, Fe_3O_4?

Problems

1. Tabulate the titer of $0.1275 N$ $KMnO_4$ in terms of: Fe, FeO, Fe_2O_3, Fe_3O_4, H_2S, H_2O_2, $Na_2C_2O_4$, SO_2, Sb_2O_3, HNO_2, Sn, UO_2SO_4, V_2O_5.

2. Find the normalities of the following solutions of $KMnO_4$:
(a) 31.38 ml $KMnO_4$ is equivalent to 0.1947 g As_2O_3.
(b) 41.47 ml $KMnO_4$ is equivalent to 0.2500 g Fe.

(c) 35.62 ml $KMnO_4$ is equivalent to a ⅕ aliquot of 1.5000 g $Na_2C_2O_4$ dissolved in 250.0 ml.   *Ans.* (a) 0.1255 *N.*

3. What is the normality of a 0.300-molar $KMnO_4$ solution under each of the following circumstances:

(a) It is used as an oxidizing agent in strongly acid solution.

(b) It is used as an oxidizing agent in basic solution.

(c) It is used as an oxidizing agent in a reaction whereby it is converted to MnO_4^{-2}.   *Ans.* (a) 1.500 *N*; (b) 0.900 *N*; (c) 0.300 *N.*

4. How many milligrams of $H_2C_2O_4 \cdot 2H_2O$ are required to decolorize 50.00 ml 0.0400 *M* $KMnO_4$?   *Ans.* 630.4 mg.

5. How many milliliters of 0.1000 *N* NaOH are needed to neutralize 35.00 ml KHC_2O_4 solution if 35.00 ml of the oxalate solution are equivalent to 40.00 ml 0.1000 *N* $KMnO_4$?

6. How many milliliters of 0.1500 *N* $KMnO_4$ are needed to titrate 1.5968 g Fe_2O_3?   *Ans.* 133.31 ml.

7. How many milligrams of As_2O_3 will react with the amount of $KMnO_4$ required in problem 6?   *Ans.* 988.9 mg.

8. How many grams per liter of potassium permanganate should be used to prepare a solution equivalent to a potassium dichromate solution containing 4.2060 g per liter?

9. 0.2030 g 99.0 per cent pure $Na_2C_2O_4$ was titrated beyond the end point by 34.60 ml $KMnO_4$ solution. 4.50 ml $FeSO_4$ solution was added for back titration, and 0.40 ml $KMnO_4$ was used to establish the end point. Find the normalities and molarities of both solutions if 1.000 ml $KMnO_4$ is equivalent to 0.900 ml $FeSO_4$.

10. If the iron in a 0.1500-g sample of ore is reduced and subsequently requires 15.03 ml 0.1000 *N* $KMnO_4$ for titration, what is the composition of the ore expressed as a percentage of: Fe, FeO, Fe_2O_3?

11. What is the percentage purity of a sample of sodium oxalate if a 0.2203-g sample requires 29.30 ml $KMnO_4$ of which 1.000 ml is equivalent to 0.006023 g Fe?   *Ans.* 96.06%.

12. Calculate the percentage of iron in a sample from the following data:

Weight of sample	0.5862 g
Volume 0.1105 *N* permanganate	32.43 ml
Volume 0.0942 *N* ferrous sulfate	
used for back titration	3.42 ml

13. The precipitated CaC_2O_4 from a 0.4207-g sample of limestone was dissolved in H_2SO_4 and required 43.08 ml 0.0958 *N* $KMnO_4$ for titration. Find the percentages of calcium as CaO and $CaCO_3$.

14. Calcium in blood is often determined by precipitating it as the oxalate, dissolving the precipitate in sulfuric acid, and titrating with standard permanganate. Find the milligrams of calcium per 100 ml of blood from the following:

5.00 ml of blood is diluted to 50.00 ml; 10.00 ml of this solution requires 1.15 ml 0.0100 *N* $KMnO_4$.   *Ans.* 23.05 mg.

15. A student unknown is a mixture of $NaAsO_2$ and NaCl. What is the

percentage of NaCl in a sample if 0.3500 g of sample require 40.13 ml 0.1105 N $KMnO_4$? What is the percentage of As_2O_3 in the sample?

16. What volume of 0.1155 N $FeSO_4$ solution should be added to a 0.5000-g sample of $K_2Cr_2O_7$ so that at least 10 ml 0.0956 N $KMnO_4$ will be used for back titration?   *Ans.* 96.55 ml.

17. By what number should the iron titer of a $KMnO_4$ solution be multiplied to give the antimony titer of the same solution?   *Ans.* 1.090.

18. What size sample of $Na_2C_2O_4$ should be taken for standardization of 0.01 N $KMnO_4$? Should individual samples of standard be used?

19. Suggest a range of sample weights for an iron ore which is approximately 60 per cent Fe_2O_3 and is to be analyzed with 0.2000 N $KMnO_4$.
Ans. 0.8–1.1 g.

20. If 10.00 ml commercial hydrogen peroxide, density 1.010, requires 36.82 ml 0.1200 N permanganate for titration, calculate the percentage by weight of H_2O_2 in the solution.

21. A sample of pyrolusite which weighs 0.4000 g is treated with 50.00 ml 0.2000 N $H_2C_2O_4$; 10.55 ml 0.1152 N $KMnO_4$ is required for back titration. What is the percentage purity of the ore?   *Ans.* 95.48%.

22. Find the percentage of available oxygen in a pyrolusite sample from the following data:
Normality $KMnO_4$, 0.0953
50.00 ml $NaAsO_2$ solution = 45.50 ml $KMnO_4$ solution
A sample of 0.5000 g is treated with 50.00 ml $NaAsO_2$ and requires 15.00 ml $KMnO_4$ for back titration.

23. Molybdenum in the $+6$ oxidation state is reduced to the $+3$ state in a Jones reductor. In this state it is added to a ferric ion solution, and an amount of Fe^{+2} is liberated which is equivalent to the molybdenum which is reoxidized to the $+6$ state. What volume of 0.1126 N $KMnO_4$ is needed to titrate the Fe^{+2} liberated by 0.2000 g 99.9 per cent pure Na_2MoO_4? How many milligrams of Fe^{+3} were consumed by the molybdenum?
Ans. 25.85 ml $KMnO_4$; 162.5 mg Fe^{+3}.

24. 1.500 g pure Pb_3O_4 is treated with 50.00 ml 0.1056 N $H_2C_2O_4$ solution. How many milliliters of $KMnO_4$ whose As_2O_3 titer is 5.000 mg are needed for back titration?

$$Pb_3O_4 + H_2C_2O_4 + 3SO_4^{-2} + 6H^+ = 3PbSO_4 + 2CO_2 + 4H_2O$$

25. In the Volhard method manganese is determined by titration with standard permanganate solution, according to the reaction,

$$2MnO_4^- + 3Mn^{+2} + 2H_2O = 5MnO_2 + 4H^+$$

(In practice manganese dioxide is not precipitated, but a zinc salt is added to the solution, and a precipitate of basic zinc manganite is obtained. The manganese, however, is in the tetravalent state.)
A 1.1246-g sample of manganese ore was dissolved and the solution diluted to 250 ml; 50.00 ml of this solution required 45.15 ml permanganate solution, 1.000 ml of which is equivalent to 8.423 mg sodium oxalate. Calculate the percentage of manganese in the sample (see note p. 290).   *Ans.* 41.58%.

Note that the permanganate solution is standardized in acid solution, against sodium oxalate, but in the Volhard process permanganate is used in basic solution, where it undergoes a valence-number change of three.

26. A 0.2643-g sample of stibnite (Sb_2S_3) is brought into solution, the sulfide removed, and the antimony, which is in the trivalent condition, titrated to the pentavalent condition by standard permanganate solution, 25.23 ml 0.1042 N permanganate being used. Find the percentage of antimony in the ore.

27. What should be the normality of a $KMnO_4$ solution so that the number of milliliters used in titration divided by two gives the percentage of iron in a 0.3000-g sample of ore? What volume of 0.1108 N $KMnO_4$ solution is required to produce exactly 1 liter of this solution?

Ans. 0.02686 N; 242.4 ml.

28. What weight of sample should be taken for analysis so that each 1.000 ml 0.2000 N $KMnO_4$ solution represents 0.500 per cent H_2O_2 in the sample?

29. What weight of material must be taken for analysis so that each 1.000 ml 0.2000 N $KMnO_4$ represents 2.000 per cent Fe_3O_4 in the sample?

CHAPTER 15

Iodimetry and Iodometry

Iodine is a weak oxidizing agent, and iodide ion a fairly strong reducing agent since the half-cell emf for the reaction,

$$2I^- = I_2 + 2e^-$$

is -0.54 volt.

In the development of analytical chemistry several factors led to an early utilization of iodine methods. They are: (1) a sensitive indicator, starch, has long been known, and (2) a reagent, sodium thiosulfate, is available for quantitative titration of free iodine. Because of these conditions a large number of analytical methods has been developed about the iodine–iodide ion equilibrium.

There are two types of iodine methods, direct and indirect. In the first, known as the iodimetric method, iodine solutions are used for titration of reducing agents which can be quantitatively oxidized at the equivalence point. The number of such reactions is limited because iodine is itself a weak oxidizing agent; further, most of these determinations can today be better done by some other method than by use of iodine. Thiosulfate alone, of the common reducing agents, must be determined by oxidation with iodine. It happens that the stronger oxidizing agents give side reactions with thiosulfate, whereas iodine oxidizes it quantitatively to tetrathionate ion, $S_4O_6^{-2}$. Some of the oxidations which can well be done by iodine are represented in the following equations:

$$H_2S + I_2 = 2H^+ + S + 2I^-$$

$$H_2SO_3 + I_2 + H_2O = SO_4^{-2} + 2I^- + 4H^+$$

$$H_2AsO_3^- + I_2 + H_2O = H_2AsO_4^- + 2I^- + 2H^+$$

$$H(SbO)C_4H_4O_6 + I_2 + H_2O = H(SbO_2)C_4H_4O_6 + 2I^- + 2H^+$$

$$Sn^{+2} + I_2 = Sn^{+4} + 2I^-$$

$$2S_2O_3^{-2} + I_2 = S_4O_6^{-2} + 2I^-$$

In indirect or iodometric methods the oxidizing agent which is to be analyzed is treated with an excess of iodide ion under suitable conditions. Iodine is liberated quantitatively, and it is titrated by a standard solution of sodium thiosulfate or arsenious acid. These methods can be used for the analysis of almost any strong oxidizing agent; consequently, there are many more applications than there are for iodimetric methods. The following equations indicate some that are widely used:

$$2MnO_4^- + 16H^+ + 10I^- = 2Mn^{+2} + 5I_2 + 8H_2O$$

$$Cr_2O_7^{-2} + 6I^- + 14H^+ = 2Cr^{+3} + 3I_2 + 7H_2O$$

$$BrO_3^- + 6H^+ + 6I^- = Br^- + 3I_2 + 3H_2O$$

$$IO_3^- + 5I^- + 6H^+ = 3I_2 + 3H_2O$$

$$ClO_3^- + 6H^+ + 6I^- = Cl^- + 3I_2 + 3H_2O$$

$$2HNO_2 + 2I^- + 2H^+ = 2NO + I_2 + 2H_2O$$

$$2Cu^{+2} + 4I^- = 2CuI + I_2$$

$$HOCl + H^+ + 2I^- = Cl^- + H_2O + I_2$$

Indicators

In both iodimetric and iodometric methods the end point is based on the presence of free iodine. This can be detected by starch or by extraction of iodine from water by an immiscible organic solvent.

Starch. Starch does not dissolve as a true solution, but it is readily dispersed as a colloid. With iodine it gives a deep blue color which is formed on the surface of the colloid particles when iodine is adsorbed and is discharged when the iodine is reduced to iodide ion. The color change is reversible and may be approached from either direction.

The presence of iodide ion is necessary for development of a good color. Anything that tends to coagulate the colloidal starch or to interfere with adsorption of iodine by the colloid will cause indicator interference. In general, heat and organic materials such as alcohol are to be avoided.

When dispersed in water, starch is very susceptible to bacterial action which causes its decomposition and makes the solution ineffective as an indicator. A solution prepared without any preservative will in general not be good for more than 2 or 3 days. Addition of a preservative inhibits this bacterial action, and it is possible to make preparations that can be stored for weeks or months. Recommended methods for preserving are addition of (1) a little mercuric iodide, (2) 1 g boric acid per 100 ml, (3) 0.1 g furoic acid per 100 ml, or (4) saturating the solution with KCl.

In the titration of iodine the starch should not be added until just prior to the end point, when the iodine color is beginning to fade. When the starch is added to a high concentration of iodine, there is a tendency to formation of a reddish color which is not readily discharged at the end point. When the titration is made with iodine, the starch may be added at any time, since iodine is not in excess until the end point is reached.

Procedure.   Preparation of Starch Indicator. Stir 1–2 g soluble starch with cold water to make a thick paste. Slowly pour this paste into 100 ml of boiling water in which 1 g boric acid crystals has been dissolved. Continue boiling for 1 minute, then cool, and store in a stoppered bottle. Discard the preparation when it becomes cloudy or when it begins to give a reddish color with iodine.

Detection of End Point by Extraction. In certain titrations, where conditions are not suitable for the use of starch indicator, the end point may be determined by an extraction method. A non-miscible organic liquid, such as carbon tetrachloride or carbon disulfide, is added to the titration solution. The heavier organic liquid forms a layer at the bottom; because of the greater solubility of iodine in the organic solvent, the concentration of the iodine is much greater in the organic than in the aqueous layer, and the color is more intense. The end point of the titration is marked by the disappearance of the iodine color in the bottom layer. After each addition of reagent, the solution is shaken and then allowed to stand until the organic layer has settled out. This method compares favorably in sensitivity with the starch indicator method, but titrations made with an extraction end point require more time than those made with the starch indicator.

Chemical Reactions and Equilibria

Before studying specific determinations by iodine methods, we should understand the chemistry of the processes and the factors which affect the complex equilibria which are encountered in many reactions.

Reactions of Thiosulfate. The reaction of thiosulfate ion with iodine in acid or neutral solutions is given by the equation,

$$2S_2O_3^{-2} + I_2 = S_4O_6^{-2} + 2I^-$$

This reaction is essentially complete at the equivalence point. In basic solutions the oxidation may proceed further, according to the equation,

$$S_2O_3^{-2} + 4I_2 + 10OH^- = 2SO_4^{-2} + 8I^- + 5H_2O$$

This reaction is not quantitative and, consequently, the titration of iodine by thiosulfate must always be made in acid or neutral solution.

In strongly acid solutions thiosulfate decomposes according to the equation,

$$S_2O_3^{-2} + 2H^+ = H_2S_2O_3 = S + H_2O + SO_2$$

It may be recalled that this reaction is the basis for a qualitative test for thiosulfate ion.

The equivalent weight of thiosulfate is a mole or **248.2** g if the pentahydrate is used. This hydrate is not suitable for preparation of standard solutions because of its water of crystallization. It is customary to make an approximate $0.1 N$ solution and to standardize by comparison with some primary standard.

Iodine. Iodine is sparingly soluble in water but dissolves readily in potassium iodide solutions because of formation of the complex I_3^- ion,

$$I_2 + I^- = I_3^-$$

The equilibrium constant is

$$K_e = \frac{[I_3^-]}{[I_2][I^-]} = 570$$

A large excess of iodide ion is used to shift the equilibrium to the right. Even with high concentrations of iodide ion, the vapor pressure of iodine over the solution is appreciable, and care must be taken to prevent loss of iodine by vaporization.

Iodine reacts with water, just as do other halogens, according to the equation,

$$I_2 + H_2O = H^+ + I^- + HIO$$

Acids tend to shift the equilibrium to the left and bases, to the right. Consequently, iodine titrations are never made in strongly basic solutions.

Light accelerates the hydrolysis of iodine by causing decomposition of hypoiodous acid:

$$2HIO = 2H^+ + 2I^- + O_2$$

Standard solutions of iodine should be preserved in dark bottles or kept inside the desk to protect them from direct sunlight.

Iodide ion in acid solution may be oxidized by air:

$$4I^- + 4H^+ + O_2 = 2I_2 + 2H_2O$$

Because of this it is necessary to guard against access of air in iodometric determinations where the oxidizing agent is treated with KI in strongly acid solutions. The presence of certain ions accelerates this atmospheric oxidation of iodide ion. For example, nitrites give the reaction,

$$2HNO_2 + 2I^- + 2H^+ = 2NO + I_2 + 2H_2O$$

This is followed by oxidation of NO to NO_2, which yields more HNO_2 and causes further oxidation. Certain metal ions, such as cuprous, can also accelerate the oxidation.

Arsenites. The emf of the couple,

$$H_2AsO_3^- + H_2O = H_2AsO_4^- + 2H^+ + 2e^-$$

is -0.56 volt, or about the same as that of iodine. Nevertheless, by proper choice of conditions the equilibrium can be so displaced that the reaction,

$$I_2 + H_2AsO_3^- + H_2O = 2I^- + H_2AsO_4^- + 2H^+$$

proceeds quantitatively to the right. This is effected by removal of hydronium ion as it is formed, by reaction with bicarbonate ion. The titration is carried out in a buffered nearly neutral solution. Enough bicarbonate or other buffer ion must be present to keep the pH near 8. If the buffer is exhausted, the hydronium ion concentration rises and the reaction does not go to completion.

In the titration of trivalent antimony with iodine, there is a similar condition and the solution must be buffered and nearly neutral. Basic antimony salts will precipitate in neutral solutions unless a substance is added which will form a soluble complex ion with trivalent antimony. Tartrates may be used for this purpose.

Standard solutions of sodium arsenite (arsenious acid) are often used instead of thiosulfate for titration of iodine. These have the advantage that they keep better than thiosulfate solutions and that a standard solution may be prepared by dissolving a known weight of arsenious oxide and making up to definite volume.

Effect of pH. Since the hydronium ion of the solution has such a variety of effects in determinations that involve the reactions of iodine, it is desirable to summarize in one place the various kinds of effects that may be encountered:

1. In iodometry it is often necessary to employ a high concentration of acid for the reaction of KI with the oxidizing constituent of the sample. This is always the case for samples which contain oxygen salts, such as permanganates, chlorates, and chromates, since hydronium ion is one of the reactants with such oxidizing agents. Subsequent titration of iodine with thiosulfate or arsenite cannot be carried out in solutions of such high acidity. Therefore, after reduction of the oxidizing constituent by iodide ion, it is customary to dilute or buffer the solution before making the iodine titration.

2. Standard iodine solutions must be neutral or acidic. Basic solutions are not stable.

3. The reaction of thiosulfate with iodine must be carried out in neutral or acid solution, never in basic solution.

4. The titration of trivalent arsenic or antimony with iodine must be done in buffered solutions of pH 5–9.

5. Solutions of KI must be protected against atmospheric oxidation when in strongly acid media.

Preparation and Standardization of Solutions

The solutions needed will depend on the samples to be analyzed. If only iodometric procedures are employed, a standard solution of sodium thiosulfate will suffice. If iodimetric analyses are required, it is necessary to have a standard solution of iodine. In the latter case it is also advisable to prepare a solution of sodium arsenite which serves both as the primary standard for the iodine solution and as a reagent for back titration when this is needed. Iodine of known strength may be directly prepared, but this is not recommended since solutions do not keep well, and frequent restandardization is required.

Potassium iodide is required in all iodine analysis methods. A good-quality analyzed-grade reagent should be used, and every lot should be tested to see that it is free of iodates. If iodates are present, they react with iodide ion when the solution is acidified, releasing iodine according to the equation:

$$IO_3^- + 5I^- + 6H^+ = 3I_2 + 3H_2O$$

Test for iodates as follows: Dissolve 0.5 g KI in 10 ml of water, add 1 ml 6 N sulfuric acid and 2 ml of starch solution. Presence of iodates is shown by immediate appearance of a blue color. Disregard a blue color which forms slowly; this is due to air oxidation of KI.

Procedure. Preparation of 0.1 N Iodine Solution. Weigh out 12 g iodate-free potassium iodide crystals (note) on the trip scale, and dissolve in 10 ml of water in a small beaker. Add 6.5 g resublimed iodine to the KI solution, and stir until it is all dissolved. Transfer the solution to a glass-stoppered bottle, and dilute to approximately 500 ml.

Note. Directions are for preparation of 500 ml of solution, which is enough for standardization and for analysis of one unknown. If a larger volume of solution is required, use proportionally larger quantities of reagents.

Procedure. Preparation of Standard Arsenious Oxide (Sodium Arsenite) Solution. Accurately weigh out a 4.5- to 5-g portion of dry primary-standard-grade arsenious oxide, and transfer to a 400-ml beaker. Add 5–10 g reagent-grade NaOH pellets and 10 ml of water. Stir, adding a little water if necessary, until all the sodium hydroxide and arsenious oxide have dissolved. Dilute to 100 ml, and neutralize with 6 N HCl, using methyl orange or methyl red indicator. When the solution is distinctly acid, add about 5 g NaHCO₃.

Transfer the solution *quantitatively* to a 1-liter volumetric flask, dilute to volume, and mix well. From the weight and purity of As₂O₃ taken, compute the exact normality. The equivalent weight is As₂O₃/4 or 49.45 g.

Procedure. Preparation of Approximately 0.1 N Thiosulfate Solution. Clean a 1-liter glass-stoppered bottle with cleaning solution and *rinse thoroughly*. Boil 1 liter of distilled water, and while still hot transfer to the cleaned bottle. Allow to cool. Add 25 g reagent-grade sodium thiosulfate crystals, Na₂S₂O₃·5H₂O, and 2–3 g borax crystals (note). Stopper the bottle, and shake until all the solid is dissolved.

Note. Na₂S₂O₃ solutions may decompose on standing as is evidenced by the appearance of milky colloidal sulfur. The precaution of using boiled water which is free of dissolved O_2 and CO_2 and of bacteria, as well as the addition of borax as preservative, leads to the preparation of a fairly stable solution.

Standardization

Arsenious oxide is recommended as the primary standard for iodine solutions. It can be obtained in primary standard grade of known assay value from the Bureau of Standards or from chemical supply sources. A further advantage (beside its ready availability) is that it is also a good standard for permanganate, thereby interrelating the oxidation equivalents of two important redox reagents. By indirectly comparing a thiosulfate solution with the same iodine solution that is standardized with arsenious oxide, we can also use this standard for thiosulfate.

A known weight of pure As_2O_3 is dissolved in NaOH

$$H_2O + As_2O_3 + 2OH^- = 2H_2AsO_3^-$$

After the solid is dissolved, the excess NaOH is neutralized by HCl and the solution is buffered by adding an excess of $NaHCO_3$. In a buffered solution the $H_2AsO_3^-$ ion reacts quantitatively with I_2.

$$I_2 + H_2AsO_3^- + H_2O = 2I^- + H_2AsO_4^- + 2H^+$$

Since each atom of arsenic changes from $+3$ to $+5$, the gram equivalent weight of As_2O_3 is $\frac{1}{4}$ mole.

Thiosulfate solutions can be directly standardized by titration against a standard iodine solution. If an iodine solution is not available, the thiosulfate solution can be standardized indirectly against known amounts of strong oxidizing agents by treating the oxidizing agent with an excess of KI (under proper conditions of acidity) and titrating the liberated iodine immediately with the thiosulfate. Since each milliequivalent of oxidant used releases a milliequivalent of I_2, the number of milliequivalents of $Na_2S_2O_3$ is equal to the milliequivalents of oxidizing agent used as standard. The most widely used primary standards for thiosulfate solutions are KIO_3, $K_2Cr_2O_7$, and copper wire (which as made for electrical use is sufficiently pure to serve as a primary standard). Standard solutions of $KMnO_4$ or $K_2Cr_2O_7$ may be used as secondary standards for thiosulfate solutions, by adding a known volume to excess KI and titrating the liberated I_2 with the thiosulfate.

Care must be taken in standardization of thiosulfate by strong oxidizing agents to avoid two likely sources of error.

1. The KI reagent used must be free of iodate, which if present

will release I_2 when the solution is acidified, as discussed on page 297. A test must be run on each new batch of KI before it can be trusted.

2. After the known amount of oxidant is added to the KI solution, there is danger of loss of iodine vapor from the flask during the subsequent titration by thiosulfate. The titration should be made immediately and with only enough swirling of the solution to give proper mixing as thiosulfate is added. It is advisable to perform the operation in a relatively large conical flask, say one that holds 500 ml.

Procedure. Standardization with Arsenious Oxide. Dry about 2 g primary-standard-grade arsenious oxide at 110°C. Weigh accurately three portions of about 0.2 g each and transfer to 250-ml conical flasks. Add 10–15 ml 1 N NaOH (note 1) and warm, to dissolve the sample. Cool the flask in tap water, add two drops of methyl red indicator, and neutralize with 1 N HCl (note 2). Add the HCl from a pipet or dropper and avoid an excess of more than three or four drops. Add about 5 g powdered $NaHCO_3$ and swirl the solution until the $NaHCO_3$ is all dissolved (note 3). Dilute the solution to 100 ml, add 3 ml of starch indicator, and titrate with iodine solution to the appearance of a deep-blue color. Determine a blank (note 4). From the weight and purity of the arsenious oxide and the net volume of iodine used, compute the normality.

Measure out three 25-ml portions of the thiosulfate solution, with a calibrated pipet. Add starch indicator and titrate each with the standardized iodine solution. Compute the normality of the thiosulfate.

Notes. 1. Prepare by dissolving 2 g reagent-grade pellets in 50 ml of distilled water.

2. Prepare by diluting 8 ml of concentrated acid to 100 ml.

3. If the end point was not overrun by a large amount in neutralization of the NaOH with HCl, 5 g $NaHCO_3$ is sufficient to neutralize all excess acid and provide a reserve amount to act as a buffer in the iodine titration. As previously shown, hydronium ion is liberated in the titration, and a buffer is needed to react with it.

4. For the blank add 1–1.5 g potassium iodide to 100 ml of water, then add five drops of hydrochloric acid, 5 g sodium bicarbonate, and 3 ml of starch solution. Titrate with iodine solution to a permanent blue color. Potassium iodide is added in the blank so that it will contain the amount ordinarily present in an analysis (because of the use of approximately 40 ml of iodine reagent which contains potassium iodide).

Errors. The chief source of error in this titration is failure to use sufficient bicarbonate to neutralize all the excess acid. If there is insufficient bicarbonate present, the end point fades badly, and too much iodine is added.

Standardization of Thiosulfate by Dichromate. This exercise illustrates the iodometric method for analysis of strong oxidizing agents which must be reduced in solutions of high acidity. A known weight of pure dichromate crystals is treated with excess potassium iodide in sulfuric acid solution, and the liberated iodine is titrated with the solution to be standardized. The reactions are:

$$Cr_2O_7^{-2} + 14H^+ + 6I^- = 3I_2 + 2Cr^{+3} + 7H_2O$$

$$3I_2 + 6S_2O_3^{-2} = 6I^- + 3S_4O_6^{-2}$$

Thus one molecule of $K_2Cr_2O_7$ requires six $S_2O_3^{-2}$ ions, and the equivalent weight of potassium dichromate is 49.04 g.

Reagent-grade potassium dichromate is usually of sufficient purity for student requirements. A small portion should be ground and dried by heating in an evaporating dish at 175–200°C for 30 minutes (heat in a sand bath, with a thermometer bulb imbedded in the powder).

Procedure. Accurately weigh out three portions of about 0.2 g of the dry powdered crystals. Transfer to a 500-ml conical flask, and dissolve in 100 ml 2 N sulfuric acid (note 1).

Add to the solution in small portions and with constant swirling 2 g powdered sodium bicarbonate (note 2), and then pour in a solution of 10 g iodate-free potassium iodide dissolved in 10 ml of water. Swirl to mix. Cover the flask with a watch glass, and allow to stand 5–10 minutes (note 3). Dilute to 350 ml (note 4), and titrate with thiosulfate solution until the brown color of iodine begins to fade; then add 3–5 ml of starch solution and complete the titration. At the end point the blue color of starch-iodine disappears, leaving the green color of chromic ion.

From the weight of dichromate used and the volume of thiosulfate, compute the normality of the latter.

Notes. 1. Prepare by adding 30 ml concentrated H_2SO_4 to 500 ml of water.

2. The carbon dioxide evolved displaces air, thereby preventing air oxidation of iodide ion in the acid solution. Swirl gently to mix, but avoid vigorous agitation. Use the specified weight of $NaHCO_3$; too much will exhaust the acid.

3. Time must be allowed for complete reaction. Longer standing is undesirable because of the possibility of air oxidation.

4. The final concentration of acid is about 0.4 N.

Errors. In addition to ordinary volumetric errors, the following may affect this titration:

1. Oxidation of iodide ion by air.

2. Insufficient acidity, which causes incomplete reduction of dichromate by iodide ion.

3. Loss of iodine vapor. The flask should be kept closed except during the titration. Avoid vigorous shaking.

4. Premature addition of starch. See the discussion of the use of starch.

5. Presence of iodate in the potassium iodide. If the test outlined on page 297 shows that iodate is present, a blank should be determined and a correction made for the volume of thiosulfate consumed in the blank.

Standardization of Thiosulfate by Copper. The reaction of copper with iodide ion and the iodometric determination of copper are discussed in the following sections. Commercial copper wire and sheet are of high purity and are often used as primary standards for thiosulfate solutions. The chief source of error is the end point, which is difficult to obtain accurately until some experience is gained.

Procedure. Weigh accurately three 0.3-g portions of copper wire (note 1). Place each portion in a 500-ml conical flask and add 1.5 ml concentrated nitric acid and 5 ml of water. Set the flask on the steam bath and allow to stand until the copper has all dissolved (note 2); then dilute to 30–50 ml.

Titrate with thiosulfate solution, following the procedure of page 305.

From the weight of copper and volume of thiosulfate, compute the normality of the latter.

Notes. 1. The equivalent weight of copper in this reaction is its atomic weight, 63.54 g.

2. Do not leave on the bath after the copper has dissolved, since, if the acid evaporates, the copper nitrate may not readily redissolve.

Errors. See discussion on page 305.

Analyses

Cr in Soluble Chromate. Obtain the sample in a stoppered weighing bottle. Dry in the oven. Weigh out accurately three 0.5- to 1-g samples as directed by the instructor, and transfer to 500-ml conical flasks. Dissolve in 100 ml 2 N sulfuric acid. Weigh out 2 g powdered $NaHCO_3$ and add this in small portions to the acid solution, swirling this gently during the addition. Dissolve 10 g potassium iodide in 15–20 ml of water and pour the solution into the flask. Swirl gently to mix but do not agitate unduly. Cover the flask with a watch glass, and allow to stand 10 minutes. Dilute to 350 ml and titrate with thiosulfate solution until the iodine color begins to fade; then add starch indicator and complete the titration.

Report the percentage of Cr in the sample. The equivalent weight is Cr/3.

Read the discussion and notes on page 300 for possible sources of error in the determination.

Sb in Soluble Antimony Sample. Obtain the sample in a stoppered weighing bottle, and dry in the oven. Weigh out accurately three 1-g samples (note 1), and transfer to 250-ml Erlenmeyer flasks. Dissolve in a solution of 5 ml concentrated HCl in 50 ml of water. Add 3 g powdered tartaric acid (note 2), and carefully neutralize with 1 N NaOH (note 3), using methyl red or methyl orange indicator. If the end point is overrun, bring the solution back to the acid side with a few drops of HCl, taking care to avoid any large excess. When the solution is barely acid, add 5 g powdered $NaHCO_3$, and swirl until the bicarbonate is dissolved. Add 3–5 ml of starch solution, and titrate immediately with a standard iodine solution. Compute the percentage of Sb in the sample. The equivalent weight is Sb/2.

Notes. 1. Based on Sb percentages of 15–25.
2. Tartaric acid forms the complex $H(SbO)C_4H_4O_6$ which is soluble. This prevents precipitation of basic antimony salts when the solution is neutralized. If tartar emetic is used as student sample (see instructor), it is not necessary to add tartaric acid.
3. Prepare by dissolving 4 g reagent-grade NaOH in 100 ml of water.

Antimony in Stibnite. The analysis of stibnite, an ore which consists chiefly of antimony sulfide along with a little silica, is a good example of a practical iodimetric determination. If arsenic and iron are present, it is necessary to use a complicated procedure for the separation of antimony, but in the absence of interfering substances it is sufficient to bring the ore into solution and titrate the trivalent antimony without any separations.

The ore is disintegrated in concentrated hydrochloric acid and the solution heated to expel hydrogen sulfide. Potassium chloride is added prior to the acid treatment to prevent loss of antimony trichloride by volatilization. Excess chloride ion gives the complex $SbCl_4^-$ ion which is not volatile. After removal of hydrogen sulfide, the solution is diluted, neutralized, and buffered by addition of sodium bicarbonate. Trivalent antimony is titrated in the buffered solution with iodine. Precipitation of basic antimony salts is prevented by addition of tartaric acid, which forms the complex $H(SbO)C_4H_4O_6$.

Procedure. Dry the sample at 110°C. Weigh out accurately three 0.5- to 3-g samples (consult instructor as to approximate percentage of antimony), and transfer to dry 500-ml conical flasks. Add about 0.3 g potassium chloride crystals, which have been ground to a fine powder. Pour 10 ml hydrochloric acid over the sample and set on the steam bath for disintegration of the sulfide. At the end of 15 minutes the disintegration should be complete; the residue (silica) should be white.

When decomposition is complete, add 3 g powdered tartaric acid, and continue heating on the steam bath for 10–15 minutes. Dilute to 100 ml by slowly adding water from a pipet. If a red precipitate of antimony sulfide appears, stop the addition of water, and place the flask on the steam bath until the precipitate redissolves and all hydrogen sulfide is expelled from the solution. If a white precipitate of basic antimony salts appears, discard the determination. Such a precipitate once formed is difficult to redissolve.

Add a drop of methyl orange or methyl red indicator (or a strip of litmus paper), and neutralize the solution by slow addition of 6 N NaOH solution (note 1). Try to avoid a large excess of NaOH. Add 6 N HCl solution (note 2) until the solution is distinctly acid, but avoid a large excess. Now add about 5 g powdered $NaHCO_3$, and swirl until it is dissolved. At this point the acid should be all neutralized, and sufficient bicarbonate should be present to buffer the solution during the subsequent titration.

Titrate with standard iodine solution, using 3–5 ml of starch indicator. From the volume and normality of iodine used compute the percentage of antimony in the ore. The equivalent weight is Sb/2 or 60.88 g.

Notes. 1. Prepare by adding 6 g reagent-grade pellets to 25 ml of water.
2. Prepare by adding 10 ml concentrated HCl to 10 ml of water.

Errors. The chief source of error is due to insufficient bicarbonate to buffer the solution completely. When this occurs, the end point fades badly.

Copper. The two chemical methods most widely used for copper determinations are electrolytic deposition and iodometric titration. The latter is based on the reaction,

$$2Cu^{+2} + 4I^- = 2CuI + I_2$$

This reaction goes to completion because cuprous iodide is precipitated. The iodine liberated is titrated, without removal of the CuI precipitate, with standard thiosulfate solution.

The reactions are carried out in acid solution but strongly acid solutions cannot be used because air oxidation of iodide ion (catalyzed by cuprous ion) is favored. The copper analysis is best done at a pH of about 3–4; a buffered acetic acid solution is ordinarily employed. The chief interfering substances are oxides of nitrogen and the elements such as trivalent or pentavalent arsenic and antimony, which react with iodine or with iodide ion. Since copper samples are usually dissolved in nitric acid, special precautions must be taken to remove oxides of nitrogen. This may be done by evaporation in the presence of sulfuric acid until all nitrates are expelled, or more simply by using only a *slight excess* of nitric acid and removing the oxides of nitrogen by adding to the solution a little urea which reacts rapidly and quantitatively with nitrous acid. Use of too great an excess of urea causes the end point of the titration to be uncertain.

The presence of a precipitate of cuprous iodide causes some difficulty in recognition of the end point. This difficulty is for the most part avoided if starch is not added until practically all the iodine has reacted. With a little experience this point is readily recognized by the fading of the iodine color. When starch is added prematurely, it is adsorbed on the surface of the precipitate and is not readily decolorized. The end point may be made sharper by adding 1–2 g ammonium or potassium thiocyanate just prior to the end point. Thiocyanate ion is adsorbed by cuprous iodide and displaces adsorbed iodine.

Analysis of Copper Sample. The sample for analysis is either a weighed portion of pure copper or an analyzed sample of copper carbonate or oxide. Prepare the solution according to the appropriate one of the following instructions.

Preparation of Solution from Analyzed Copper Sample. Weigh out three portions of proper size (see instructor) and put each in a 500-ml conical flask. Add 5 ml of water and heat to near boiling; then add 6 N HNO_3, 1 ml at a time, until the sample is dissolved. Avoid an excess of more than 2 ml of acid. When the sample is all dissolved, rinse down the walls of the flask and add water to make the volume about 50 ml.

Preparation of Solution from Weighed Copper Sample. Obtain the sample in a 250-ml volumetric flask that is calibrated relative to a 50-ml pipet. The sample is a weighed portion of pure copper to which 7–8 ml concentrated HNO_3 has been added. Add about 10 ml of water and warm on a steam bath until the sample is completely dissolved. Cool, dilute to volume, and mix thoroughly. Pipet four 50-ml aliquot portions into 500-ml conical flasks.

Procedure. Titration of Copper Solution. Heat the solution to boiling and add 1 g urea dissolved in 5 ml of water. Boil for 1 minute. Cool in tap water. Add $4 N$ ammonia solution (prepare by adding 1 volume of concentrated solution to 3 volumes of water) dropwise from a pipet until a precipitate of cupric hydroxide forms or until the solution just begins to show the deep-blue color of the $Cu(NH_3)_4^{+2}$ complex ion (note 1). Add 5 ml glacial acetic acid. If the solution is warm, cool to room temperature. Add 5 g iodate-free potassium iodide dissolved in 10 ml of water. Cover the flask with a watch glass and allow to stand for 2 minutes. Titrate with standard thiosulfate solution until the iodine color begins to fade. Add 3 ml of starch indicator and continue the titration until the deep-blue color changes to gray. Add 2 g powdered ammonium or potassium thiocyanate crystals and swirl the solution gently until the crystals dissolve. The solution should turn back to the starch-iodine blue color as adsorbed iodine is displaced from the CuI particles. Complete the titration to a sharp color change.

If a weighed copper sample was used, compute the weight of copper in each aliquot portion, and report the total weight received. If an analyzed sample was used, report the percentage of copper.

Notes. 1. If any appreciable concentration of ammonium salts is present, the hydroxide ion concentration is so repressed that the cupric hydroxide precipitate does not form. The use of too much nitric acid or urea may lead to this condition. If too much ammonia is used, it is neutralized by the acetic acid, leading to formation of an ammonium acetate buffer. This condition causes poor end points. Careful control of pH is important.

2. It often happens that students spoil the first sample by failure to recognize the proper point at which to add starch. Use of four samples is specified so that three good values may be obtained, even if the first is spoiled.

3. A blank should be run on the thiocyanate to establish the absence of reducing agent impurities (often sulfides) in the reagent which might react with iodine. Dissolve 2 g of reagent in a small amount of water, add 5 ml of starch indicator, and determine the volume of standard I_2 solution to produce the blue color.

Errors. In addition to the usual volumetric errors, the following may occur:

1. Presence of oxides of nitrogen, leading to excess oxidation of iodide ion.

2. Incorrect pH, due to use of too much nitric acid, urea, or ammonia.

3. Difficulty in determination of end point.
4. Air oxidation, because of too slow titration.
5. Loss of iodine from the solution, due to vigorous swirling.

Copper Ore. Copper ores usually contain iron, arsenic, and anti-mony. All these can cause interference in the iodometric determina-tion of copper unless the copper is isolated or unless the titration is carried out under conditions that will prevent interference. The copper can be isolated by dissolving the sample in nitric acid, convert-ing to a sulfuric acid solution, and precipitating the copper with zinc or aluminum. The precipitated copper is then dissolved in nitric acid and determined according to the procedures given previously. In the Park method[1] no separations are made. Interference from ferric iron is prevented by addition of ammonium bifluoride, which causes forma-tion of the stable complex, $FeF_6{}^{-3}$. Arsenic and antimony are oxi-dized to the pentavalent condition by the procedure used for bringing the sample into solution. In this form they cause no interference provided the pH of the solution is carefully regulated at a value of about 3.5. A phthalate buffer serves for pH regulation.

Procedure. Weigh out sufficient sample to furnish about 0.2 g Cu (instructor), and transfer to a 150-ml beaker. Add 20 ml nitric acid, and heat on the steam bath until all copper is in solution. Evaporate on steam bath to a volume of about 5 ml. Add 5 ml of urea solution (1 g in 25 ml), and heat for a few minutes; then dilute to 30 ml, place the beaker on a wire gauze, and bring the solution to a boil (avoid spattering). Filter while hot, and wash the residue well with hot dilute nitric acid (1 : 100), catching the washings with the filtrate in a wide-mouth conical flask. Place the filtrate on the steam bath, and evaporate to about 30 ml. Cool and add 6 N ammonium hydroxide to precipitate the iron. When a precipitate begins to appear, add the hydroxide a drop at a time, testing after each addition for the smell of ammonia above the solution. When ammonia is faintly detectable by smell, after residual gas is blown away from above the surface, the iron is all precipitated. There should not be sufficient excess ammonia at this point to give the deep-blue color of the cupric ammonia com-plex. Add exactly 2.0 g ammonium bifluoride, and swirl until it is dissolved; then add 1.0 g potassium acid phthalate, and swirl until dissolved. Add 3.0 g potassium iodide, and swirl gently until dissolved. Titrate at once with standard thiosulfate solution, as directed in the preceding exercise. Report the percentage of copper in the ore.

[1] B. Park, *Ind. Eng. Chem., Anal. Ed.,* **3,** 77 (1931).

Iodide Ion. The amount of iodide ion in a sample can be determined by quantitative oxidation to iodine and titration of the iodine with thiosulfate. It is, of course, necessary to remove all excess of oxidizing agent before the thiosulfate titration. This limits the choice of oxidizing agents that can be used. Sodium nitrite in dilute sulfuric acid solution is a convenient oxidizer; the excess of nitrite can be removed by urea, as described in the iodometric method for copper. Bromine or chlorine cannot be used unless care is taken to oxidize all the iodine to iodic acid, since otherwise there is uncertainty about the equivalents of iodine formed. When the amount of iodine in a sample is very small, an acid solution may be treated with bromine and boiled until all bromine is expelled. This oxidizes iodine quantitatively to iodic acid. The sample is now treated with an excess of potassium iodide, and iodine is liberated. The liberated iodine is titrated with standard thiosulfate. This method has the advantage that each equivalent of iodide ion originally present gives six equivalents of titratable iodine. The reactions are:

$$I^- + 3Br_2 + 3H_2O = IO_3^- + 6Br^- + 6H^+$$

$$IO_3^- + 5I^- + 6H^+ = 3I_2 + 3H_2O$$

This method is often used for the determination of iodine in biological samples.

Questions

1. Write ionic equations to represent all the reactions involved in the following; include reactions used to dissolve materials, etc.
   (a) The iodimetric determination of sulfur present as H_2S.
   (b) The standardization of I_2 solution with As_2O_3.
   (c) The standardization of $Na_2S_2O_3$ solution with Cu.

2. Why have so many volumetric redox methods been developed which involve iodine?

3. What errors arise in iodometric procedures from:
   (a) Too low a pH at the time of titration?
   (b) Addition of starch at the beginning of the titration?
   (c) Use of partially decomposed starch solution?

4. Why is borax added to thiosulfate solutions? What other materials might be used instead of borax?

5. Why is starch often an unsatisfactory indicator if a solution contains a high concentration of electrolytes?

6. Describe a method whereby a thiosulfate solution could be standardized by pure iron wire.

7. In the standardization of thiosulfate with permanganate, dichromate, etc., why is the standard added to a strongly acid solution? Why is the solution later diluted before titration with thiosulfate? Why cannot thiosulfate be directly titrated against such oxidizing agents as permanganate?

8. Why must the iodimetric determination of As^{+3} or Sb^{+3} be made in buffered solutions? What pH is employed?

9. Predict the effects of the following errors on the specified determinations:

(a) Sodium carbonate is not added in the standardization of $Na_2S_2O_3$ with $K_2Cr_2O_7$.

(b) Urea is not added in the analysis for Cu.

(c) Atmospheric oxidation takes place during the standardization of $Na_2S_2O_3$ with Cu.

(d) The KI used in an analysis of a soluble chromate sample contains a small amount of KIO_3.

10. In the stibnite determination what would be the effect on the result if:

(a) The H_2S were not all boiled out of solution?

(b) The $NaHCO_3$ were not added?

11. Why is it advisable, in indirect iodine determinations, to prepare *four* samples, instead of the three samples ordinarily employed for other volumetric analyses?

12. List the errors which may result from the incorrect use of starch solution.

13. List the errors common to all indirect analyses involving iodine.

14. Describe a method for standardization of a thiosulfate solution by indirect comparison with a standard sodium hydroxide solution. Give equations for each reaction involved. What indicator should be used for each titration?

Problems

1. Express the strength of a $0.0900\,N$ iodine solution in terms of the titer for the following substances: As, As_2O_3, As_2O_5, H_2S, H_2SO_3, S, CdS.

Ans. 3.371 mg As.

2. What is the titer of $0.1000\,N$ thiosulfate solution for each of the following substances: I, Cu, Fe, $K_2Cr_2O_7$, Cr_2O_3, $KMnO_4$?

3. How much of the substance in column B will react with the stated amount of reagent in column A?

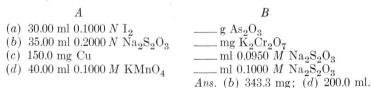

A	B
(a) 30.00 ml 0.1000 N I_2	____ g As_2O_3
(b) 35.00 ml 0.2000 N $Na_2S_2O_3$	____ mg $K_2Cr_2O_7$
(c) 150.0 mg Cu	____ ml 0.0950 M $Na_2S_2O_3$
(d) 40.00 ml 0.1000 M $KMnO_4$	____ ml 0.1000 M $Na_2S_2O_3$

Ans. (b) 343.3 mg; (d) 200.0 ml.

4. Calculate the normalities of the following thiosulfate solutions:

(a) 31.67 ml $Na_2S_2O_3$ is equivalent to 0.3500 g pure I_2.

(b) 35.15 ml $Na_2S_2O_3$ is equivalent to 0.2500 g electrolytic Cu.

(c) 41.22 ml $Na_2S_2O_3$ is equivalent to 37.84 ml I_2; 35.62 ml I_2 is equivalent to 0.2750 g As_2O_3.

(d) 29.08 ml $Na_2S_2O_3$ is equivalent to 0.1000 g $KBrO_3$.

(e) 36.52 ml $Na_2S_2O_3$ is equivalent to 0.2982 g KIO_3.

Ans. (a) 0.0871 N; (e) 0.2289 N.

5. A solution of thiosulfate is standardized with pure potassium dichromate. From the following data calculate the normality of the solution.

Buret readings: after	41.73 ml
before	0.51 ml
Weight of sample plus container	16.3852 g
Weight of container	16.1731 g
Buret readings for the blank: after	0.15 ml
before	0.07 ml

6. 1.000 ml $K_2Cr_2O_7$ is equivalent to 0.005585 g Fe; 20.00 ml $K_2Cr_2O_7$ liberates the iodine which is titrated by 32.46 ml $Na_2S_2O_3$. What is the normality of the latter solution? *Ans.* 0.06161 N.

7. Find the copper titer of a thiosulfate solution from the following data:

40.16 ml $KMnO_4$ is equivalent to 200.0 mg As_2O_3.

35.79 ml $KMnO_4$ liberates iodine which is titrated by 41.07 ml $Na_2S_2O_3$.

8. To a mixture of 97.31 ml 0.1096 N I_2 and 97.21 ml 0.1098 N $Na_2S_2O_3$ a drop of starch is added. Is the mixture blue or colorless?

9. How many milligrams of As_2O_3 will react with the I_2 equivalent to 35.00 ml $Na_2S_2O_3$ solution if 40.00 ml $Na_2S_2O_3$ solution is equivalent to 3.000 mM metallic copper? *Ans.* 129.8 mg.

10. What range of weights of $K_2Cr_2O_7$ as a primary standard should be used for the standardization of a 0.0500 N $Na_2S_2O_3$ solution? Should individual samples or aliquot portions be used? *Ans.* 75–120 ml.

11. How many grams of $KIO_3 \cdot HIO_3$ (potassium biniodate) are contained in 500 ml 0.1000 N solution? *Ans.* 1.625 g.

12. A soluble chromate is treated with excess KI in the presence of strong acid; after dilution 35.00 ml 0.1600 N $Na_2S_2O_3$ is required to titrate the liberated I_2. What is the weight of Cr in the sample?

13. Calculate the percentage of CuO in an ore from the following:

Weight of sample	0.4217 g
Volume of $Na_2S_2O_3$	35.16 ml
41.22 ml $Na_2S_2O_3$ is equivalent to 0.2121 g $K_2Cr_2O_7$	

Ans. 69.59%.

14. Find the percentage of Cu_2O in an ore from the following:

30.10 ml I_2 is equivalent to 0.2475 g As_2O_3.

50.00 ml $Na_2S_2O_3$ is equivalent to 30.10 ml I_2.

0.6092 g ore requires 30.46 ml $Na_2S_2O_3$. *Ans.* 35.81%.

15. What would be a satisfactory range of sample weights for the analysis,

by titration with 0.1250 N I_2, of an unknown which is approximately 15 per cent As_2O_3?

16. 18.00 ml 0.1000 M $KMnO_4$ reacts with excess KI to liberate a quantity of I_2 which in turn requires 45.00 ml $Na_2S_2O_3$ for titration. 30.00 ml thiosulfate solution is needed to titrate the iodine liberated by a copper sample. Calculate the milligrams of copper in the sample.

17. The As_2O_3 titer of an I_2 solution is 5.100 mg per milliliter. Find the percentage purity of a stibnite ore which required 27.50 ml I_2 to titrate a 0.5000-g sample. *Ans.* 48.18%.

18. What is the percentage purity of a sample of tartar emetic, $K(SbO)C_4H_4O_6$, if 0.5063 g requires 38.63 ml of an I_2 solution for titration in the presence of $NaHCO_3$. The As_2O_3 titer of the I_2 is 1.000 mg per milliliter.

19. The reaction between IO_3^-, I^-, and H^+ to form I_2 and the reaction of the latter with $S_2O_3^{-2}$ may be used to standardize an acid. A weighed sample of iodate is dissolved in a solution which contains an excess of KI and $Na_2S_2O_3$. The solution is titrated with the acid until all the iodate is used up and H^+ is in excess as shown by methyl orange indicator. In such a standardization 32.56 ml of an acid was needed to titrate a solution containing 0.5455 g pure KIO_3. What is the normality of the acid? *Ans.* 0.4697 N.

20. A 0.2352-g sample of pyrolusite (MnO_2) is heated with excess hydrochloric acid, and the distillate is passed into excess potassium iodide solution. The iodine liberated requires 47.82 ml 0.1123 N thiosulfate solution for titration. Calculate the percentage of manganese dioxide in the pyrolusite. Write equations for the reactions. *Ans.* 99.25%.

21. An iron sample contains 0.0500 per cent of S, present as sulfide. If a 1.000-g sample is analyzed iodimetrically, what concentration of iodine solution should be used for the volume of solution to be 30 ml or greater? *Ans.* 0.001040 N.

22. Lead may be estimated volumetrically by electrolyzing to convert the lead to the dioxide PbO_2, dissolving the dioxide in an acid solution of potassium iodide, and titrating the liberated iodine with thiosulfate; 0.5670 g of an ore required 36.23 ml 0.0998 N thiosulfate for titration of the liberated iodine. Calculate the percentage of lead in the ore.

$$PbO_2 + H^+ + I^- = Pb^{+2} + I_2 + H_2O \quad \text{(unbalanced)}$$

23. Lead can be determined by precipitating it as $PbCrO_4$, dissolving in acid, adding excess KI, and titrating with $Na_2S_2O_3$. What volume of 0.1400 N $Na_2S_2O_3$ would be required to titrate the iodine liberated by the lead sample of problem 22? Often the acidified chromate solution is directly titrated with $FeSO_4$ solution. What volume of 0.1250 N $FeSO_4$ solution would be required to titrate the sample of problem 22? *Ans.* 38.74 ml $Na_2S_2O_3$; 43.39 ml $FeSO_4$.

24. Calculate the normality of a $Na_2S_2O_3$ solution from the following:
Normality NaOH = 0.1000
21.29 ml NaOH is equivalent to 32.15 ml KHC_2O_4.

32.15 ml KHC_2O_4 is equivalent to 54.60 ml $KMnO_4$.

27.30 ml $KMnO_4$ is equivalent to 21.29 ml $Na_2S_2O_3$.

25. A 3.565-g sample of bleaching powder is shaken thoroughly with water and diluted to 500 ml. 25.00 ml of this requires 29.62 ml 0.1042 N $Na_2S_2O_3$ to titrate the I_2 liberated when the bleaching powder solution is mixed with KI. Find the percentage of available chlorine in the bleaching powder.

Ans. 61.40%.

26. How many milliliters of a solution which contains 1.500 g 99.5 per cent As_2O_3 in 250 ml are needed to directly titrate a 50.00-ml aliquot of the bleaching powder solution of problem 25?

27. Small amounts of gold are estimated by an iodometric procedure whereby the iodine producing reaction is:

$$AuCl_3 + 3KI = AuI + I_2 + 3KCl$$

What volume of $N/1000$ $Na_2S_2O_3$ is needed to titrate the iodine liberated by the gold in 1.000-g ore which contains 0.030 per cent Au? *Ans.* 3.042 ml.

28. 0.5000 g of a sample of KI is added to 50.00 ml 0.1000 N KIO_3 which is acidified; the mixture is boiled until the I_2 is expelled. The solution is cooled, and 2–3 g KI is added. The remaining KIO_3 liberates I_2 which is titrated with $Na_2S_2O_3$. It requires 21.21 ml 0.0956 N $Na_2S_2O_3$ to titrate. Find the percentage purity of the KI.

29. The Hinman method for the volumetric determination of sulfate involves the addition of an acidified solution of $BaCrO_4$ to the unknown. After the $BaSO_4$ has precipitated, the solution is made basic, and the excess $BaCrO_4$ precipitates. One mole of CrO_4^{-2} is left in solution for each mole of SO_4^{-2} precipitated. The precipitate is removed by filtration, and the solution is acidified, treated with KI, and titrated with $Na_2S_2O_3$. How many milliliters of 0.1108 N $Na_2S_2O_3$ are needed to titrate a 0.1500-g sample of $(NH_4)_2SO_4$ treated as described? *Ans.* 31.45 ml.

30. What weight of steel must be taken for analysis so that each 1.000 ml 0.0100 N I_2 solution represents 0.0100 per cent S in the sample?

CHAPTER 16

Oxidation by
Ceric and Dichromate Ions

Prior to the discovery of internal redox indicators, some thirty years ago, potassium permanganate was the only strong oxidizing agent in general use. The color change from purple to colorless when the permanganate ion is reduced makes this reagent able to serve as its own indicator, as we have seen in Chapter 14. As a result of this situation we find, even today, that a large fraction of the oxidation titration methods in the standard reference works of analytical chemistry are based on $KMnO_4$. We have seen, however, that there are several objections to the use of permanganate. The solution must be boiled, to destroy reducing agents in the water, and then filtered before standardization—a somewhat inconvenient operation when preparing a large quantity of reagent. After standardization the solution may undergo changes in concentration, and frequent restandardization is needed. The solution must be protected from light. Permanganate is such a strong oxidizing agent that, when titrations are made in HCl solution, it is necessary to take special precautions to prevent oxidation of chloride ion, an undesirable side reaction.

Since the development of internal redox indicators that can be used with any oxidizing agent, there has been a persistent search for oxidizing agents to replace permanganate. Today we have a number of such agents. Chief among these is tetravalent cerium, which has in sulfuric acid solution about the same oxidizing power as permanganate. Its solutions are very stable and can be kept for years without re-standardization. It can be used for most of the determinations that formerly required permanganate, such as determinations of iron, arsenic, antimony, oxalates, and many others. The initial cost of ceric solutions is relatively high, as compared with permanganate, but the cost is more than offset by the saving of technician's time in preparation and standardization of the solutions. Other reagents of less general applicability include potassium dichromate and potassium bromate or iodate.

Cerate Oxidimetry

Tetravalent cerium dissolves in strongly acid solutions, forming complex anions whose formulas in solution are not definitely established. The strength of ceric solutions as oxidizing agents depends on the acid present. The oxidizing power in HCl solutions is less than that of MnO_4^- ion, in H_2SO_4 about the same as permanganate, and in HNO_3 and $HClO_4$ considerably higher than permanganate. Thus, by proper choice of media, we can vary the oxidation potential to fit the needs of specific determinations—a feature that MnO_4^- does not have. Ordinarily the standard solution for cerate titrations is made up in H_2SO_4 and is referred to as a solution of ceric sulfate. Ferroin is used as redox indicator for titrations with ceric sulfate.

Since we do not know the coordination states of Ce^{+3} and Ce^{+4} ions in solution, we represent the reduction by the simplified equation

$$Ce^{+4} + e^- = Ce^{+3}$$

Preparation of Ceric Sulfate Solution. Several ceric salts are available for analytical use. Most widely used are ceric ammonium sulfate and ceric hydroxide, which can be obtained from all supply houses. Also available are prepared cerium salt solutions in H_2SO_4, which can be purchased ready for standardization.

Procedure. Slowly pour 60 ml 18 M H_2SO_4 (concentrated) into 300 ml of water in a beaker, with constant stirring. Weigh out approximately $\frac{1}{10}$ mole of the soluble ceric salt provided (use the formula printed on the label to compute the mole weight). Add the salt to the hot acid solution and stir until dissolved. Cool, dilute to 1 liter, and transfer to a glass-stoppered bottle. Allow to stand overnight before standardization. If any precipitate appears (probably ceric phosphate), filter through a porous crucible before standardization.

Standardization. The primary standards recommended for permanganate solutions are also used for standardizing ceric sulfate solutions. Ferrous ion is oxidized rapidly and quantitatively by ceric sulfate, but the reactions with arsenious oxide or sodium oxalate are too slow for titrations unless catalysts are added to speed up the reactions. Osmium tetroxide has been found to be a good catalyst for oxidation of As_2O_3 and ICl a good catalyst for oxidation of oxalates. Only a trace of each catalyst is required. Apparently the catalyst oxidizes the reducing agent, then is itself reoxidized by the ceric ion, thus serving as an electron carrier.

Procedure. Arsenious Oxide. Dissolve 20 g reagent-grade NaOH in 80 ml of water. Weigh accurately three 0.2-g portions of dried primary-standard-grade arsenious oxide, putting each portion into a 250-ml conical flask. Add 10 ml of the NaOH solution and gently swirl the liquid until the solid is dissolved. Add 75 ml of water, 25 ml $3\ M\ H_2SO_4$, two drops of 0.01 M osmium tetroxide solution (note 1), and two drops of Ferroin indicator. Titrate with the ceric sulfate solution to a sharp color change (note 2). Run a blank. Calculate the normality of the ceric sulfate solution. The equivalent weight of arsenious oxide is $\frac{1}{4}$ mole.

Notes. 1. Prepare the solution by dissolving 0.25 g osmium tetroxide (perosmic acid) in 100 ml 0.1 M H_2SO_4. Keep in a dark-glass dropping bottle.

Care must be used in making up the solution because of the high toxicity of osmium salts. Open the sealed ampule containing the volatile oxide under a hood. Remove sample by a spatula and weigh on a trip scale under the hood. Transfer the weighed portion to the bottle containing the sulfuric acid. The diluted solution is not particularly dangerous to handle.

2. The end point is a sharp color change from pink to pale blue.

Procedure. Iron Wire. Weigh accurately three 0.22-g portions of primary-standard-grade iron wire which is clean and free of rust. Put each portion into a 500-ml conical flask and dissolve the wire in 10 ml HCl, warming on a steam bath until solution is complete. While still hot, add stannous chloride solution (note 1), a drop at a time, until the color of the iron solution changes from the yellow of the ferric chloride complex to the pale green of ferrous ion in solution. The first drop of $SnCl_2$ may complete the reduction. Add just one drop of $SnCl_2$ after the color change. Cool the solution to room temperature and add 20 ml mercuric chloride solution (note 2) all at once. Swirl

from the total volume used. Compute the ratio $\dfrac{\text{ml thiocyanate}}{\text{ml silver}}$.

Accurately weigh a 1.1- to 1.2-g sample of pure dry sodium chloride, transfer to a 250-ml volumetric flask, dissolve, and dilute to volume. Withdraw, by a pipet, $\frac{1}{5}$ aliquot portions and introduce into 250-ml conical flasks. Add 5 ml of the nitric acid solution, and then add silver nitrate solution from a buret until it is in excess (note 3). Add 1–3 ml nitrobenzene (note 4) and 1 ml of the indicator solution, shake, and then titrate with thiocyanate. From the volume of thiocyanate used compute the volume of silver nitrate in excess, and deduct this from the total volume to obtain the net volume required for precipitation of the chloride. From the net volume and the weight of chloride sample employed compute the Cl^- titer of the solution.

Notes. 1. Care must be taken to avoid the addition of too little thiocyanate. Near the end point the reagent must be added dropwise and the solution thoroughly shaken before the next addition.

2. The end point color is due to the formation of the complex $Fe(SCN)_6^{-3}$ ion.

3. Near the equivalence point the precipitate will coagulate. After this occurs, shake well, allow to stand a moment, and add a little silver nitrate to the clear supernatant liquid. If precipitation is complete, no marked cloudiness will occur. Continue until it is certain that silver nitrate is in excess.

4. Optionally, instead of adding nitrobenzene, filter the solution through a filtering crucible, wash with a little water, and titrate the clear filtrate with thiocyanate solution. This method has the advantage that at the end point only a small amount of precipitate is in the solution.

Analysis of Chloride Samples. Prepare a solution of the sample. Analyze aliquot portions, following the procedures employed for standardization. Report the percentage of chlorine in the sample. Alternatively, weigh out three 0.25- to 1.0-g samples as directed. Dissolve each, add excess standard $AgNO_3$, and complete the titration with standard KSCN, as in the standardization.

Errors and Interfering Substances. The usual volumetric errors are encountered. In addition, the beginner is prone to undertitrate on adding thiocyanate. The necessity for removal of silver chloride has been emphasized. Other substances that form insoluble silver salts must be absent, as must bivalent mercury which forms a stable complex ion with thiocyanate ion. Nitrous acid and oxides of nitrogen must be boiled off, since they attack thiocyanic acid, causing a premature red color in the solution. The amount of excess silver nitrate should be kept small, thereby reducing the volume required in back titration.

Procedure.   Analysis of Silver Alloy. Prepare solutions as directed above. Standardize the thiocyanate solution as follows. Accurately weigh 0.3–0.4 g pure silver sheet or foil, and dissolve in an Erlenmeyer flask in 5–10 ml 1 : 1 nitric acid. Boil the solution until oxides of nitrogen are expelled. Dilute to 100 ml, add ferric indicator, and titrate with thiocyanate, as directed in the preceding exercise. Compute the Ag^+ titer of the thiocyanate.

In the analysis of an alloy proceed just as in the standardization. Report the percentage of silver.

Determination of Chloride by the Mohr Method

The theory of this method has been given on page 181. The titration is made in neutral solution to an end point which is shown by the formation of a colored precipitate of silver chromate.

Procedure. Standardization. Accurately weigh a 1.1- to 1.2-g sample of pure dry sodium chloride, transfer to a 250-ml volumetric flask, and dilute to volume. Withdraw one-fifth aliquot portions by a pipet, and deliver into 250-ml porcelain casseroles (note). Add 2 ml 0.1 M potassium chromate solution, and titrate, without further dilution, with silver nitrate solution prepared according to directions on page 324. The end point is marked by the first permanent appearance of a colored precipitate of silver chromate. Compute the Cl^- titer of the silver nitrate solution.

Note. The end point for this titration is more readily observed in a casserole than in a flask. Stir the solution with a short glass stirring rod.

Analysis of Soluble Chloride. Accurately weigh 1.5–2 g, and make up to a volume of 250 ml in a calibrated flask. Titrate aliquot portions (note) according to the procedure used in standardization. Calculate the percentage chlorine in the sample.

Note. Test the solution of the sample with litmus paper after the aliquot portion is measured. If it is acid, add chloride-free sodium bicarbonate until the solution shows a neutral reaction, or neutralize, using phenolphthalein indicator, with chloride-free base. If the sample is strongly basic, neutralize with chloride-free nitric acid to the phenolphthalein end point.

Errors and Interfering Substances. All the errors common to volumetric determinations may be encountered. The solution must be nearly neutral. The chief difficulty in using this method lies in the technique of determining the end point. Interfering cations are those that are colored, such as copper and nickel; those that give precipitates with chromates, such as lead and barium; and those that precipitate as basic chlorides in neutral solution, such as bismuth and tin. Interfering anions are those whose silver salts are insoluble in neutral solution, such as phosphate, arsenate, and sulfide.

Other Applications. The Mohr method may be used for the determination of bromides but is not satisfactory for iodides and thiocyanates because adsorption by the precipitate causes an unsatisfactory end point. Since the introduction of adsorption indicators, the Mohr method is not widely used, even for chlorides, for the end point is difficult to detect, and the titration is tedious. The method, however, is capable of high accuracy in the hands of an experienced analyst.

Complex Ion Titrations

Formation of complexes between ions of the sample and of the titrating reagent may be used in volumetric analyses in much the same way as we have utilized formation of precipitates in the methods of the preceding section.

A widely used application of complex ion formation in titrations is the Liebig method for determination of cyanide ion by titration with silver ion. The reaction is

$$Ag^+ + 2CN^- = Ag(CN)_2{}^-$$

The dicyanoargentate complex is so stable that the concentration of silver ion remains very low until all the cyanide ion of the solution is used up, which occurs at the equivalence point. Adding another drop of reagent, containing silver ion, causes a large increase in the concentration of Ag^+ in the solution and a precipitate of the slightly soluble $Ag[Ag(CN)_2]$ appears, which is taken as the end point. In practice it is found that a better end point is obtained by adding ammonia and iodide ion to the solution. The ammonia prevents precipitation of $Ag[Ag(CN)_2]$, by formation of the $Ag(NH_3)_2{}^+$ complex, but AgI is so insoluble that it precipitates, even in the presence of NH_3, with one drop of reagent in excess of the equivalent amount.

Determination of Hardness of Water. A useful application of complex ion titration is the determination of calcium and magnesium in water. The hardness of water is measured in parts per million (ppm), or milligrams per liter, expressed in terms of calcium as $CaCO_3$. Actually the hardness is due both to calcium and magnesium salts, but the two are determined together in the titration.

The complexing agent used for the titration is a salt of the tetraprotic acid, ethylenediaminetetraacetic acid (EDTA) which forms very stable

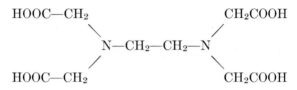

complex ions with metal ions. Its salts used for this purpose are marketed under the trade names of Versene, Triton B, Complexone II, and Sequestrene. The one used in this determination is the disodium salt, with the formula $Na_2H_2Y\cdot2H_2O$, where Y is the tetravalent anion of EDTA. When Ca^{+2} is treated with H_2Y^{-2}, a very stable complex is formed

$$Ca^{+2} + H_2Y^{-2} = CaY^{-2} + 2H^+$$

Magnesium ion forms a similar complex, MgY^{-2}, which is far less stable than the calcium complex.

When a sample containing calcium and magnesium ions is titrated with a solution of EDTA, the calcium ions are first tied up as CaY^{-2}. As more reagent is added and the calcium ions are all combined as complex, the magnesium ions form MgY^{-2}. The desired end point of the titration is the point at which *all* the calcium and magnesium ions of the solution have combined with the complexing agent. It has been found that the indicator Eriochromeschwartz T forms a colored complex with magnesium ion and that this complex is less stable than the MgY^{-2} complex. Consequently, when EDTA is added to a solution containing the magnesium-indicator complex, we have the reaction

(1) $$\underset{\text{Wine red}}{MgIn^-} + H_2Y^{-2} = MgY^{-2} + \underset{\text{blue}}{HIn^{-2}} + H^+$$

and a color change is observed when the last of the indicator ion is displaced from its complex with magnesium ion. Since all calcium solutions do not contain magnesium ion, it is customary to prepare the EDTA solution with a small amount of magnesium ion. When added to a solution containing calcium ion and indicator, the magnesium complex first loses Y^{-4} to calcium

(2) $$MgY^{-2} + Ca^{+2} = CaY^{-2} + Mg^{+2}$$

and the free Mg^{+2} ion reacts with the indicator

(3) $$Mg^{+2} + HIn^{-2} = MgIn^- + H^+$$

After all the calcium ion is complexed, reaction (1) occurs and a color change is observed.

Procedure. Prepare a standard calcium solution by dissolving an accurately weighed 0.5-g portion of pure $CaCO_3$ (note 1) in the minimum amount of $1\ M$ HCl and diluting to 500 ml in a volumetric flask. Calculate the milligrams of $CaCO_3$ equivalent to 1 ml of solution. Prepare the reagent solution by dissolving 4 g of the disodium salt of EDTA and 0.1 g $MgCl_2 \cdot 6H_2O$ in water and diluting to 1 liter.

Standardize the EDTA solution by comparison with the standard calcium solution. Pipet a 25-ml sample of calcium solution into a conical flask. Add 10–15 ml ammonia buffer (note 2) of pH 10 and four or five drops of indicator (note 3). Titrate with the EDTA solution. The end point is a color change from wine red to a clear blue.

Compute the weight of $CaCO_3$ equivalent to 1 ml of the EDTA solution, the $CaCO_3$ titer.

Obtain an unknown calcium solution and titrate a 25-ml sample, just as in the standardization. From the volume of EDTA used and the $CaCO_3$ titer, compute the weight of $CaCO_3$ in the sample, in milligrams. Multiply this value by 40 to obtain the weight of $CaCO_3$ per liter of the solution. This weight in milligrams is the parts per million of $CaCO_3$.

Notes. 1. Crystals of Iceland spar or calcite are recommended as standards.

2. Prepare the buffer solution by adding 6.75 g NH_4Cl to 57 ml concentrated ammonia solution and diluting to 100 ml. The pH is slightly above 10.

It is essential to make the titration at a pH near 10. If the solution is more acidic than this, the indicator exists as H_2In^- instead of HIn^{-2} and does not give the desired color change. If the solution is more basic, magnesium precipitates as $Mg(OH)_2$.

3. Prepare the indicator solution by dissolving 0.5 g analytical reagent-grade Eriochromeschwartz T in 100 ml of alcohol. The solution does not keep well (about 1 month), and each bottle should be dated when prepared.

Questions

1. Is the strength of a silver nitrate solution as determined by the Mohr method greater than, equal to, or less than the strength as determined by an adsorption indicator titration, pure sodium chloride being used as the primary standard in each case?

2. When a solution of hydrochloric acid is to be standardized by the

Mohr method, the acid is first neutralized. What indicator should be used? What reagent would you employ for the neutralization?

3. Is it necessary to neutralize a hydrochloric acid solution if it is to be standardized by silver nitrate titration by the adsorption indicator method? If the solution is neutralized, what indicator should be employed? What would be the effect on the apparent molarity if the neutralization were not carried out?

4. What relation must exist between the charge on the indicator ion and the charge on the titrating ion when an adsorption indicator is used?

5. Why is dextrin used in the chloride determination with dichlorofluorescein as indicator?

6. In a Volhard analysis for Cl^- the AgCl was not removed by filtration before the back titration with SCN^-. Was the determined percentage of chloride too high or too low?

7. Explain how I^- can serve as indicator in the titration of CN^- by Ag^+.

8. Why is it not required to filter off the silver iodide in the determination of iodine by the Volhard procedure?

9. Devise a procedure for determining the silver present in a sample of photographic emulsion which contains silver, bromine, and organic matter.

Problems

1. What is the molarity of an NH_4SCN solution, 25.75 ml of which is required to titrate 215.3 mg 99.9 per cent pure Ag after the metal has been dissolved in HNO_3?   *Ans.* 0.0774 *M*.

2. How many grams of $AgNO_3$ should be present per liter of solution so that each milliliter is equivalent to 1.000 mg Cl^-?   *Ans.* 4.791 g.

3. How many milliliters of 0.1500 *M* $AgNO_3$ are needed to titrate:
   (*a*) 4.000 mM $BaCl_2 \cdot 2H_2O$?
   (*b*) 0.5000 g KCN?
   (*c*) 35.50 ml 0.1571 *M* KSCN?   *Ans.* (*a*) 53.33 ml.

4. A sample of silver alloy is dissolved in nitric acid and the silver determined by a Volhard titration. If 38.62 ml 0.1175 *M* KSCN solution is required for the titration, how many grams of silver are in the sample?

5. How many milliliters of 0.1200 *M* KSCN will be needed in the back titration after the addition of 30.00 ml 0.1100 *M* $AgNO_3$ solution to 0.3655 g KI dissolved in 100 ml of water?   *Ans.* 9.15 ml.

6. An iodide sample which weighs 3.000 g is analyzed by the Volhard process; 49.50 ml 0.2000 *M* $AgNO_3$ is added, and 6.50 ml 0.1000 *M* KSCN is needed to titrate the excess silver. Calculate the percentage of iodide in the sample.   *Ans.* 39.13%.

7. A sample contains NaBr and inert material. A ⅕ aliquot is dissolved in water and 50.00 ml 0.1050 M AgNO$_3$ is added; 8.00 ml KSCN solution is required to produce the red color with the ferric alum indicator. Calculate the grams of NaBr present if 1.000 ml AgNO$_3$ is equivalent to 1.250 ml KSCN.   *Ans.* 2.355 g.

8. 40.00 ml 0.1000 M AgNO$_3$ is added to 0.3300 g of a sample which contains 1.300 mM BaCl$_2$·2H$_2$O and impurities inert to AgNO$_3$. How many milliliters of 0.1200 M KSCN solution will be required for back titration?

9. From the following data determine the percentage of chloride in a sample:

Standardization:	Weight NaCl	0.2500 g
	Volume AgNO$_3$	63.46 ml
	Volume KSCN	3.79 ml

37.90 ml KSCN is equivalent to 50.00 ml AgNO$_3$.

Analysis:	Weight of sample	0.6250 g
	Volume AgNO$_3$	35.00 ml
	Volume KSCN	7.58 ml

Ans. 10.37%.

10. In a cyanide determination 35.66 ml silver nitrate solution is needed to titrate to the appearance of turbidity. If 1.000 ml AgNO$_3$ = 0.002875 g KCl, how many grams of cyanide, computed as NaCN, are present in the sample?   *Ans.* 0.1348 g.

11. A small amount of KI is added to a solution of NaCN. 35.00 ml 0.1200 M AgNO$_3$ is required to produce a precipitate of AgI. Find the milligrams of NaCN present.   *Ans.* 411.8 mg.

12. Mercury can be complexed as Hg(SCN)$_2$ and determined by a procedure similar to the Volhard process. What is the percentage of mercury in an alloy, 0.7500 g of which required 45.10 ml KSCN solution whose Ag titer is 9.542 mg per milliliter?

13. A 0.5216-g sample of a zinc ore was brought into solution and the zinc separated and titrated with ferrocyanide solution, 23.21 ml being required; 32.00 ml of this solution was required for 0.2613 g pure zinc oxide in the standardization. Find the percentage of zinc in the ore. The reaction is

$$2K_4Fe(CN)_6 + 3ZnCl_2 = K_2Zn_3[Fe(CN)_6]_2 + 6KCl$$

Ans. 29.19%.

14. In an aqueous solution of 0.2100 g of a mixture containing potassium cyanide and chloride, 14.56 ml 0.1000 M silver nitrate was required to produce a faint permanent turbidity; 30.00 ml more of the silver nitrate was added (an excess) and the precipitate of AgCl and AgCN filtered off from the solution. The filtrate and washings were titrated with 13.06 ml 0.1000 M thiocyanate solution. Calculate the percentages of KCl and KCN in the sample.

15. Nickel is determined volumetrically by the addition of a known amount of cyanide to an ammoniacal solution of the sample. The excess of cyanide is titrated by standard silver solution. Potassium iodide is often added as an indicator. (Silver iodide precipitates as soon as there is any silver in excess.)

$$Ni^{+2} + 4CN^- = Ni(CN)_4^{-2}$$
$$Ag^+ + 2CN^- = Ag(CN)_2^-$$
$$Ag^+ + I^- = AgI$$

How much potassium cyanide should be added to a solution of 0.2500 g of nickel ore containing 20.00 per cent nickel in order that at least 15.00 ml 0.1100 M silver nitrate will be used in the back titration? *Ans.* 436.8 mg.

16. If 5.0 mM sodium chloride is dissolved in 150 ml of water, calculate the concentration of chloride ion after the addition of the following volumes of 0.1000 M silver nitrate solution: 0, 10, 20, 30, 40, 45, 48, 49, 49.3, 49.5, 49.9, 50, 55. Plot the negative logarithm of chloride ion concentration (pCl) against the milliliters of silver nitrate added. Compare this curve with the titration curves of Chapter 7.

17. The solubility product for silver chloride is 10^{-10} and for silver thiocyanate 10^{-12}. If the concentration of thiocyanate ion which is needed to affect the ferric ion indicator is $10^{-5} M$, what concentration of chloride ion will be in equilibrium at the end point in a solution which is in contact with both silver chloride and silver thiocyanate? What volume of 0.1 M thiocyanate solution will need to be added to 100 ml of solution in order to bring this amount of chloride into solution, according to the equation,

$$AgCl + SCN^- = AgSCN + Cl^-$$

If the stoichiometric volume of thiocyanate required in the Volhard determination of chloride is 15.00 ml, what is the percentage of error introduced if the silver chloride is not removed by filtration?

18. Calculate the theoretical error in a Mohr titration in which 50.00 ml 0.1 M silver nitrate solution is needed at the equivalence point, and the concentration of chromate ion is $10^{-4} M$. Assume a final volume of 100 ml. *Ans.* 0.2%.

19. What should be the concentration of chromate ion so that there will be no theoretical error in the determination of Br^- by the Mohr method? Is it possible to use such a concentration?

20. Sulfate may be determined by titration with standard lead nitrate solution. Potassium iodide is used as a spot-test indicator. What concentration of iodide ion should be used in order that the spot test show a turbidity at the stoichiometric point in a sulfate titration? Use the solubility product values in the Appendix. *Ans.* $10^{-2} M$.

21. 0.4985 g $CaCO_3$ is dissolved in HCl and diluted to 500 ml. A 25-ml aliquot requires 23.62 EDTA solution for titration. 100.0 ml of a water sample require 30.13 ml EDTA solution for titration. What is the total hardness of the water expressed in parts per million of $CaCO_3$?

22. Which of the following pipets—10 ml, 20 ml, 25 ml, 50 ml, 100 ml— should be used to measure the sample taken for titration of a water whose total hardness is approximately 900 ppm $CaCO_3$ if the EDTA solution available has a $CaCO_3$ titer of 0.8385 mg? *Ans.* 25 ml.

PART FOUR—GRAVIMETRIC

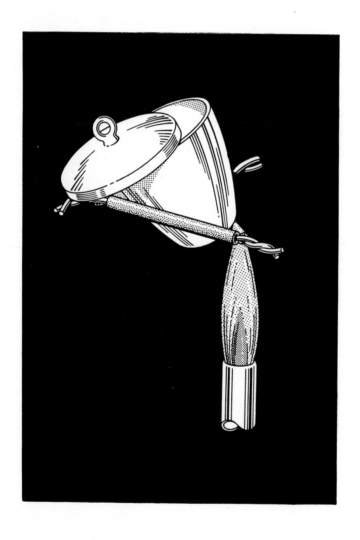

ANALYSIS BY PRECIPITATION

CHAPTER **18**

General Principles

Analyses in which a given substance is isolated from the sample and weighed in some pure form are known in general as gravimetric analyses. Isolation may be carried out by precipitating the sought substance in an insoluble form, by depositing a pure metal by electrolysis or by converting the substance to a gas which is absorbed in a suitable reagent. Of the three, precipitation is by far the most widely used. We shall in this and following chapters consider the principles and methods of precipitation analysis. This is the oldest, the most fundamental, the most versatile, and, we might add, the most unpopular method of analysis. Precipitation methods are laborious and fraught with difficulties. Nevertheless they can be used where other methods are not applicable.

In all precipitation analyses we seek to achieve the following:

1. The sought substance is completely removed from the remainder of the sample, i.e., precipitation is complete.

2. The precipitate as weighed is a pure substance of definitely known chemical composition.

The conditions developed for the various precipitation analyses are those that have been found to best achieve these conditions. In order to understand them, we must first study the general properties of precipitates.

337

Physical Properties of Precipitates

Precipitation. The phenomenon of precipitation, so familiar from our experience in qualitative analysis, is in reality one of the most complicated processes we deal with, even though the reaction is apparently very simple. We mix two solutions, containing respectively the anion and cation of an insoluble substance, and observe that shortly after mixing a precipitate appears. Usually this is instantaneous, but on occasion supersaturated solutions may remain for some time without forming a precipitate, unless a seed crystal is added. When a precipitate does appear, it may be in a variety of forms—as very large single crystals, very small crystals, large curds, or a slimy gelatinous mass.

Whatever the form, all crystals have one feature in common. The ions or molecules are packed closely together in an orderly array that is repeated throughout the crystal. A good representation of the arrangement of the ions in a simple crystal can be made by filling a box with marbles; the marbles assume the regular arrangement that gives the closest packing. If the crystals are composed of ions, such as AgCl, the two ions are of different size but are so arranged that every positive ion is surrounded by negative ions and vice versa.

It is truly remarkable that a billion billion ions (or more) which are randomly dispersed in a few hundred milliliters of solution can within a few seconds become arranged in the orderly planes of a crystal. The process is initiated when a few ions clump together, to start the nucleus of the crystal. After nuclei are formed, more ions strike the surface and adhere, always in the regular pattern characteristic of that crystal, thus causing the crystal to grow. In a given solution this growth continues until there is equilibrium between the ions in solution and the solid phase. At equilibrium the rate of deposition of ions on the surface is just equal to the rate at which ions leave the surface to go into solution. These two processes continue as long as the crystal remains in contact with the mother liquor.

Size of Crystals. The nature of the precipitate when equilibrium is attained depends on the properties of the specific crystal species present and on the conditions under which the reagents containing the ions of the crystal are mixed.

1. If mixing is rapid and a very large number of nuclei are formed

at the start of crystallization, each particle may be of colloidal size when equilibrium is reached. A colloidal particle is so small that gravitational forces do not overcome the Brownian movement caused by bombardment with solvent molecules. Thus, instead of settling out as a precipitate, the particles may remain in suspension indefinitely. Colloidal particles have, in general, a diameter less than 2000 angstrom units (an angstrom unit, A, is 10^{-8} cm). They are too small to be seen by the naked eye, and many are too small to be observed by an ordinary microscope. They pass through the pores of the usual filter material. A precipitate that remains dispersed as a colloid is useless for analytical purposes.

2. The colloid first formed may coagulate to give a precipitate. Some colloids are readily coagulated, to form a mass of curds. A familiar example is the AgCl precipitate as ordinarily obtained. The particles we observe settling out of solution are not individual crystals but are composed of a large number of colloidal particles rather loosely held together. Other colloids may agglomerate to form a gelatinous mass, a semi-liquid state coherent enough to settle out of solution but without any fixed shape or size. Such a coagulated mass is known as a *gel*. It holds large amounts of the liquid from which the precipitation is made. Precipitates of ferric and aluminum hydroxides, known as hydrous oxides, are gels. The individual colloidal particles making up the gel are very small indeed; if they were not agglomerated as a gel, it would be impossible to filter out these particles.

3. The colloidal particles first formed may grow into large crystals that eventually precipitate. If precautions are taken to start precipitation slowly, so that an excessive number of nuclei is not formed, then each particle grows to a relatively large size and we have a crystalline precipitate that settles out of solution readily and is retained by ordinary filters. The speed of crystallization, and consequently the number of particles formed in a given system, can be controlled to some extent by the manner in which the reagents are mixed. In general, the number of nuclei formed is a function of the amount by which the concentrations of the ions in the solution exceed the equilibrium value at the time precipitation begins. If precipitation is made to occur very slowly, we can, in the extreme case, obtain a single crystal, as in the slow evaporation of a salt solution. We do not seek single crystals in analyses but often we do employ the principle of slow deposition in order to obtain crystals of such size that filtration is made easy. To accomplish this objective we attempt to start the precipitation under conditions of relatively large solubility

and then decrease the solubility to obtain further precipitation on the surface of the crystals first formed.

For example, if we precipitate CaC_2O_4 by adding a soluble oxalate to a solution of a calcium salt, we obtain a mass of very small crystals. But if we mix $C_2O_4^{-2}$ and the Ca^{+2} ions in an acid solution, in which CaC_2O_4 is soluble, and then slowly neutralize the acid by adding ammonia solution, we obtain large crystals. Better still, if urea is added to the acid solution containing Ca^{+2} and $HC_2O_4^{-}$ ions and the solution boiled, the urea gradually hydrolyzes to give ammonia

$$CO(NH_2)_2 + H_2O = 2NH_3 + CO_2$$

and this NH_3 neutralizes the acid

$$NH_3 + HC_2O_4^{-} = NH_4^{+} + C_2O_4^{-2}$$

As this occurs the concentration of $C_2O_4^{-2}$ is slowly increased, and a very slow and uniform precipitation of CaC_2O_4 occurs. Precipitations of this type, wherein the precipitant is gradually formed in the solution, are known as "homogeneous precipitations."

When precipitates are extremely insoluble, the analyst has but limited control of the conditions of precipitation, because as soon as the reagents are mixed the initial degree of supersaturation in the solution is large. The usual procedure is to mix warm dilute solutions of the ion slowly, with vigorous stirring. The reagent is added dropwise, from a buret or pipet, and advantage is taken of the increased solubility of most salts with rise in temperature. Even so, with some precipitates, such as $BaSO_4$, it is difficult to obtain large enough particles for good filtration.

Solubility. The solubility of a given crystal is the result of two opposing forces. On the one hand is the attraction of water molecules for the ions of the crystal. Opposed to this are the forces holding the ions together in the crystal, the lattice forces. These vary widely from one crystal to another. Thus, silver ion dissolves to the extent of several moles per liter from the surface of a crystal of $AgNO_3$, whereas the solubility of AgI crystals is less than 10^{-8} moles per liter. The difference is that the lattice forces in AgI are stronger than those in $AgNO_3$.

Effect of Foreign Ions on Solubility. The analyst can control to some extent the forces tending to pull ions of a crystal into solution and thereby alter the solubility. For most crystals a rise in temperature increases solubility, perhaps by weakening the lattice forces.

This effect is small, however, in comparison with effects obtained by introducing foreign ions into the solution.

1. *Effect of Common Ion.* We have seen in the discussion of the solubility product constant that if an excess of one ion is added to a saturated solution of a salt, the concentration of the other ion is decreased to maintain constancy of the ion product. Advantage is often taken of this relation in precipitations, to reduce the amount of a given ion left in the solution. Thus, if AgCl is precipitated by adding NaCl to a solution of $AgNO_3$, an excess of Cl^- reduces the concentration of Ag^+ remaining in solution. A common ion is sometimes used in the wash water, to reduce solubility losses.

2. *Removal of Ions by Reactions.* In a saturated solution of an ionic salt MA we have the equilibrium

$$MA(s) = M^+ + A^-$$

Anything that reacts with either the M^+ or the A^- ion displaces the equilibrium to the right, i.e., increases the solubility. If A^- is the ion of a weak acid, the addition of hydronium ion will reduce the concentration of A^- ion, thereby increasing the solubility of the salt. If A^- is the OH^- ion, NH_4^+ will react with it, thereby increasing the solubility. An example is the solubility of $Mg(OH)_2$ in solutions of ammonium salts. Any substance that forms a complex with the M^+ ion will increase the solubility of MA. A familiar example is the solubility of AgCl in NH_3 solutions.

3. *Effect of Salts That Do Not React with Ions of the Precipitate.* The solubility of sparingly soluble substances is usually increased to some extent by the presence of neutral salts, i.e., salts that do not react with an ion of the precipitate. A discussion of this effect, which is due to attractions of ions of opposite charge for one another, is beyond the scope of an elementary text. The Debye-Hückel theory, treated in all physical chemistry texts, affords a mathematical explanation in terms of the "ionic strength" of the solution and the relations of the activities or effective concentrations of ions to their formalities. For present purposes it is sufficient to state that the effect of neutral salts in increasing solubilities is not of great significance in practical precipitation analyses, for most of the precipitates employed are so slightly soluble that solubility losses in either pure water or salt solutions are among the least of the errors encountered.

Effect of Particle Size upon Solubility. Ordinarily we think of the solubility of a given substance as constant. This is true when we are dealing with precipitates but is not true for the very small particles of colloids. When the particle size is less than 1 micron (10^{-3} mm in diameter), there is an appreciable increase in solubility with decrease in particle size. The effect is attributed to surface effects in very small particles, and as yet there is no wholly satisfactory explanation for it.

Digestion. The increased solubility of small particles causes a phenomenon known as "Ostwald ripening" of precipitates. When a mixture of large and small crystals is *digested* with the mother liquor (heated near the boiling point in contact with the saturated solution), it is found that the small crystals dissolve and their ions redeposit on the surfaces of the larger crystals. This effect is often beneficial in making the precipitate easier to filter and wash, for it eliminates the small particles that might clog the pores of the filter.

Digestion of precipitates causes the crystals to become more regular in shape, in addition to increasing the particle size. Freshly formed crystals are often of quite irregular shape, much like feathers or snowflakes, with very large surface areas. Such formations occur because ions from solution tend to deposit at points or protuberances on the surface, where fresh solution first comes into contact with the surface. On digestion the irregular needles redissolve and the ions redeposit on the major crystal faces, to create a more compact crystal with much smaller surface area. There may also be deposition of solid between adjacent small particles, to cement them together into a single larger particle. This effect is perhaps as important or more so than the dissolving and redeposition of small crystals, in the enhancement of filterability of precipitates by digestion.

It should not be inferred from the preceding discussion that digestion is beneficial for all types of precipitates. Precipitates that are made up of particles of colloidal size only, such as the coagulated curds of AgCl or the gels formed by hydrous oxides of iron and aluminum, do not appear to be altered by digestion and ordinarily are heated only long enough to effect coagulation.

Purity of Precipitates

The major source of error in gravimetric analyses is the presence of impurities in the precipitate. It is difficult to prevent ions other than the ones that make up the lattice from coming down with the precipitate. Even the complete removal of water from a precipitate is often a difficult problem.

Coprecipitation. A dragging down of normally soluble ions with an insoluble substance is known as *coprecipitation*. This can occur in three ways: (1) by incorporation of foreign ions in the crystal lattice, to form a solid solution, (2) by surface adsorption of foreign ions and (3) by occlusion.

Formation of solid solutions is not usually a source of difficulty in analyses, for such mixed crystals are formed only when there are present two species having the same charge and approximately the same ionic radius. Ordinarily the unwanted ions of the solution are quite unlike those of the precipitate and will not fit into the lattice. At times the chemist may take advantage of mixed crystals, as in the separation of very small quantities of artificially prepared radioactive substances from a solution. A carrier ion is added to the solution in amounts sufficient to give a precipitate. This carrier ion is the same species as the radioactive substance or is an ion of like size and charge. When it is precipitated, the sought radioactive ion is incorporated into the crystals and dragged down, even though present in such small quantity that its precipitation alone is not possible.

Before discussing contamination of precipitates by adsorption of foreign substances at surfaces, we must consider briefly the properties of surfaces and the nature of adsorption on surfaces. The ions, atoms, or molecules that make up a crystal are held together by attractive forces. In the case of ionic crystals, such as the precipitates we deal with in analyses, the forces are of two kinds (1) the van der Waals attractions between adjacent atoms and (2) the electrostatic forces between ions of opposite charge.

The attractive forces of a given ion inside a crystal are satisfied—that is, the ion is surrounded by all the neighbors it can hold. A surface ion, however, has only neighbors in the same plane and beneath that plane. It therefore has a residual attractive force that can hold other ions or molecules coming into contact with the surface. Every surface holds foreign material to some extent except when it is in a vacuum. This phenomenon is known as *adsorption*.

In crystals of filterable size the amounts of impurities held at the surface are usually negligibly small, because only a small fraction of the total number of ions in the crystal lie in the surface. In the very small particles of colloids, however, a large fraction of the total ions of the particle may lie in the surface—the smaller the particle the larger the fraction of surface ions. The amount of adsorption by colloidal particles is consequently much greater than for the same mass of material in the form of larger crystals.

Charge and Coagulation of Colloids. It will be recalled that formation of a crystal starts with a small nucleus and that the nucleus grows by deposition of positive and negative ions onto the surface. If more of one kind of ion deposits than of the other, this leads to a charge on the particle. To illustrate, consider the precipitation of AgCl by adding Ag^+ to a solution of Cl^- ions. After nuclei begin to form, they grow by adding silver and chloride ions. Before the equivalence point is reached, the solution contains an excess of chloride ions; after deposition of all the silver ions in the solution, additional chloride ions will deposit to start another layer. This leads to a negative charge on the particles. The existence of such a charge can be shown by placing electrodes connected to the positive and negative terminals of a battery in the solution. It is observed that all the particles migrate toward the positive electrode. If the particles are formed in the presence of excess silver ion, then it is found that the particles are positively charged.

In the situation just described we have shown that a charge is gained when a particle has more of one type of its lattice ions than of the other type. Foreign ions may also be adsorbed at a surface, thereby giving the particle a charge. It is found that all colloidal particles in aqueous dispersion do have a charge, either by an excess of lattice ions or by adsorption of foreign ions. Since electric charges of like sign repel one another, the charge on colloidal particles tends to prevent collisions with one with another, thus keeping the particles from coalescing. If we wish to coagulate a colloid, it is necessary to neutralize the effects of the charges by causing the particles to adsorb ions of charge whose sign is opposite that of the particle. If, for example, AgCl is precipitated from NaCl solution, we first observe a colloidal dispersion of very fine particles. As the equivalence point is approached, the concentration of silver ion in the solution rises and the excess negative charge, from chloride ions, is partially neutralized by adsorption of positively charged silver ions. At this point of low-charge density on the surface of the particles, they no longer repel one another but rather can collide and coalesce to form a curdy precipitate which settles out of the solution.

It is not essential that the ion which neutralizes the surface charge and permits the particles to coagulate be an ion of the precipitate, as it is in the precipitation of silver chloride. Every charged particle tends to attract ions of opposite charge, and ions other than those of the lattice may be adsorbed to neutralize the charge. This may be demonstrated experimentally by adding standard solutions of electrolytes to identical portions of a colloidal solution and noting the volumes required to effect coagulation. Results of a study made with negatively charged arsenic trisulfide sol are given in Table 18–1. It will be noted that ions of the same positive charge have roughly the same coagulating value, but that divalent ions are much more effective than monovalent ions and that trivalent ions are of still

TABLE 18–1. Coagulating Values of Electrolytes for Negatively Charged Arsenic Trisulfide Sol*

Electrolyte	Millimoles per liter
LiCl	58
NaCl	51
KNO_3	50
KCl	50
NH_4Cl	42
HCl	31
$MgCl_2$	0.72
$MgSO_4$	0.81
$CaCl_2$	0.65
$SrCl_2$	0.64
$BaCl_2$	0.69
$AlCl_3$	0.093
$Al(NO_3)_3$	0.095
$Ce(NO_3)_3$	0.080

* Values taken from H. B. Weiser, *Colloid Chemistry*, Wiley, second edition, 1949, page 253.

higher effectiveness than divalent ions. This shows that (1) the higher valence ions are more strongly adsorbed than those of lower valence and (2) it is not necessary that the coagulating ion be adsorbed in an amount sufficient to neutralize the charge of the particle completely. This is shown by the fact that less than 0.1 mM of trivalent aluminum ion coagulates the same amount of colloid as 50 mM of monovalent sodium or potassium ion, a ratio of 1 : 500. Since ions of higher valence are so strongly adsorbed by colloidal particles, it is necessary to avoid the presence of such ions during precipitations, to prevent their inclusion with the precipitate.

In other studies of the effects of ions in coagulating colloids it is found that the ions forming the most insoluble compounds with ions of a crystal are the ions most strongly adsorbed by that crystal. This generalization is known as the rule of Paneth-Fajans-Hahn.

The foregoing treatment gives only part of the story. In discussing the charges on colloid particles, we have considered only the charges that are tightly held to the surface. We must also consider another effect, the condition of the solution in the immediate vicinity of the charged particle. The charges of the adsorbed ions on the surface attract ions of opposite charge, so that the layer of solution surrounding the particle contains an excess of those ions whose charge is opposite the charge on the particle. These loosely held ions in the solution are known as *counter ions*.

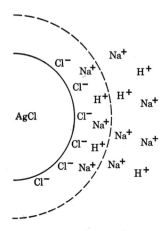

A schematic representation of a charged AgCl particle is shown in Fig. 18–1. This particle was formed in a solution containing chloride ions, and the lattice contains more Cl^- ions than Ag^+ ions. Thus the particle is negatively charged. The negative Cl^- ions of the lattice hold some of the positive ions of the solution tightly to the surface, but there are fewer adsorbed positive ions on the particle than Cl^- ions, since the particle as a whole carries a negative charge. In our diagram the lattice surface is indicated by the solid line, and the adsorbed Cl^- and the posi-

FIGURE 18–1. Adsorption at the surface of a colloidal particle of AgCl. The particle is formed in excess chloride ion.

tive ions by the portion within the dotted line. The layer of loosely held counter ions is indicated by the portion of the drawing outside the dotted line.

When the negatively charged particle of Fig. 18–1 migrates under the influence of an electric field (a phenomenon known as electrophoresis), the portion within the dotted line goes to the positive pole and the counter ions migrate in the opposite direction.

With this picture in mind, let us consider what happens when a colloid is coagulated by addition of ions having a charge opposite that of the particle. If Ag^+ ions are added to the solution in contact with the particle of Fig. 18–1, they fit into the lattice positions adjoining the Cl^- ions. When enough Ag^+ ions have been added to partially neutralize the charge of the particles, they may collide with one another and coalesce, thus forming a precipitate.

If an electrolyte containing a non-lattice positive ion is added to the solution, this ion is attracted into the layer of counter ions. The effect of the counter ion layer is a partial neutralization of the particle charge; if the layer contains some ions of high-charge density, such as the $+2$ and $+3$ valence cations, the effect may be to neutralize the charge enough so that particles can collide and coalesce.

When this happens, the precipitate drags down a layer of solution which contains the counter ions, and this leads to contamination of the precipitate by foreign ions.

Peptization of Precipitates. When a curdy precipitate of coagulated AgCl particles is washed with water, it is observed that a milky filtrate comes through the filter. This milkiness is caused by redispersion of the coagulated colloid particles, a process known as *peptization*. Peptization results from removal of the counter ions, which dissolve in the wash water. As they are removed and are no longer effective in neutralizing the charges on the particles, the charges repel one another and the particles disperse.

We must remove the counter ions, insofar as possible, to prevent contamination of the final precipitate. At the same time an error is introduced if much of the precipitate peptizes and passes through the filter. The solution to the problem is to use in the wash liquid ions which will replace the counter ions, but which are volatile and will be removed when the precipitate is dried. A dilute solution of HNO_3 is used for washing AgCl precipitates, so that the Na^+ counter ions are replaced by H^+ ions. When the precipitate is dried, HNO_3 is driven off, leaving a quite pure precipitate for weighing.

Effect of Concentration on Adsorption. When adsorption takes place from solutions, the amount adsorbed per unit area of surface is related to the concentration of ions in solution. A curve showing the amount adsorbed at constant temperature as a function of concentration is known as an adsorption isotherm. A typical isotherm for adsorption from solution is shown in Fig. 18–2. This isotherm shows that at low concentrations the amount of adsorption increases practically linearly with the concentration, but at higher concentrations the isotherm tends to level out. The amount adsorbed on a given sample can be minimized by keeping the concentration of the adsorbed ions in the solution to as low a value as possible.

Occlusion. Impurities other than those held to the surface by adsorption may be mechanically held inside capillary pores within a crystal particle. This type of coprecipitation is known as occlusion. It may occur when a crystal particle is an agglomerate of small colloidal particles that have coalesced. When this occurs, there are small pores between the individual particles of the aggregate; these pores hold mother liquor—water and the foreign ions of the solution. If the pores remain open, most of the mother liquor is removed in washing, but sometimes in ripening of the precipitate there is further crystallization in such a way as to seal off some of the pores and trap the entrained liquid inside the particle. The water of such trapped liquid is later vaporized if the precipitate is strongly ignited, but the dissolved ions remain within the pore as an impurity.

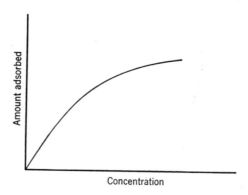

FIGURE 18–2. Effect of concentration on amount adsorbed.

The Operations of Precipitation Analyses

The techniques of the operations used in precipitation analyses have been described in Chapter 4. We are now in a position to consider the reasons why the various operations are done as they are, keeping in mind that the objective is to make the separations as complete as possible and to obtain the precipitate in as pure a condition as possible.

Dissolution of Sample. Most analyses start with a small sample of a solid material, which should represent the average composition of the gross lot from which it was taken. The first step in the analysis is to put this sample into solution. In the simple exercises of an elementary course the sample is usually dissolved (as received) in the appropriate solvent, e.g., water, acid, or base. In more advanced analyses, such as minerals or alloys, it is often found that some component of the sample is left as a residue when the bulk of the sample goes into solution in the selected solvent. Advantage may be taken of this behavior to isolate the insoluble component while dissolving the remainder of the sample. Thus, in the analysis of a mineral which contains a silicate along with carbonates and other salts, we find that SiO_2 is insoluble in the HCl solution that dissolves all the other ions present. Silica is determined by drying and weighing the residue from dissolution of the sample. Another example is the residue of SnO_2 obtained when a brass sample is dissolved in HNO_3. This too can be dried and weighed, to obtain the tin analysis for the sample.

Precipitation. The precipitants used in gravimetric procedures have been selected to give precipitates of low solubility which filter well and are reasonably pure. Even so, careful attention to detail is required to obtain a good analysis. There is a reason for every step in the procedure. The objective is to start the formation of the precipitate under conditions of maximum solubility, so as to obtain a relatively small number of nuclei, and then to complete the process under such conditions that the initial nuclei grow to large and well-formed crystals which are readily filterable. Conditions are arranged to minimize the concentrations of foreign ions that may be coprecipitated and thus contaminate the final product. Each analysis has its own special problems, and no set of conditions is universally applicable. We find, however, that the following general principles are to some degree applicable to most determinations.

1. Precipitation is usually made from hot solutions. Most crystals are appreciably more soluble at $100°C$ than at room temperature, and consequently the initial degree of supersaturation is less in the hot solution than it would be in a cold one. This leads to formation of the minimum number of initial nuclei. Further advantages are that heat aids in the coagulation of colloidal particles, that the amount of adsorption is decreased by increase in temperature, and that the speed of exchange of ions between solid and solution is increased. Since the solubility is usually increased with rise in temperature, it is necessary in most analyses to cool the solution to room temperature before filtration.

2. Dilute solutions are used, for both the sample and the added reagent. The advantages are (a) precipitation is initiated at a low degree of supersaturation, so as to obtain a relatively small number of initial nuclei, and (b) the concentrations of foreign ions that might be adsorbed by the surface are kept to a low value.

3. When the precipitate is a salt of a weak acid, and therefore soluble in a strong acid, it is advantageous to mix the reagents in an acid solution and then to bring about a slow precipitation by slow neutralization of the acid. This can be done either by adding ammonia solution or by the process known as homogeneous precipitation, as described on page 340.

4. Reprecipitation is often employed for precipitates that are readily

dissolved in dilute solutions of strong acids, e.g., hydroxides or salts of weak acids. The precipitate first formed is filtered and partially washed and then dissolved in a dilute acid solution. A second precipitation is obtained by neutralizing the acid. This process takes advantage of the relation between concentrations of ions in solution and their adsorption on the surface of a precipitate. The first precipitate may hold a relatively large amount of adsorbed impurities, which may be ions of high concentration in the mother liquor. These ions are present at much lower concentration in the solution resulting when the first precipitate is dissolved, since only the amounts previously adsorbed are now present. Consequently the amount adsorbed when the second precipitate is formed is much lower than for the first precipitation.

Reprecipitation is of course not applicable to many substances whose precipitates are not readily dissolved. It is widely used for oxalates, phosphates, carbonates, and hydroxides.

Digestion. After precipitation is completed, the precipitate is allowed to stand in contact with the mother liquor for periods of 1 to several hours, usually at a temperature near the boiling point of the liquid. This permits Ostwald ripening of the precipitate. Advantages are (1) increase in the particle size, (2) decrease in surface area and the amount of adsorbed material, and (3) improvement in filterability as the crystals become larger and more regular in form.

Filtration. The method of filtration used has no effect on the purity or properties of the precipitate. These properties, however, do influence the choice of method. Reducible precipitates, such as AgCl, cannot be filtered by paper, which must be burned away and would thereby cause reduction of the precipitate during ignition. Precipitates which have large and well-formed crystals can be filtered either by paper or a suitable porous crucible, with or without suction. Gelatinous precipitates must be filtered by paper, without suction. Suction causes trouble by pulling small particles into the pores of the paper and causing it to clog. An open-texture paper, with large pores, should be used for filtration of the gels.

Washing. The precipitate is washed for removal of adhering mother liquor and soluble surface impurities. The wash liquid is selected on the basis of the following.

1. The wash liquid should not leave a nonvolatile residue, i.e., should contain no salts that are not vaporized when the precipitate is dried.

It may contain HCl, HNO$_3$, or ammonium salts, but not such salts as NaCl.

2. If the solubility of the precipitate is appreciable, the wash liquid may contain a common ion which will minimize the amount of precipitate that goes into solution. This ion should be the ion of the precipitant, not of the substance sought in the analysis. For example, HCl could not be used in the wash liquid for a precipitate of AgCl in a chloride analysis; instead of washing out excess silver ion, it would cause further precipitation with it. However, HCl could be used in a silver analysis.

3. An electrolyte is added to the wash liquid used for precipitates of agglomerated colloids, to prevent peptization as the counter ions are removed.

4. Hot wash liquid is used whenever the solubility of the precipitate permits. The viscosity of water is decreased at elevated temperatures, and it flows through the filter more rapidly. Also, hot liquids dissolve soluble impurities more rapidly than cold.

5. Washing is done by addition of the liquid in small portions, each of which is thoroughly drained before the next is added. This is discussed in Chapter 4.

Drying and Ignition. The wet precipitate of an analysis must be thoroughly dried before it can be weighed as a pure substance. Water may be bound by precipitates in several ways: (1) loosely adhering, (2) adsorbed on the surface, (3) occluded within the granules as the particles are formed, and (4) as water of hydration, with a definite chemical formula. Loosely adhering water (excess wash liquid) is removed by drying at a temperature slightly above 100°C, or in some analyses by washing the precipitate with a volatile liquid such as acetone, which removes the water and is itself removed by air drying. Most precipitates we deal with need more drastic treatment. Drying at 110–150°C will remove adsorbed water from some, but others may require ignition at high temperatures. The hydrous oxides of iron and aluminum are examples of precipitates that hold water very tenaciously. After ignition, care must be used in weighing such precipitates, especially when the room air is at high relative humidity. Only rarely are precipitates weighed as definite hydrates, because it is difficult to insure that all loose water is removed without the risk of partially decomposing the hydrate. Usually hydrates are decomposed by ignition at elevated temperatures and the precipitate weighed in the anhydrous condition. There are some exceptions, however. An example is MgNH$_4$PO$_4\cdot$6H$_2$O, which can be dried by washing successively with alcohol and ether and then drawing air through the filter until the ether is removed.

When the precipitate as received has a definite composition that is stable at the temperature used to remove water, the only function of ignition is this removal of water. In many analyses ignition also decomposes the precipitate, by driving off volatile oxides such as CO_2. Decomposition can be either helpful or harmful. Care must be taken to avoid decomposition of AgCl or $BaSO_4$ (which at high temperature is slowly decomposed to BaO and SO_3), for a partial decomposition would invalidate a determination. On the other hand we seek decomposition of the precipitate obtained when calcium is determined by precipitation of $CaC_2O_4 \cdot H_2O$. If this precipitate is heated to remove adhering water, there is partial decomposition of the hydrate, leaving a residue of indefinite composition. The hydrate may be decomposed by ignition at higher temperature, leaving a residue of anhydrous calcium oxalate; but if the temperature is not carefully controlled, this compound is partially decomposed to carbonate. Usually the precipitate is ignited at a carefully controlled temperature of 500–600°C, which gives a residue of $CaCO_3$ or else is ignited above 900° to CaO.

In summary, then, for each specific analysis we employ drying or ignition conditions that assure the purest final product. Detailed directions are given with each procedure, which must be followed with care.

When filtration is done with paper, it must be burned away during the ignition, as described in Chapter 4. Care must be taken in this operation to prevent possible reduction of many precipitates by carbon from the paper or the carbon monoxide formed as the carbon burns. A frequent student error is partial reduction of a barium sulfate precipitate by the reaction

$$BaSO_4 + 4C = BaS + 4CO$$

This makes the result of the analysis low, since a mole of BaS weighs less than a mole of $BaSO_4$, the form in which we assume all the barium to be in our final computations.

Errors of Precipitation Analyses

In conclusion, we may summarize the likely sources of error in precipitation analyses. These should be studied carefully and kept in mind in every analysis. These errors may be grouped into two main classes: (1) incomplete separation of the sought ions from the sample solution and (2) failure to obtain a pure precipitate of definite composition for the final weighing. In general, little difficulty is encountered in the first class. The precipitates used in analyses are usually so insoluble that, barring a careless failure to add sufficient reagent to precipitate all the sample, the amount left in the mother liquor is negligibly small.

Inability to obtain a completely pure precipitate is a problem in almost all precipitation analyses. Many factors may cause difficulties.

1. Coprecipitation of foreign ions.

2. Postprecipitation of sparingly soluble substances after the initial precipitate has been formed. An example is precipitation of magnesium oxalate on the surface of a calcium oxalate precipitate, under conditions at which magnesium oxalate alone would not precipitate. This postprecipitation is caused by the high concentration of oxalate ions on the surface, as illustrated in Fig. 18–1 for chloride ions on the surface of an AgCl precipitate.

3. Insufficient washing to remove soluble adsorbed salts. A test should always be made for completeness of washing; usually we test for chloride or sulfate ions, impurities that are readily identified.

4. Incomplete ignition leading to retention of water.

5. Overignition causing decomposition of some precipitates.

6. Reduction by carbon of the paper.

7. Incomplete combustion of the filter paper, leaving a residue of carbon which adds to the weight observed.

8. Adsorption of water or carbon dioxide by the ignited precipitate after ignition.

With respect to the errors arising from the purity of the precipitate, it can be noted that the following usually lead to high results: postprecipitation, insufficient washing, incomplete ignition, and errors in weighing due to adsorption of water or carbon dioxide. Reduction or decomposition during ignition most generally leads to low results.

Errors which arise from coprecipitation and wrong composition of precipitate must be considered for each case because they can lead to high or low results or even occasionally will not affect the results.

In the case of coprecipitation, we must consider two things: (1) what is coprecipitated and (2) to what does the coprecipitated material ignite? Let us assume that in a sulfate analysis barium nitrate is coprecipitated with the $BaSO_4$. Actually the barium nitrate should not have precipitated at all, and, hence, it is extra material which ignites to BaO and is weighed along with $BaSO_4$, and the error arising from coprecipitation leads to high results. If, however, in a barium analysis, barium nitrate were coprecipitated, the results are different. In this case the barium ion combined with nitrate ion should be combined with sulfate ion. In the ignited product the BaO weighs much less than the $BaSO_4$ which it should be, and errors from coprecipitation lead to low results.

In an iron analysis $Fe_2(SO_4)_3$ may be coprecipitated with the $Fe(OH)_3$. On ignition both of these substances are converted to Fe_2O_3, and, hence, no error arises from this coprecipitation.

Finally, before leaving the subject, may we say that, despite the potential sources of error in precipitation analyses, such analyses are capable of giving very accurate results. Most of our accepted atomic weights are based on gravimetric analyses, but analyses in which all possible precautions were taken to avoid the numerous types of errors we have discussed. As for the work of this course, if instructions are carefully followed, good results will be obtained.

Questions

1. Differentiate between a solution of sodium sulfide and a colloidal dispersion of arsenic trisulfide with respect to:

(a) Size of the dispersed particles.

(b) Charge on the particles.

(c) Filterability of the dispersed material.

(d) Effect of the addition of NH_4Cl.

2. Discuss the effects of the following on the amount of adsorption by a precipitate.

(a) Total surface area.

(b) Temperature.

(c) Concentration of the adsorbed ion.

(d) Solubility of the salt corresponding to the ions adsorbed primarily and secondarily.

(e) Valence of the secondarily adsorbed ion.

3. Hydrous ferric oxide is crystalline in internal arrangement. On what evidence is this statement based?

4. What is the origin of the electric charge found on the surface of colloidal particles?

5. If barium sulfate is precipitated by addition of barium chloride solution to a sulfuric acid solution, what charge would you expect to find on the particles while they are in the colloidal condition? What counter ion might be effective in neutralizing this charge? What substance would you expect to find dragged down with the precipitate because of the adsorption of the counter ion on coagulation of the sol?

6. Explain why silver chloride sol is coagulated just before the equivalence point is reached in the addition of silver nitrate to sodium chloride.

7. How may the addition of an electrolyte to the wash water prevent peptization of a precipitate?

8. Could dichlorofluorescein (Chapter 17) serve as indicator for the titration of silver nitrate by sodium chloride? Explain, on the basis of the discussion given in this chapter.

9. A sample of Na_2SO_4 has appreciable amounts of NaCl and $NaNO_3$ impurities. In the analysis for the sulfate content by precipitation as $BaSO_4$, which of the following are advisable to prevent contamination of the precipitate by $Ba(NO_3)_2$?

  (a) Reprecipitation.

  (b) Several evaporations of sample solution to dryness with HCl before precipitation.

  (c) Addition of the unknown to $BaCl_2$ rather than the reverse.

10. Explain briefly how each of the following tends to minimize coprecipitation:

  (a) Digestion or aging.

  (b) Reprecipitation.

11. Give two reasons why most precipitations are conducted in hot dilute solution.

12. Give two reasons why a volatile electrolyte with an ion common to the precipitate is used in the wash water.

13. Draw a picture which represents the various adsorption layers on a colloidal particle of $BaSO_4$ which is formed by adding H_2SO_4 to a solution of $BaCl_2$.

14. Explain why slow addition of reagent will give a more filterable precipitate than rapid addition. Why is it not important to add reagent slowly in the precipitation of ferric hydroxide?

15. Why are many precipitations made by mixing the reagents in an acid solution and then slowly neutralizing the acid?

16. What is the purpose of reprecipitation? Under what conditions can reprecipitation be employed?

17. List the routine methods for minimization of coprecipitation.

18. What is the pupose of ignition of a precipitate? What factors govern the choice of ignition temperature?

Problems

1. How many spherical colloidal oil particles 20 A in radius can be made from a spherical oil drop whose radius is 2 microns? The density of the oil is 0.9 g per milliliter. What is the percentage increase in surface?

Ans. 10^9 particles; 105% increase.

2. If the molecular weight of the oil in problem 1 is 150, how many molecules are contained in the particle of 20 A radius?

3. 0.5000 g Fe^{+3} is precipitated as hydrated ferric oxide. During ignition 90 per cent of the iron is converted to Fe_2O_3, and the remainder is present as Fe_3O_4. What does the ignited precipitate weigh? What should it have weighed if all the iron were in its proper form?

CHAPTER 19

Typical Precipitation Analyses

The determinations of this chapter have been selected as representative of the more common types of precipitates encountered in gravimetric analyses; they also illustrate the most widely used methods of filtration and ignition.

Chlorides

Chlorides are precipitated as the slightly soluble silver chloride, AgCl, by treating a dilute solution of the sample with a solution of silver nitrate. The precipitate has low solubility, is easily obtained in a condition nearly free of impurities, is readily filtered and washed, and does not require ignition. For these reasons it is frequently chosen as the first exercise in gravimetric analysis.

Properties of Silver Chloride Precipitates

Silver chloride belongs to the class of curdy or agglomerated precipitates. When the reagents are first mixed, a colloidal suspension is obtained which coagulates readily, leaving a clear supernatant liquid. The coagulated mass forms in large lumps which are readily separated by filtration. The precipitate is not greatly contaminated by foreign substances, especially when it is formed by the addition of silver nitrate to chloride solutions. Silver chloride is readily peptized on washing by pure water, but the addition of a little nitric acid to the wash water prevents this. The solubility is about 2 mg per liter at 25°C and 21 mg per liter at 100°C in pure water, but it is much less in the presence of excess silver nitrate because of the common-ion effect.

Silver chloride cannot be ignited in the presence of organic matter because at elevated temperatures it is readily reduced. Furthermore, a high temperature cannot be employed in ignition because the melting point is 460°C; on heating above this point there is considerable loss by volatilization. Therefore it is customary to filter through a filtering crucible and to dry at 110–130°C. In the most precise work the precipitate is fused by carefully controlled heating in an electric oven to remove the last traces of moisture.

An outstanding characteristic of silver halide precipitates is their sensitivity to light. Direct sunlight, or even diffuse light, causes a darkening of the precipitate, from decomposition into the elements. The halogen formed by the decomposition escapes, leaving silver dispersed within the precipitate. In the analysis of silver this decomposition leads to low results, but in the analysis of chlorine the result is high if decomposition occurs before filtration because the liberated chlorine reacts with silver nitrate solution to form additional precipitate. Decomposition after filtration causes relatively little error, since only the comparatively light chlorine atom escapes and the heavy silver atom remains.

Analysis of Soluble Chloride

Procedure. Wash three filtering crucibles, and bring to constant weight by drying at 110–130°C. Use either fritted-glass or porcelain crucibles.

The unknown substance is a mixture of soluble chlorides, which contains no interfering ions. Dry the sample at 100–120°C, and weigh three portions of 0.3–0.6 g (note 1). Transfer each portion to a 250-ml beaker, and dissolve in 100–150 ml water to which 1–2 ml nitric acid has been added (note 2). Calculate the weight of silver nitrate necessary for precipitation, assuming that the sample is sodium chloride. Prepare a solution containing approximately 20 mg silver nitrate per milliliter, and add the calculated amount of this solution, plus an excess of about 10 per cent, to each portion of sample. Add the silver nitrate dropwise from a pipet, keeping the solution well stirred during the addition. After addition is completed, heat the solution nearly to boiling, and keep at this temperature until the precipitate has coagulated. Allow the precipitate to settle, and then test for complete precipitation by adding a few drops of silver nitrate solution to the supernatant liquid. Continue the addition of silver nitrate, in small portions, until precipitation is complete. Cover the beaker with a watch glass, and allow it to stand in the desk (note 3) for 1–2 hours before filtration.

Use suction filtration, as described in Chapter 4. Carefully pour the clear supernatant liquid through the crucible, retaining as much of the solid in the beaker as possible. Again test for complete precipitation by adding a few drops of silver nitrate solution to the filtrate. Wash with 0.01 N nitric acid as follows. Add 20–25 ml wash solution to the precipitate, stir well, allow the particles to settle, and decant the supernatant liquid through the filter. Repeat the washing by decantation three or four times; then transfer the precipitate to the crucible. Use a stirring rod equipped with a policeman to rub all particles of precipitate from the walls of the beaker, and wash the particles into the crucible. Fill the crucible half full of wash solution, and allow it to drain completely. Repeat the process two or three times; then remove the funnel from the filter flask, insert a test tube in the flask as shown in Fig. 4–10, page 73, and collect a 5-ml sample of wash solution. Test this for completeness of washing by adding a little hydrochloric acid. If no appreciable cloudiness forms, washing may be assumed to be complete. If it is not complete, continue the washing until a satisfactory test is obtained. Remove the crucible from the funnel, place it in a small beaker (properly labeled), and dry in an oven at 110–130°C for 2–4 hours. Cool in a desiccator, and weigh; then redry, and reweigh. Repeat until a weight constant within 0.3 mg is obtained. The second period of drying usually gives a weight that agrees with the first one. From the weight of silver chloride obtained, calculate the weight of chlorine and the percentage in the sample.

Notes. 1. The weight of the final precipitate should be 0.4–0.5 g. Consult instructor.

2. Nitric acid is added to the solution as an aid in coagulating the precipitate and in order to prevent precipitation of silver hydroxide, carbonate, etc.

3. As protection from light.

4. Add a few drops of silver nitrate solution to a nitric acid solution similar to that employed for solution of the sample. If any cloudiness appears, obtain chloride-free reagents.

Errors. The chief sources of error are the moisture retained on ignition and the decomposition of the precipitate under the action of light. Adsorption of foreign ions is not a source of great error, nor are the losses due to solubility. Interfering substances are the following:

1. Anions whose silver salts are insoluble in nitric acid, such as bromide, iodide, thiocyanate, sulfide.

2. Heavy metals whose chlorides are insoluble, in particular, mercurous and lead ions.

3. Substances that form complex ions with silver. If the solution is properly acidified, these substances cause no difficulty.

Other Applications. The procedure may be employed with little modification for other determinations. The most widely used of these are:

1. *Bromine and Iodine.* Because these ions are readily oxidized, it is important to use less nitric acid than in the chloride precipitation and to wash the precipitate nearly free of nitric acid before drying.

2. *The Oxyhalogen Acids and Their Salts.* These may be determined by reducing them to the halogen ion and precipitating with silver. The reduction may be made by treating a nitric acid solution of the sample with chloride-free sodium nitrite.

3. *Organic Halogen Compounds.* These may be analyzed if the halogen is first converted to the ion by fusion with sodium peroxide in a bomb or by heating with fuming nitric acid in a closed tube (Carius method).

4. *Silver.* Silver ion is determined gravimetrically by precipitating it with hydrochloric acid. A large excess of acid must be avoided, since silver chloride is appreciably soluble in hydrochloric acid because $AgCl_2{}^-$ complex ions form.

Gravimetric Standardization of Hydrochloric Acid

Gravimetric standardization of hydrochloric acid, by precipitation of silver chloride, is a convenient and accurate method which has the advantage of being independent of the purity of any primary standard. The disadvantage of the method lies in the fact that the determination does not give the amount of hydrochloric acid but the amount of chloride ion. If the acid is contaminated by ammonium chloride, this leads to error.

Procedure. Measure out accurately, from a calibrated pipet, a 50-ml portion of the acid. Add a drop of methyl red indicator, and neutralize with $0.1 N$ sodium hydroxide solution (note). Add 1 ml nitric acid to the neutral solution, and determine the amount of chloride present according to the previous procedure. From the weight of precipitate obtained, calculate the number of milliequivalents of chloride ion in the solution and the normality of the solution.

Note. Test the sodium hydroxide for chloride by acidifying a 5-ml portion with nitric acid and adding silver nitrate. Heat. If a perceptible precipitate appears, the sodium hydroxide is unsuitable for use.

Iron

Iron or steel samples are seldom analyzed by a gravimetric method because of the accuracy and convenience of volumetric and instrumental methods. The gravimetric methods illustrated in this section are nevertheless very important, because in mineral analyses the separation of iron (and related substances) is always made by precipitation. These methods are also valuable in student training because they illustrate the technique of analysis with gelatinous precipitates. In the following section directions are given for iron analyses only, but the operations employed for filtration, washing, ignition, and so on, are applicable to all the other substances (such as Al^{+3}, Cr^{+3}, Ti^{+4}, and Zr^{+4}) that precipitate as hydrous oxides on the addition of ammonia.

When iron, in the ferric condition, is treated with ammonia solution, it is quantitatively precipitated as hydrous ferric oxide according to the reaction:

$$2Fe^{+3} + 6NH_3 + xH_2O = Fe_2O_3{\cdot}yH_2O + 6NH_4{}^+$$

The precipitate is of indefinite composition because of the water which is more or less tightly bound with it. Such precipitates are known as *hydrous oxides,* rather than hydroxides, although the two terms are commonly used interchangeably, even though the precipitate cannot be represented accurately by the formula $Fe(OH)_3$. On ignition the precipitate yields ferric oxide, in which form it is weighed.

Properties of Hydrous Ferric Oxide

Hydrous ferric oxide is a very insoluble substance. Precipitation is quantitative, even from slightly acid solutions. The precipitate first forms as a dispersed phase, but on heating in the presence of electrolytes it coagulates to a gelatinous mass which settles out of suspension. Prolonged boiling tends to break up the aggregates and cause the precipitate to become slimy. If the precipitate is well coagulated, it may be readily filtered and washed, provided a porous paper is employed and washing is done by decantation. The use of suction is not recommended, for suction drags particles of the gel into the pores of the filtering medium, clogging it so that no liquid can be drawn through. The solubility of the hydrous oxide is so small that precipitation and washing can be carried out with hot solutions without any danger of loss. Electrolytes, usually ammonium salts, are added to the wash liquid to prevent dispersion of the coagulated gel.

As might be expected from their colloidal character, the hydrous oxides have a great tendency to adsorb other ions present. If precipitation is made from basic solution, the primarily adsorbed ion is the hydroxide ion, and this readily holds by secondary adsorption positive ions which may be present. If there is a large excess of ammonium ion in the precipitating and wash solutions, the adsorption of other positive ions can be kept at a minimum, and, since ammonium salts volatilize on ignition of the precipitate, little harm is caused by the adsorption. As might be expected from the general nature of adsorption, divalent ions are more strongly adsorbed than monovalent ones. Since the precipitates are readily dissolved by addition of acids, it is customary to employ reprecipitation in order to minimize the amount of coprecipitation.

The ignition of the hydrous oxides requires rather high temperatures in order to drive off water. Ferric oxide is made completely anhydrous by heating at 1000°C, but most of the water is expelled by heating at lower temperatures. Aluminum oxide requires still higher temperatures. If the precipitate contains only iron, the final ignition may be made in a porcelain crucible over a Meker burner, but, if the precipitate is one of combined oxides, such as obtained in mineral analysis, it is advisable to ignite for a short time with a blast lamp. Iron oxide is readily reduced to the magnetic oxide Fe_3O_4 by carbon or reducing gases from the filter paper, and it is necessary to permit free access of air while burning off the paper. Should reduction occur, the product may be reoxidized by prolonged ignition, or by careful addition of a little concentrated nitric acid, evaporation, and reignition.

Analysis of Iron Solution

The samples used as student exercises for gravimetric iron analyses are usually (1) solutions of weighed iron samples, or (2) analyzed samples of ferrous ammonium sulfate. The former has the advantage that each student can be given an individual sample of accurately known weight of iron, with added constituents that make a double precipitation essential.

Procedure. Before beginning the analysis, clean three porcelain crucibles (properly identified) and bring to constant weight by ignition at the highest temperature of a Meker burner (or in the muffle furnace if one is provided). The second heating should give a weight in agreement with the first one.

Obtain the sample of weighed iron wire (note 1) in a 250-ml volumetric flask. Dilute to volume, mix well, and measure out three 50-ml aliquot portions from a pipet which has been calibrated with reference to the flask. Deliver the samples into 400- to 600-ml beakers. Do not begin the analysis unless a 2- to 3-hour period is available for work, since the filtration once started cannot be interrupted (note 2).

★ Heat the sample solution to boiling, and add 1–2 ml concentrated nitric acid, a drop at a time, from a pipet (note 3). Continue boiling for a minute in order to expel oxides of nitrogen. Dilute the solution to 200–300 ml, heat to boiling, and slowly add, from a pipet, filtered $6N$ ammonia (note 4), stirring constantly and heating during the addition. After a precipitate begins to form, add the base in small portions until the odor of ammonia persists in the steam above the boiling solution. Continue to boil for about a minute after precipitation is complete, then remove the flame, allow the precipitate to settle, and test for complete precipitation by adding a few more drops of ammonia. Note the color of the coagulated precipitate (note 5). Decant (note 6) the clear supernatant liquid through an open-texture filter paper (Whatman no. 41 or an equivalent grade), taking care to retain the bulk of the precipitate in the beaker. Test the filtrate for completeness of precipitation by adding a little ammonia solution. Add to the beaker holding the bulk of the precipitate 25–30 ml ammonium nitrate wash solution (note 7), heat, and decant the hot solution through the filter paper. Wash twice in this manner.

Carefully remove the filter paper from the funnel, taking care that no precipitate is lost, and place it in the beaker containing the bulk of precipitate. Pour in about 30 ml $3N$ hydrochloric acid, and warm gently. Break up the filter paper with a stirring rod. All the precipitate should dissolve readily in the hot acid solution. After solution is complete, dilute to 100 ml and precipitate by addition of ammonia solution, as before. Filter through an open-texture ashless paper, wash twice by decantation, and then transfer all the precipitate to the filter. Wipe the beaker with a small piece of dry ashless filter paper to remove the last traces of precipitate, and add the filter paper to the precipitate in the funnel. Wash the precipitate in the funnel with hot ammonium nitrate solution until a 5-ml portion of wash water gives only a faint cloudiness when acidified and treated with silver nitrate solution (note 8). If only a faint cloudiness appears, washing may be assumed to be complete. Allow the filter paper to drain thoroughly, fold over the edges, transfer to a previously ignited and weighed porcelain crucible, and ignite according to the directions in Chapter 4. Complete the ignition by heating for 30 minutes at the highest temperature of the Meker burner (note 9). Cool the crucible in a desiccator for 20–25 minutes and weigh. Reignite for 20-minute periods until the weight is constant within 0.3 mg. Compute the weight of iron received.

Notes. 1. The sample is 0.7–1.0 g iron wire dissolved in 10–20 ml hydrochloric acid. Usually some inert impurity, such as sodium sulfate, is added to the flask, to make the analytical conditions comparable to those when ferrous sulfate samples are used.

2. The gelatinous precipitate obtained must not be allowed to dry on the paper before washing is completed. On drying, the precipitate breaks up into lumps, and in this condition it cannot be washed.

3. Ferrous ion is oxidized according to the reaction,

$$3Fe^{+2} + NO_3^- + 4H^+ = 3Fe^{+3} + NO + 2H_2O$$

The solution darkens at first, owing to formation of the complex, $FeSO_4 \cdot NO$. As oxidation proceeds, the color changes to the bright yellow of the ferric ion. Bromine water may be substituted for nitric acid if desired. Add 10–15 ml saturated bromine water and boil until all bromine is expelled.

4. Filtered ammonia must be used in order to prevent the introduction of silica, which is often present as a suspension in alkaline solutions. If the ammonia solution is clear, filtration is not needed.

5. The precipitate should be reddish brown in color. If a dark-green precipitate is obtained, indicating ferrous hydroxide, dissolve the precipitate in hydrochloric acid, reoxidize with nitric acid, and reprecipitate. The precipitation of ferrous hydroxide is not complete in the presence of ammonium salts.

6. In the filtration of gelatinous precipitates it is important to use every precaution to insure a rapid flow. Read the directions of Chapter 4. Use an open-texture filter paper. Make certain that the paper fits the funnel tightly. Filter while hot.

7. The wash solution contains 8–10 g ammonium nitrate per liter. Prepare by dissolving solid ammonium nitrate or by adding ammonia to a solution of 10 ml nitric acid per liter of water until it is neutral to litmus paper.

8. Chloride ion in small concentration is not harmful; its presence is used here as a convenient test for completeness of washing.

9. A blast lamp should be used for ignition if the usual burners do not give a flame sufficiently hot to heat the bottom of the crucible to bright redness. The crucibles must be brought to constant weight at the same temperature as that used for ignition of the precipitate. If a blast lamp is used, the precipitate need be ignited for only 10- to 15-minute periods.

Errors. The interfering substances are the anions which yield precipitates in basic solution with ferric ion, such as phosphate and arsenate, and all the metals except those of the alkaline earth and alkali group. If barium, calcium, and strontium are present, distilled ammonia must be used to avoid the presence of carbonate, which would precipitate these metals as carbonates. Sufficient ammonium salts must be present to prevent the precipitation of magnesium if this is present. The errors due to incomplete oxidation and to drying of the precipitate have been mentioned. Incomplete washing leads to high results. Reduction is very likely to occur during ignition of the precipitate. Reduction during ignition and incomplete ignition are probably the most frequent errors of beginning students.

Other Applications. Aluminum, trivalent chromium, zirconium, titanium, and trivalent manganese are precipitated as hydrous oxides by an excess of ammonia. The more amphoteric ones, aluminum and chromium, are completely precipitated only in solutions of carefully controlled pH, since, if much excess ammonia is added, the solubility is increased. (See determination of mixed oxides in limestone analysis.) Precipitation of manganese is complete only if it is first oxidized by treatment with bromine or persulfate.

Analysis of Ferrous Ammonium Sulfate

The samples are analyzed mixtures of ferrous ammonium sulfate with inert salts.

Procedure. Obtain the sample in a stoppered weighing bottle. Do not dry. Weigh out into 400–600-ml beakers three samples of 0.8–1 g each. Moisten with 5 ml concentrated hydrochloric acid, add 50 ml water, and heat until the sample is dissolved. Follow the procedure on page 363 for the iron determination, beginning with the star.

Sulfur

One of the most widely used (and at the same time least exact) gravimetric analyses is the determination of sulfur or barium ion by precipitation as barium sulfate. The ions react quantitatively according to the equation,

$$Ba^{+2} + SO_4{}^{-2} = BaSO_4$$

giving a precipitate whose solubility is only 3–4 mg per liter in pure water. Unfortunately, however, the precipitate is generally quite impure, and the accuracy of the analysis is never very great.

Properties of Barium Sulfate Precipitates

Barium sulfate crystals are usually so very finely divided that the analyst must use care in order to obtain a filterable product. Precipitation from concentrated and cold solutions gives such a fine-grained product that it passes through any obtainable filter paper, but, if precipitation is carried out from hot dilute solutions and a period of digestion is allowed, a coarser precipitate can be obtained which is readily separated by filtration.

The solubility of barium sulfate is slight and is decreased by excess of barium or sulfate ion. It is increased, however, by the presence of hydronium ion. This is due to repression of the concentration of sulfate ion by formation of the bisulfate ion $HSO_4{}^-$. In the presence of $1\,N$ hydrochloric acid the solubility is about thirty times greater than in pure water. Nevertheless, it is customary to carry out precipitation in weakly acid solutions $(0.01–0.05\,N)$, because the precipitates so obtained not only are less contaminated than those from neutral solutions but also consist of larger crystals and are therefore more easily filtered. The presence of acid also prevents the precipitation of barium carbonate, barium phosphate, and other barium salts that are insoluble in neutral solution.

The coprecipitation of other salts with barium sulfate has been widely studied. Almost every constituent of the solution may be dragged down, even though the precipitate is formed under the most favorable conditions, that is, by the addition of barium chloride to sulfuric acid solutions. The chlorides and sulfates of the univalent cations are the least adsorbed foreign substances. Nitrate and chlorate ion, if present, are somewhat adsorbed, as barium salts. The bivalent metals whose sulfates are soluble cause little more difficulty than the univalent metals, but the presence of any substances that have slightly soluble sulfates must be avoided. The only trivalent metals likely to be present are iron, aluminum, and chromium. Of these, iron is coprecipitated to a very large extent, usually as a basic sulfate, and chromium and aluminum are coprecipitated to a smaller extent. Chromium, however, should be avoided because it may form a complex with sulfate ion.

Reprecipitation cannot be employed to obtain purer precipitates of barium sulfate, because no solvent is available in which the precipitate can be dissolved. Coprecipitation of certain substances is best prevented by removal of the offending ion, either by precipitation or by chemical reaction. Thus ferric ion, which is readily coprecipitated, may be reduced to ferrous ion, or it may be removed by precipitation as the hydroxide before precipitation of barium sulfate. Since not all substances can be removed (for example, the alkali and ammonium ions), their coprecipitation may only be minimized by proper conditions of formation and digestion of the precipitate.

Coprecipitation may be such as to make a result either high or low. In the determination of sulfate the adsorption of barium chloride leads to a high result, but the adsorption of sulfuric acid (either as the free acid or as barium acid sulfate) leads to a low result because sulfuric acid is volatilized during ignition. For the same reason the adsorption of ferric sulfate gives low results. Adsorption of barium chloride leads to low results in the determination of barium, but all foreign sulfates cause high results. Why?

The ignition of barium sulfate presents some difficulties. If filter paper is used, it must be carefully burned away, with plentiful access of air, since the sulfate may be reduced by carbon

$$BaSO_4 + 4C = BaS + 4CO$$

A filtering crucible may advantageously be used for this determination, thereby eliminating any reduction of the precipitate. The precipitate must be ignited at a temperature of 700–800°C to remove water.

Determination of Sulfur in a Soluble Sulfate

The samples ordinarily employed are mixtures of alkali sulfates and chlorides. They are free of the substances which are most extensively coprecipitated.

Procedure. Weigh out three 0.5-g portions (note 1) of the dried sample (note 2), and transfer to 400–600-ml beakers. Dissolve in a little water, add 1 ml hydrochloride acid, and dilute to 200–300 ml. Heat to boiling, and, with constant stirring and continued heating, add barium chloride solution (note 3) dropwise from a buret which is mounted in a slanting position with the tip above the precipitation beaker. After 15–20 ml have been added, interrupt the process, allow the precipitate to settle, and test for completeness of precipitation by adding a few more drops of barium chloride. If a further precipitate forms, add another 5 ml of reagent and then test again for completeness of precipitation. Continue in this manner until an excess of reagent is present. Cover the beaker, and set it on the steam bath for an hour (note 4). The precipitate should be coarse enough to settle readily after stirring, and the supernatant liquid should be clear.

Complete the determination by one of the following procedures. If an electric muffle furnace is available, a filtering crucible is recommended to avoid the danger of reducing the precipitate.

Filter Paper Method. Select an ashless filter paper according to the manufacturer's recommendations for barium sulfate precipitates (Whatman 42 or equivalent). Decant the hot supernatant liquid through the filter paper, and test the filtrate for complete precipitation. If no more precipitate forms, reject the filtrate (note 5) and replace the beaker beneath the funnel. Transfer the precipitate to the filter and wash with hot water until a test for chloride ion shows that little is present. After washing is complete, fold down the edges of the filter paper, and transfer the moist precipitate to a previously ignited and weighed porcelain crucible. Carefully burn off the filter paper, and ignite at the highest temperature of the Meker burner for 15–20 minutes after the paper is completely removed. Cool in a desiccator, and weigh. Repeat, with 15-minute ignitions, until the weight is constant within 0.5 mg. Compute the percentage of sulfur as SO_3 in the sample.

Filtering Crucible. A porous-porcelain crucible is used. Sintered glass is not satisfactory because of the ignition temperature required.

Prepare the crucible and bring to constant weight by ignition in a muffle furnace at 700–800°C. Filter and wash in the usual manner, and dry in an oven at 110°C. Ignite at 700–800° for 30-minute periods until constant weight is attained.

Report the percentage of sulfur as SO_3.

Notes. 1. The amount of sample is chosen to yield 0.5–1.0 g of precipitate.

2. If hydrated sulfate samples are employed as student unknowns, they should not be dried before weighing.

3. Prepare an approximately 5 per cent solution by dissolving 5 g $BaCl_2 \cdot 2H_2O$ in 100 ml of water.

4. One hour is a minimum period of digestion. A longer time is not harmful.

5. The bulk of the filtrate is removed before washing in order that only a small volume need be refiltered, should any precipitate go through the paper. Carefully examine the washings for suspended matter. If any is found, refilter through the same paper.

Errors. The determination of sulfur is subject to the following errors.

1. Loss by solubility because of strongly acid solution.

2. Impurity of precipitate. See preliminary discussion.

3. Reduction during ignition. This is a very frequent error in student analyses if filter paper is used.

4. Interfering substances (see preliminary discussion).

5. Loss of solid during filtration. The fineness of barium sulfate crystals often causes some particles to pass through the filter paper.

Other Applications. The method is applicable for the analysis of barium and for all sulfur compounds which can be quantitatively oxidized to sulfates. Organic sulfur compounds may be oxidized by fusion with sodium peroxide in a Parr bomb, or by treatment with fuming nitric acid in a sealed tube. Lead and strontium may be determined gravimetrically as the sulfates if the solubility of the precipitate is diminished by the addition of alcohol to the solution. The gravimetric method is not recommended for standardization of sulfuric acid solutions because of the errors due to coprecipitation.

Phosphorus, Magnesium

This method can be used for determination of either phosphorus or magnesium. It is based on precipitation of the complex salt, magnesium ammonium phosphate hexahydrate, $MgNH_4PO_4 \cdot 6H_2O$, from an ammoniacal solution. The precipitate may be dried at room temperature and weighed as the hexahydrate, or, by ignition it may be converted to the pyrophosphate $Mg_2P_2O_7$ and weighed in this form.

Properties of the Precipitate

Precipitation is carried out by slow neutralization of an acid solution which contains magnesium, ammonium, and acid phosphate ions. While acid is in excess, no precipitate is formed, but on neutralization the concentration of the phosphate ion is increased, and a supersaturated solution is obtained from which the hexahydrate slowly precipitates. After addition of ammonia, the solution must be allowed to stand for several hours before filtration, in order that precipitation may be complete.

The equilibria involved in the formation of the precipitate are complex. Application of the solubility product principle gives the expression,

$$[Mg^{+2}][NH_4^+][PO_4^{-3}] = K_{sp}$$

Since the third ionization of phosphoric acid,

$$HPO_4^{-2} + H_2O = PO_4^{-3} + H_3O^+$$

is very slight, the concentration of phosphate ion in an aqueous solution depends on the concentration of hydronium ion. The solubility of the precipitate is therefore highly dependent on the pH of the solution. In pure water at room temperature it is quite soluble. Addition of ammonium chloride increases the solubility rather than decreases it (as we might expect from the K_{sp} relation) because ammonium chloride gives by hydrolysis an acid solution which represses the concentration of phosphate ion. In ammonia solution, however, the concentration of phosphate ion is increased, and the precipitate is less soluble. Precipitation is therefore carried out in an ammoniacal solution. Some ammonium ion must be present, both for formation of the precipitate and to prevent precipitation of magnesium hydroxide (since ammonium ion decreases the hydroxide ion concentration).

It is difficult to obtain a precipitate of the exact composition represented by the formula $MgNH_4PO_4$. Unless the concentrations of the various ions and the pH of the solution are held within narrow limits, the precipitate will be contaminated by phosphates of other composition, such as $MgHPO_4$ and $Mg_3(PO_4)_2$. The best results are found if precipitation is carried out by slow neutralization of an acid solution, but even that will not insure a precipitate of the correct composition and it is usually necessary to use a double precipitation. The precipitate first obtained, of uncertain composition, is dissolved in acid, and the final precipitation is made from this solution under carefully determined conditions.

The solution must be free of all metal ions other than those of magnesium, the alkalies, and ammonium. Salts of the alkali metals are adsorbed to a marked extent in the first precipitation, but a double precipitation will give accurate values except in the presence of potassium, which tends to form mixed crystals of the composition $Mg(NH_4,K)PO_4$. If potassium is present, three or more precipitations may be needed. In the determination of phosphorus from samples containing metal ions other than the alkalies, a preliminary separation is made by precipitation with ammonium molybdate. In magnesium determinations other metals are first removed by precipitation, just as in the procedures of qualitative analysis.

The hexahydrate is rather stable at room temperature, and Worsham[1] has suggested drying by alcohol and ether and weighing in this form. The method has been tested in student analyses by the present authors. Although drying at room temperature is not in general an advisable procedure, because there is no criterion about completeness, in this determination it is often advantageously employed because of the difficulties attendant on ignition of the precipitate. When this method

[1] W. A. Worsham, Jr., dissertation, Columbia University, 1923.

is used, the precipitate must not at any time be warmed, since at temperatures of 40–60°C the hexahydrate loses water and forms the monohydrate.

Hoffman and Lundell[2] have studied the ignition of the precipitate and recommend a temperature of 1100°C. At 1000°C constant weight is attained only slowly, and at 1200°C the pyrophosphate slowly decomposes. Student results accurate enough for most determinations may be obtained by ignition in a filtering crucible at 900–1000°C. Ignition with filter paper has the serious disadvantage that the precipitate is easily reduced by the paper. Correct results can be obtained only by very slow charring of the filter paper and subsequent low-temperature heating, with plentiful access of air until the carbon of the paper is entirely consumed.

Analysis of Soluble Phosphate

Dry the sample and weigh out three portions of 0.5–0.6 g each, into 400-ml beakers. Dissolve in 15–20 ml 6 N HCl and dilute to about 125 ml. Place the three beakers in a dish of ice and add to each a few drops of methyl red indicator and 10 ml magnesia mixture (note 1) for each 100 mg of P_2O_5 present (ask instructor for approximate percentage of P_2O_5 in the sample and use this value to compute the weight present).

Procedure. Precipitation. Slowly add concentrated ammonia solution (filter first if the solution has any suspended matter) until the indicator shows a color change. Stir constantly while adding the ammonia, but avoid striking the side of the beaker with the rod (note 2). This causes scratched places. When a precipitate starts to form, add 5 ml more ammonia and allow the solution to stand in ice water for 4 hours (note 3), replenishing the ice from time to time.

Reprecipitate as follows. Filter the cold solution through a retentive filter paper, keeping the precipitate in the beaker; then wash once by decantation with 25 ml 1 N ammonia which has been chilled. Discard the filtrate and washings. Place beneath the funnel the beaker which contains the precipitate, and dissolve the precipitate by slowly pouring 50 ml hot 1 N hydrochloric acid through the filter paper, taking care to wet all portions of the paper repeatedly. Wash the paper with a little hot water, catching all washings in the beaker which contains the sample. Discard the paper. Dilute the solution

---

[2] J. I. Hoffman and G. E. F. Lundell, *J. Research, Nat. Bur. Standards,* **5,** 279 (1930).

of the redissolved sample to 100 ml, add 2 ml magnesia mixture, and again precipitate as above. After standing, filter and ignite according to one of the following methods. From the weight of precipitate calculate the weight of phosphoric anhydride (P_2O_5) in the sample and its percentage.

A. As Hexahydrate. Prepare three filtering crucibles, but instead of drying in an oven pass through the crucibles two 15-ml portions of 95 per cent alcohol followed by two 15-ml portions of ether. Continue to draw air through the crucible 5 minutes after the last portion of ether is added, then remove the crucible, dry the outside with a clean cambric towel, and let it stand in the balance case for 20 minutes before weighing. After weighing, repeat the washing and reweigh. The second weight should agree within 0.4 mg. Place the weighed crucible in its holder, apply suction, and decant the well-cooled supernatant liquid from the precipitate. Wash once by decantation with cold 1 N ammonia, then transfer the precipitate to the crucible, and complete the washing, sucking the precipitate dry after each addition of liquid. The total volume of wash liquid need not exceed 50–75 ml. When washing is complete, as shown by a test for chlorides, dry the precipitate by passing alcohol and ether through the filter, as in the preparation of the crucible. Every detail in drying and weighing the precipitate should be exactly like the same operation in the preparation of the crucible in order to cancel errors. Weigh as the hexahydrate, $MgNH_4PO_4·6H_2O$. A second weighing is unnecessary.

B. Ignition in Filtering Crucible. Prepare three filtering crucibles, and ignite to constant weight at 900–1000°C in a muffle furnace. Filter the sample, and wash with ammonia, as directed previously; then dry the precipitate in an oven and ignite for 30-minute periods until constant weight is obtained. The final form of the precipitate is as the pyrophosphate $Mg_2P_2O_7$.

C. Filter Paper. Decant the solution above the sample through a small ashless paper, and wash the residue twice with cold 1 N ammonia by decantation; then transfer the precipitate to the paper, and continue washing until a test shows absence of chlorides. Use as little wash liquid as possible. Suction may be used advantageously for filtration. See page 71. Allow the paper to drain thoroughly, transfer it to a previously ignited and weighed porcelain crucible, and very slowly dry and char the filter paper, leaving the crucible uncovered. When no more smoke escapes, set the crucible in a slanting position, and oxidize the carbon by a very small flame. When the carbon is oxidized, complete the ignition as usual, heating with a blast lamp for 30-minute periods until a constant weight is obtained.

Notes. 1. Preparation: To 50 g $MgCl_2 \cdot 6H_2O$, dissolved in 500 ml of water, add 100 g NH_4Cl and a slight excess of NH_4OH. Allow to stand overnight, and filter if cloudy. Make slightly acid with HCl, and dilute to 1000 ml.

2. Crystals adhere very firmly to any scratched places on the wall of the beaker.

3. Standing overnight at room temperature is also satisfactory.

Errors. The most important ones in this determination are those mentioned in the introductory section. They are:

1. Loss by solubility. If directions are followed this is negligible.

2. Incorrect composition of precipitate. If a double precipitation is carefully carried out, a precipitate of correct composition may be obtained.

3. Ignition errors. See previous discussion.

4. Interfering substances. All common metal ions except those of magnesium and the alkalies must be absent. Arsenates if present are precipitated as magnesium ammonium arsenate.

Determination of Magnesium

After a solution containing magnesium ions, ammonium ion, and HPO_4^{-2} ions in proper proportions is prepared, the procedure for determination of magnesium is exactly the same as the one for phosphorus.

Procedure. Weigh out three 0.5-g portions of the dried sample, into 400-ml beakers. Add 15–20 ml of water and cover the beaker with a watch glass; then add slowly 5 ml HCl. Warm if necessary to dissolve the sample. Rinse off and remove the watch glass, and rinse down the sides of the beaker. Dilute the solution to about 125 ml. Dissolve 2 g $(NH_4)_2HPO_4$ in a little water and add to the sample. Add a few drops of methyl red indicator to each solution. Place the three beakers in a bath of ice water, and complete the determination by adding ammonia solution, reprecipitating, filtering, and igniting as directed in the procedure of page 372. Report the percentage of MgO in the sample.

Determination of Phosphoric Anhydride in a Mineral

The determination of phosphorus in a mineral is a much more complicated procedure than the simple analysis given above. The chief mineral sources of phosphorus are calcium phosphate rocks, in which calcium phosphates of varied composition are associated with large amounts of fluorine and smaller amounts of silicates, carbonates, aluminum, and magnesium. In nitric acid solution phosphates can be separated from the interfering constituents ordinarily present by precipitation as ammonium phosphomolybdate. This precipitate, which is approximately represented by the formula,

$$(NH_4)_3PO_4 \cdot 12MoO_3 \cdot H_2O$$

is not definite enough in composition to be used for a final weighing in exact analyses, but is dissolved in ammonia and the phosphorus is precipitated as magnesium ammonium phosphate, as in the analysis above.

Because of the colloidal nature of ammonium phosphomolybdate, precipitation is best carried out in warm solutions. If, however, the temperature rises much above 60°C, the nitric acid solution of molybdic acid employed as reagent decomposes, and the precipitate is contaminated by molybdic anhydride, MoO_3. Large amounts of ammonium nitrate are necessary in formation of the precipitate, and a large excess of molybdic acid reagent is needed. The precipitate is readily peptized if washed with water but remains as a gel in the presence of dilute ammonium nitrate solutions.

A number of interfering substances may cause difficulty in the precipitation of ammonium phosphomolybdate. Arsenic forms a complex molybdate and must be absent. Large amounts of sulfates, chlorides, fluorides, and organic substances retard formation of the precipitate, but these are not usually present in phosphate rocks in amounts sufficient to cause trouble. Silica must be removed before precipitation of phosphorus. Titanium, zirconium, and tin may lead to loss of phosphorus, by formation of insoluble phosphates which remain in the silica precipitate. Iron and tin may cause loss of phosphorus at the time of dissolving the phosphomolybdate precipitate in ammonia, but intereference from this source may be eliminated by adding to the ammonia solution a small amount of ammonium citrate. This, by formation of complex ions with the metal ion, causes the solution of ferric or tin phosphates.

In analyses which do not require highest accuracy the precipitate of ammonium phosphomolybdate may be dissolved in a measured volume of standard sodium hydroxide solution and the excess of base titrated with standard acid. The reaction is:

$$(NH_4)_3PO_4 \cdot 12MoO_3 + 23NaOH$$
$$= 11Na_2MoO_4 + (NH_4)_2MoO_4 + NaNH_4HPO_4 + 11H_2O$$

Procedure. Weigh out duplicate samples of 0.3–0.4 g of the well-ground and dried mineral. Transfer to a small beaker, add 15 ml 6 N nitric acid, cover with a watch glass, and heat until the sample has dissolved (note 1), replacing acid if any evaporates. After solution of the sample, set the watch glass on glass hooks which hang on the rim of the beaker, and evaporate to dryness on the steam bath. Heat in an oven for 1 hour at 110–120°C (note 2), digest with 25 ml 6 N nitric acid to dissolve the soluble portions, filter off the silica (catching the filtrate in a 250-ml beaker), and wash with a little hot water. Test for completeness of washing by neutralizing a 5-ml portion of wash liquid with ammonia. If calcium and phosphate ions are still present, a gelatinous precipitate of calcium phosphate will form. Should an appreciable precipitate form, add the tested portion to the solution and continue washing.

To the filtrate and washings, which should not exceed 100 ml in volume, add 6 N ammonia until a slight permanent precipitate is formed; then make slightly acid by addition of nitric acid until the precipitate dissolves. Avoid a large excess. Heat the solution to 60°C, stirring with a thermometer; then lower the flame so as to keep it at just this temperature, and add with stirring 75 ml of warmed molybdate reagent (note 3). Digest at this temperature for half an hour, with frequent stirring, and then allow the solution to stand 4 hours at room temperature. Filter and wash several times by decantation with ammonium nitrate (note 4) solution. Set the beaker in which the bulk of precipitate remains beneath the funnel used for filtration, and dissolve the precipitate by pouring through the filter 20 ml 5 N ammonia solution. Wash the filter with several portions of hot 0.5 N ammonia and then with hot dilute hydrochloric acid solution, catching all washings in the beaker holding the solution of the sample. The total volume should not exceed 100 ml. Add methyl red indicator to the solution, and carefully neutralize with 6 N hydrochloric acid. If this operation is not done with care, a precipitate of phosphomolybdate may come down. Should this occur, add ammonia until the precipitate is redissolved, and again neutralize. Add to the barely acid solution 15 ml of magnesia reagent, and precipitate magnesium ammonium phosphate according to directions on page 371. A double precipitation is essential because of the presence of molybdate. Weigh the precipitate according to one of the methods described, and determine the percentage of P_2O_5 in the mineral.

Notes. 1. The reaction is

$$Ca_3(PO_4)_2 + 4H^+ = 3Ca^{+2} + 2H_2PO_4^-$$

2. For dehydration of silica. See Chapter 20 for discussion.

3. Prepare the reagent as follows: Add 120 g molybdic acid (85 per cent MoO_3) to a solution of 80 ml concentrated ammonia in 40 ml of water, and stir until all has dissolved. Filter the cloudy solution, and pour the filtrate slowly and with constant agitation into a cold mixture of 400 ml concentrated nitric acid and 600 ml of water. Filter before use, since nearly always there is a precipitate of molybdic anhydride.

4. 5 g per 100 ml.

Organic Precipitants—Dimethylglyoxime

There are available many organic substances which serve as precipitants for metals not only by forming ionic compounds but also by forming coordination complexes. Some reagents are like hydrogen sulfide and serve to precipitate groups of cations. Others are quite specific such as dimethylglyoxime for nickel.

Properies of Nickel-Dimethylglyoxime Precipitate

The red color of $NiC_8H_{14}N_4O_4$ is familiar as a confirmatory test for nickel in qualitative analysis. The precipitate is extremely bulky and the amount of solid resulting from the precipitation of more than 0.1 g nickel is too large to be handled conveniently. The equation for the precipitation is:

$$Ni^{+2} + 2C_4H_8N_2O_2 \rightarrow NiC_8H_{14}N_4O_4 + 2H^+$$

Using structural formulas we have:

The precipitate is quite insoluble in dilute ammonia when ammonium salts and excess of the reagent are present. It is soluble in dilute acid and in solutions containing more than 50 per cent alcohol by volume. Palladium, gold, and ferrous iron interfere by precipitating with the reagent. Iron must be oxidized to the ferric state. Precipitation of its hydrous oxide along with those of other trivalent ions is prevented by the formation of complexes with tartaric acid. Sufficient ammonium salts must be present to buffer the ammonia solution and thus prevent the precipitation of hydroxides of bivalent ions such as zinc, magnesium, and others.

The precipitate is made more filterable by starting precipitation in hot, somewhat acid, solution and then making the mixture basic with ammonia.

Analysis of Nickel Oxide

Procedure. Bring filtering crucibles to constant weight by washing with water and drying at 120–150°C. Accurately weigh into 400-ml beakers samples of sufficient size to furnish 20–30 mg nickel (note 1). Add 20 ml nitric acid, warm gently, and finally boil until all reddish-brown fumes have ceased and the volume is reduced to approximately 5 ml (note 2). Cool, add 20 ml HCl, and evaporate to dryness. Repeat with a 10-ml portion of HCl. Bake at 110° for 1 hour. Cool, add 10 ml HCl, warm gently, and add 100 ml of hot water. Digest until all particles have been dissolved.

Add 20 ml of a solution of 10 g NH_4Cl per 100 ml of water and 10 ml of a solution of 20 g tartaric acid per 100 ml of water. Dilute to 200 ml and neutralize by slow addition of 8 M NH_3 (note 3). Just acidify with HCl and add five drops in excess.

The precipitating reagent is a solution of 1 g dimethylglyoxime in 100 ml 95 per cent alcohol. Warm the solution containing the nickel to around 70°C and add from a buret enough dimethylglyoxime reagent so that 0.5–0.7 ml is used for each milligram of nickel present (note 4). While stirring the mixture, finally add 8 M NH_3 to neutrality plus two or three drops in excess. Digest on a steam bath for 1 hour. Cool and filter. Keep the top quarter inch of the crucible surface dry since the precipitate has a tendency to creep over wet surfaces and some may be lost. Be sure to test the first portion of the filtrate for completeness of precipitation.

Wash with water at room temperature until free of chloride. Do not let the solid dry out until washing is complete. Dry for at least 1 hour in an oven at 120–150°C and weigh as $NiC_8H_{14}N_4O_4$ which contains 20.31 per cent nickel.

Notes. 1. The bulkiness of the precipitate limits the size of sample.

2. If all black particles are not disintegrated, treat with more HNO_3 and re-evaporate.

3. If a precipitate forms at this point, dissolve it in HCl and add more tartaric acid.

4. About 4 mg dimethylglyoxime is the stoichiometric quantity for the precipitation of a milligram of nickel. If too great an excess of reagent is added, the dimethylglyoxime may be thrown out of solution by the dilution of the alcoholic solution with water. Such an occurrence can be recognized by the much lighter color of the dimethylglyoxime itself.

Questions

A. Chloride

1. Why is filtering crucible used for filtration of AgCl?

2. Explain the following, and predict the effects on the analysis if the steps were omitted:
   (a) The addition of HNO_3 to the unknown solution.
   (b) The excess of $AgNO_3$ solution.
   (e) The presence of HNO_3 in the wash water.

3. List two metals and five anions which can be determined by slight modifications of the methods of this section.

4. What do you think of the determination of lead by precipitation as lead chloride?

5. An unknown contains a mixture of sodium carbonate and sodium chloride. Describe a method by which each constituent could be determined.

6. Why is *hot* 0.01 N HNO_3 solution not used to wash the AgCl?

7. How is completeness of washing tested?

8. What substances might contaminate a precipitate of silver chloride? (Consider the adsorption conditions at the time of precipitation.) If this method were used for determination of silver, what might be adsorbed?

9. Which of the following ions will interfere with a Cl^- determination: Hg^{+2}, Hg_2^{+2}, Co^{+2}, Br^-, F^-, SO_4^{-2}, NO_3^-?

10. During the determination of silver as AgCl, the following substances were coprecipitated. What effect do they have on the result if the ignition process consists of drying at 110° in the oven?
    (a) NaCl.
    (b) $AgNO_3$.

11. Why can washing be effectively done with a smaller volume of wash liquid when a filtering crucible is used than when filter paper is used?

12. Outline a method for the analysis of a sample containing both chloride and chlorate.

13. How might the chlorine content of chloroform be determined?

14. How would you determine the bromine content of a photographic emulsion?

15. If a solution of hydrochloric acid is standardized gravimetrically, the value found for the concentration may be higher than that found by volumetric standardization. Suggest possible reasons for this.

16. List three methods for standardizing a solution of hydrochloric acid.

B. Iron

1. State the reason for using an electrolyte in the wash solution. Could NH_4Cl be used? HCl? ($FeCl_3$ is volatile at ignition temperatures.)

2. Explain briefly:
   (a) Why is reprecipitation employed?
   (b) Why is the precipitate washed with ammonium nitrate solution?
   (c) Why is paper chosen as the filtering medium? Why is the use of macerated filter paper recommended?
   (d) What other ions (if present) will interfere with the determination?

3. What effect on the observed percentage of iron will each of the following errors have? How can these errors be eliminated?
   (a) Incomplete oxidation to Fe^{+3}.
   (b) SiO_2 in the ammonia.
   (c) Coprecipitation of $Fe(OH)SO_4$.
   (d) Reduction to Fe_3O_4 during the ignition.

4. Could $K_2Cr_2O_7$ be used as the agent to oxidize the iron to Fe^{+3}?

5. If Mg^{+2} is present in the solution from which Fe and Al are to be precipitated as combined hydroxides, a fairly high concentration of ammonium chloride must be present. Why?

6. Why is the gravimetric method seldom used for the determination of iron in ores? Why, in rock analyses, are iron and aluminum precipitated gravimetrically?

7. Which would you expect to be most highly coprecipitated with hydrous ferric oxide, provided the two were present in equal concentration, Ba^{+2} or K^+? Why?

8. Why is reprecipitation employed for iron but not for chlorine?

9. Show, by equations, that if sulfur, as a basic ferric sulfate, is coprecipitated with ferric hydroxide the result of the analysis is not affected.

10. What error is introduced if an unwashed ferric hydroxide precipitate is allowed to dry in the filter before washing is completed?

C. Sulfur

1. What effect would the coprecipitation of each of the following have: $Ba(NO_3)_2$, Na_2SO_4, $Fe_2(SO_4)_3$?
   (a) On a sulfur determination in which the precipitate is collected in a filtering crucible and dried in oven?
   (b) On a barium determination in which the precipitate is collected on a filter paper and ignited to red heat?

2. How would each of the following affect the results of the analysis for sulfur?
   (a) Use of hot wash water.
   (b) Precipitation in a strongly acid solution.

(c) Coprecipitation of ferric chloride (the precipitate is ignited to red heat).

(d) Reduction to BaS during the ignition.

3. Which of the following have to be absent in the determination of Ba^{+2} as $BaSO_4$: Mg^{+2}, Pb^{+2}, Na^+, NO_3^-, Cl^-, PO_4^{-3}?

4. Why is the solubility of barium sulfate increased by the presence of acids?

5. How would you analyze a sample of $BaSO_4$?

6. Outline a procedure for the analysis of PbS.

7. List the various ways of minimizing or avoiding contamination of precipitates. Which methods have been applied in the procedures of this chapter?

D. Phosphorus, Magnesium

1. Describe the various methods which may be used for ignition of a precipitate of magnesium ammonium phosphate.

2. Why must phosphorus in apatite be precipitated by molybdate reagent before it is precipitated as the magnesium salt?

3. List the most probable errors. How does each affect the result?

4. What is the effect on the result of an analysis for phosphorus if part of it precipitates as $Mg_3(PO_4)_2$ which ignites without change?

Problems

1. What is the percentage purity of a sample of $KClO_3$ which weighed 0.7440 g and yielded a precipitate of AgCl which weighed 0.8600 g?

Ans. 98.83%.

2. How many milliliters HNO_3 (density 1.42, 69.8 per cent HNO_3 by weight) are needed to oxidize the iron in 50 ml of a solution which contains 10 mg ferrous sulfate per milliliter:

(a) On the basis that the reduction product of HNO_3 is NO and that enough acid must be added to furnish both the necessary H^+ and NO_3^-?

(b) On the basis that NO is formed but that the solution is acidic and only enough HNO_3 must be added to furnish the necessary NO_3^-?

3. A sample of Mohr's salt [$FeSO_4 \cdot (NH_4)_2SO_4 \cdot 6H_2O$] and impurities which are inert to both NH_3 and $BaCl_2$ yielded a precipitate of $BaSO_4$ which weighs 0.2500 g.

(a) What weight of Fe_2O_3 could be obtained from the sample?

(b) What is the percentage of Mohr's salt present if the sample weighed 0.5000 g?

4. What volume of 0.12 M $BaCl_2$ is required to precipitate the $BaSO_4$ after the oxidation of a sample of 0.1 g of ore which is approximately 80 per cent FeS_2?

Ans. 11.1 ml.

5. What weight of pyrite (FeS_2) must be taken for analysis so that each milligram of precipitated $BaSO_4$ shall represent 0.1 per cent sulfur in the sample?   *Ans.* 137.4 mg.

6. Cobalt is sometimes precipitated with α-nitroso-β-naphthol as $Co[C_{10}H_6O(NO)]_3$ and ignited in a stream of O_2 to Co_3O_4. At other times the precipitate is ignited in H_2 and weighed as Co. What weight of cobalt would have been obtained from the same weight of sample which produced 0.2125 g C_3O_4?   *Ans.* 0.1560 g.

7. A sample of $CaCO_3$ and $FeCO_3$ contains equal portions of each by weight. Strong ignition produces CaO and Fe_2O_3. What is the weight of a 1.000-g sample after thorough ignition?   *Ans.* 0.6247 g.

8. A 0.7500-g sample of an alloy steel yielded on electrolysis 0.1532 g of a mixed deposit of Co and Ni. This is dissolved, and the precipitated nickel dimethylglyoxine ($NiC_8H_{14}N_4O_4$) weighs 0.3560 g. What are the percentages of the two metals in the alloy?   *Ans.* 9.64% Ni; 10.79% Co.

9. A student sample is prepared by mixing $Ca_3(PO_4)_2 \cdot CaCl_2$ with SiO_2. A 0.5000-g sample yields 0.1825 g $Mg_2P_2O_7$. What volume of an $(NH_4)_2C_2O_4$ solution of 40 mg per milliliter is needed to precipitate the calcium from a 1.000-g sample of the unknown?

10. What volume of NH_3 solution (density = 0.946 g/ml, 15% NH_3) is required to precipitate the iron as ferric hydroxide from a sample of $FeSO_4 \cdot (NH_4)_2SO_4 \cdot 6H_2O$ which in turn required 0.30 ml HNO_3 (density = 1.350 g/ml, 55% HNO_3) to oxidize it? Assume that the HNO_3 is reduced to NO and that the 0.30 ml furnished both the needed NO_3^- and H^+.
                                                    Ans. 0.955 ml.

11. What error in parts per thousand and in percentage of Cl present is made if 1.0 mg AgCl is mechanically lost in an analysis whereby a 0.3500-g sample produces a precipitate of AgCl which weighs 1.000 g?

12. I^- can be separated from other halides by precipitation as PdI_2 and weighed as such or reduced in a current of H_2 to Pd. A 0.5000-g sample of KI and NaCl and inert impurities gave a precipitate of mixed silver halides which weighed 0.8000 g. A 0.2000-g sample yielded 0.04250 g metallic palladium. Find the percentages of NaCl and KI in the sample.
                                        Ans. 66.10% KI; 27.18% NaCl.

13. A mixture of $KClO_4$ and $KClO_3$ can be analyzed as follows:
    (a) The mixture is ignited with NH_4Cl and AgCl is precipitated from the residual KCl from the $KClO_4$ and $KClO_3$.
    (b) The mixture is treated with SO_2 which reduces only the chlorate to chloride, and again the precipitated AgCl is weighed.
From the following data find the percentage of $KClO_4$ in a mixture:

Weight of sample	0.2610 g
Weight AgCl from step 1	0.1490 g
Weight AgCl from step 2	0.0745 g

CHAPTER 20

Analysis of Limestone

The complete analysis of a complex sample, involving the determination of several constituents, is a useful exercise in the training of analytical chemists. The operations involved in the determination of each constituent are the same as in the determination of that constituent in a simple mixture, but a complete analysis is more difficult because errors introduced at any stage of the procedure will affect all subsequent determinations made on the sample. The material almost universally chosen for the first complete analysis is a dolomitic limestone rock. This mineral is chosen both because limestone is commercially important and because this analysis is typical of all mineral analyses. The analysis is one of the easiest of the complex mineral analyses because the samples are usually brought into solution readily, whereas many other minerals require considerable skill for this operation.

General Discussion of Methods

Composition of Limestones. Hillebrand and Lundell[1] discuss mineral analysis under the general classifications of silicate and carbonate rocks. The silicate rocks contain large amounts of silicates and can be

[1] Hillebrand, Lundell, Bright, and Hoffman, *Applied Inorganic Analysis,* second edition, Wiley, 1953.

382

brought into solution only by fusion with sodium carbonate, whereas carbonate rocks are those containing relatively small amounts of silicates but large amounts of carbonates. The carbonates are, in general, more or less readily disintegrated by treatment with dilute acids. These classifications are not absolute, nor are the analytical methods peculiar to one class or the other, since after solution of a sample the analysis of all minerals is very similar. Limestone rocks contain large amounts of carbon dioxide as carbonates of calcium and magnesium, nevertheless, it is often necessary to fuse the sample in order to effect solution.

In the exact analysis of a limestone a determination is made for each constituent thought to be present; but, for many purposes, and in particular for the commercial analysis of limestones, a condensed or "proximate" analysis is made. In a proximate analysis the determinations are:

Loss on ignition.
Impure silica.
Combined oxides (Fe_2O_3, Al_2O_3, TiO_2, Mn_3O_4, and P_2O_5).
Calcium oxide.
Magnesium oxide.

All the constituents of the sample except sodium, potassium, chlorine, and sulfur are determined at some stage of the proximate analysis. The summation of the weights found in this analysis should therefore total the weight of sample used, except for the small amounts of the substances mentioned which escape detection. The term, proximate analysis, does not, therefore, indicate approximate accuracy but refers to the number of constituents measured. In the limestone analysis, as for most minerals, it is customary to express the percentages of the various susbtances as their oxides. The combined percentages for minerals which are salts of oxygen acids should therefore total nearly 100 per cent.

Loss on Ignition. When limestones are ignited at high temperatures, all the combined carbon dioxide and water are expelled and organic matter is oxidized. In addition, there are minor changes due to oxidation of sulfides and of ferrous salts, decomposition of sulfates, and volatilization of potassium chloride. Since the greater part of the loss on ignition is due to escape of carbon dioxide, the loss in weight is usually the only determination made for the carbon dioxide content of the sample.

The readiness with which carbon dioxide is expelled depends on the composition of the sample. Magnesium carbonate is decomposed at quite low temperatures. If a sample containing magnesium carbonate is rapidly heated at a high temperature, the gas may be evolved so vigorously as to carry away solid particles. Calcium carbonate requires somewhat higher temperatures for complete decomposition. If ignition is completed at a temperature above 900°C, all carbon dioxide is driven off.

In addition to expelling carbon dioxide, ignition plays an important role in the subsequent analysis. Many silicates which originally cannot be decomposed by acids are converted by ignition into calcium and magnesium salts by reaction with the oxides of these elements. These salts are then readily decomposed by treatment with dilute hydrochloric acid. The reactions which take place may be represented by the following equations (SiO_2 is used as illustration):

$$CaCO_3 = CaO + CO_2$$
$$SiO_2 + CaO = CaSiO_3$$

Solution of Sample and Determination of Silica. When decomposable silicates are treated with dilute acids, the reaction may be represented by the equation,

$$CaSiO_3 + 2H^+ + yH_2O = Ca^{+2} + SiO_2 \cdot (y + 1)H_2O$$

The hydrated silicon dioxide, or silicic acid, is in the form of a dispersed gel. When the solution is evaporated, some of the water is driven off from the gel, and insoluble silicon dioxide, still somewhat hydrated, remains as a residue. When this residue is treated with water or dilute acids, a portion of the gel again passes into a dispersed condition. The tendency toward dispersion may be decreased by *dehydrating* the silica after evaporation. Ordinarily dehydration consists in heating the residue for a time at a temperature of 100–140°C. Higher temperatures may be employed, giving still better dehydration, but their use is not advantageous because at elevated temperature there is a hydrolysis of hydrated iron salts to a basic substance not readily soluble in acids. More than one evaporation and dehydration, without an intervening filtration, does not decrease the amount of silica which passes into suspension. If, however, after the first evaporation and dehydration, the residue is extracted with acids (in order to dissolve soluble salts), the insoluble portion separated by filtration, and the solution again evaporated and dehydrated, most of the remaining silica is now recovered in an insoluble form. This is the procedure followed in analysis. In the insoluble form left after dehydration at 100–140°C, silicon dioxide still retains a large amount of water. This

highest temperature of the blast lamp (use silica triangles), allow to cool until no redness is visible, and transfer to a desiccator. Cool the crucibles for 15 minutes (note 1), and weigh accurately. Reignite for a 5-minute period, cool, and reweigh.

If the sample has gritty particles, which can be felt by rubbing a pinch of the powder between the fingers, grind thoroughly in an agate mortar (usually not required for student samples). Grind only a small portion at a time, then regrind in larger portions to assure complete mixing. Dry the well-ground sample overnight at 110–140°C, and store in a glass-stoppered weighing bottle which is fitted with a weighing spoon. Weigh out samples of 0.5–0.8 g (note 2), and transfer to crucibles which have previously been ignited and weighed. Cover the crucible and heat over a Tirrill burner, gradually increasing the size of the flame over a period of 5–10 minutes (note 3) until the bottom of the crucible is a dull red. Transfer to the blast lamp, and heat for 10 minutes at the highest temperature obtainable (note 4), directing the flame against the bottom and one side of the crucible, which is upright in its silica triangle. Cool, weigh, and reheat (note 5) for 10-minute periods until the weight is constant, or until the rate of change is approximately the rate at which the crucible loses weight. Compute the loss on ignition, (note 6), and calculate the percentage loss.

Notes. 1. Since many of the precipitates in the analysis are very hygroscopic, it is important to observe the following precautions:

(*a*) Always cool ignited crucibles for the same length of time, with the desiccator in a cool place. This length of time should be the minimum time required to bring the crucible to room temperature, since, after removal of the cover, the atmosphere in a desiccator is for some time practically the same humidity as the external atmosphere.

(*b*) Keep the crucible covered during the process of cooling and weighing.

(*c*) Make all weighings in as short a time as possible. On the second and subsequent weighings, place the proper weights on the balance pan before the crucible is removed from the desiccator. After the crucible is on the balance pan, only one or two sets of swings should be necessary to compute the exact weight.

2. The weight of sample used is governed by the amount of magnesium. Since this is often present in amounts as high as 20 per cent of MgO, it is not desirable to employ a sample larger than 0.5 g unless the approximate composition is known. In no case should the amount of magnesium greatly exceed 0.1 g of MgO. Use of small samples increases the relative error in the determination of the minor constituents (silicon and combined oxides) but nevertheless permits sufficient accuracy for these.

3. Heat should be gradually applied, to prevent too rapid decomposition of magnesium carbonate. If this occurs, there is danger of loss of sample; particles of magnesium oxide may be carried away by the escaping gas.

4. If this is the first time the blast lamp is used, ask an assistant to inspect the

adjustment. A properly adjusted lamp will bring a platinum crucible nearly to white heat, or at least to a bright-yellow color.

5. Two or three periods of heating should be sufficient to expel carbon dioxide completely. If constant weight is not obtained on the third heating, crush the sample, as directed for calcium oxide on page 394, and reignite until the weight is constant. It often happens that the sample sinters into a lump which encloses undecomposed particles, thereby preventing complete decomposition.

6. Attainment of constant weight sometimes requires such a long time that the student is in doubt about the final weight of the crucible after the prolonged ignition. It is advisable to reweigh the crucible after the sample has been removed, and to use this weight as the final weight of the crucible, in order to obtain the corrected weight of sample after ignition. After removal of the sample it is necessary to heat only to redness (for removal of surface moisture) before weighing the empty crucible.

Silica

Procedure.   Put about 10 ml of water (note 1) into a 400-ml beaker, and transfer the sample from the crucible to the beaker, taking care to prevent any loss. If any particles adhere to the crucible lid, brush them into the beaker. Remove adhering particles from the crucible by adding to it 1 ml hydrochloric acid (1 : 1), heating gently for a short time, and washing the solution into the beaker. Rinse the crucible thoroughly, while holding it over the beaker. Add to the beaker 15 ml hydrochloric acid, cover with a watch glass, and allow it to stand on the steam bath until the sample has completely disintegrated. The process may be hastened by breaking up the sintered mass of sample with a blunt stirring rod, which is left in the solution until the final filtration. When the sample has completely disintegrated, rinse off the watch glass cover, rinse down the sides of the beaker, and evaporate the solution to dryness on a steam bath. During the evaporation the beaker should be covered by a watch glass resting on bent glass hooks which hang on the rim of the beaker. When the residue is dry, allow it to cool, drench with 10 ml hydrochloric acid (note 2), and digest on the steam bath for a short time; then add 20 ml of water and bring the solution to a boil. Decant the solution through a small ashless filter paper. Add to the residue in the beaker 20-ml dilute hydrochloric acid (1 : 9) and boil for 5 minutes. Filter, this time transferring the residue to the paper. Wash with dilute hydrochloric acid (1 : 100) ten times. Fold the filter paper, and reserve for ignition.

Evaporate the filtrate and washings to dryness on the steam bath. When no fumes of acid can be detected above the hot beaker, heat the residue for 1 hour at 120–140°C on an asbestos pad on a hot plate (note 3). Cool, drench with hydrochloric acid, digest, dilute, and filter (note 4), as previously directed. Catch the filtrate and washings in a 400- to 600-ml beaker, and reserve for the following determination. Place both filter papers in a previously ignited and weighed platinum or porcelain crucible, and ignite, completing this process by 10 minutes' heating with a blast lamp. Allow a plentiful access of air while smoking off and burning the paper. Otherwise a carbide of silicon is formed. Weigh, and reignite with a blast lamp until constant weight is obtained. Report the percentage of impure silica in the sample.

Notes. 1. Too vigorous reaction may result if the ignited sample, which contains a large amount of calcium oxide, is added directly to an acid solution.

2. The sample is first drenched with concentrated acid since the salts of iron and aluminum present are more easily brought into solution in concentrated acid than in dilute acid.

3. The temperature used for baking the residue should not exceed 140°C, because at higher temperatures the oxides of iron present are rendered insoluble and are not redissolved during the digestion period. It is also claimed that, in the presence of magnesium, silica is rendered somewhat soluble by baking at elevated temperatures.

4. A fresh filter paper is used for the second filtration. If the original paper were used, some of the first residue might redissolve (disperse). A very small paper may be used for the second filtration, since the amount of silica obtained never exceeds a few milligrams.

Errors. 1. Incomplete disintegration of the sample.

2. Incomplete removal of iron and aluminum oxides from the silica. The error from this source is largely compensated by the amount of silica which passes into solution.

3. Introduction of silica from some reagent.

Combined Oxides

Procedure. Dilute the filtrate from the silica determination to 200 ml, and add sufficient hydrochloric acid to make its concentration 1–1.5 M (note 1), taking into account the amount of acid used in digestion and filtration after the second dehydration of silica. Add a few drops of methyl red indicator, heat the solution to boiling, and neutralize (note 2) with 6 N ammonia (note 2) freshly prepared from a new bottle, to avoid the presence of $(NH_4)_2CO_3$. Boil 1–3 minutes, allow the precipitate to settle, and decant the solution through a rapid ashless filter paper (Whatman no. 41). Wash by decantation three times with 20-ml portions of hot ammonium chloride solution (2 g per 100 ml). Bring each portion to a boil before filtering. Reserve the combined filtrate and washings for the next determination. Carefully remove the filter paper from the funnel, taking care that no precipitate is lost, and place it (note 3) in the beaker containing the bulk of precipitate. Add to the beaker 30 ml hot 3 N hydrochloric acid, and break up the paper with a stirring rod. When all the precipitate is dissolved, dilute to 100 ml, bring the solution to a boil, and precipitate as before. Filter through an ashless paper, catching the filtrate in the beaker which holds the first filtrate and taking care to keep most of the precipitate in the beaker. Wash twice by decantation with hot ammonium chloride solution, transfer the precipitate to the filter paper, and wash thoroughly with hot ammonium chloride solution. Make the combined filtrates and washings slightly acidic with hydrochloric acid, and place the solution on the steam bath for evaporation. Drain the paper, fold, transfer to a crucible, and ignite, finally bringing to constant weight by heating for 5–10-minute periods with a blast lamp. Report the percentage of mixed oxides, R_2O_3, in the sample.

Notes. 1. Sufficient acid should be present to yield on neutralization a concentration of ammonium salts high enough to prevent precipitation of magnesium hydroxide.

2. If the precipitate is so voluminous that the indicator end point cannot be seen, add ammonia until it can be detected by the odor in the steam above the solution; then boil until the odor of the ammonia is very faint. To make this test, blow away the steam before smelling the vapor. It is essential that the final concentration of ammonia be very small, to prevent loss of aluminum.

3. The precipitate may be dissolved by pouring acid through the filter paper, but, since the presence of macerated filter paper is advantageous in the final precipitation, the entire paper is added to the acid solution. The paper readily disintegrates to give a suspension of filter pulp.

Errors. 1. Salts of calcium and magnesium may be adsorbed in the precipitate of hydrous oxides. A double precipitation largely eliminates this error.

2. Calcium carbonate will precipitate if there is carbonate ion in the solution. Use only freshly prepared ammonia, and do not allow the alkaline solution to stand in the presence of air.

3. Precipitation of aluminum may be incomplete if much excess ammonia is used.

4. Magnesium hydroxide may precipitate if insufficient ammonium salts are present.

5. Any silica remaining in the solution is dragged down by the precipitate of hydrous oxides.

6. Incomplete ignition may cause the weight of precipitate to be too great. Aluminum hydroxide in particular requires a high temperature for complete dehydration.

7. Iron oxide may be reduced by filter paper during the ignition.

There should be a plentiful access of air during the early stages of ignition. Keep reducing gases from entering the crucible.

8. If phosphates are present in large amounts, calcium phosphate will precipitate from the ammoniacal solution.

Calcium

Evaporate the (acidified) combined filtrate and washings from the mixed oxide determination to 250 ml (note 1), add a few drops of methyl red indicator, bring the solution to a boil, and add 1–1.5 g oxalic acid or ammonium oxalate (dissolved in a little water). Add freshly prepared 3 N ammonia solution from a pipet, while continuing to boil the solution, until a precipitate of calcium oxalate begins to form. Complete the neutralization (note 2) over a period of 10 minutes by slowly adding ammonia solution from a buret mounted above the beaker. The beaker is placed on a ring stand so that it may be constantly heated. Mechanical stirring is advisable if the facilities are available. When the indicator has just turned color, allow the precipitate to settle; test for complete precipitation by adding a little ammonium oxalate solution. If no further precipitate forms, allow the solution to stand (note 3) for 1 hour. Filter, keeping the precipitate in the beaker as far as possible (note 4). Wash twice by decantation with ammonium oxalate solution (note 5). Acidify the filtrate and washings, and place on the steam bath for evaporation.

Place the beaker with the bulk of precipitate beneath the funnel, and dissolve the precipitate in 50 ml 2–3 N hot hydrochloride acid solution. Pour the acid solution through the paper, in order to dissolve any particles of precipitate which may be on the paper. Add the acid in small portions from a pipet, directing the stream about the upper edge of the paper. Wash the paper several times with hot water, catching all washings in the beaker in which the first precipitate has been dissolved. Discard the paper. Dilute the solution to 250 ml, add 0.5 g oxalic acid or ammonium oxalate, and again precipitate the calcium as before. Allow to stand 1 hour, and filter. *Add the filtrate and washings to those from the first precipitation.* The choice of filtering medium is governed by the method to be used for the final estimation. If platinum crucibles are available, filter with ashless paper (Whatman no. 40 or 42) and ignite in platinum to calcium oxide. If platinum is not available, filter in a filtering crucible and determine the calcium volumetrically according to the procedure given on page 282.

If the calcium is to be ignited to the oxide, filter with a suitable grade of quantitative paper, and wash thoroughly with ammonium oxalate solution (note 5), until a test for chlorides with a 5-ml portion gives only a slight cloudiness (note 6). Ignite in a platinum crucible. When the paper is completely oxidized, interrupt the heating and crush the precipitate with a blunt stirring rod. Carefully wipe particles of precipitate from the rod with a piece of ashless filter paper, place the paper in the crucible, and burn it. After crushing, ignite for 15 minutes with the blast lamp, using the highest temperature attainable. Cool and weigh; then reignite for 10-minute periods until constant weight is obtained. Three periods are generally required to assure the attainment of constant weight. Report the percentage of calcium oxide in the sample.

Notes. 1. The concentration of calcium should not exceed 1 mg CaO per milliliter.

2. Note that the precipitate is largely formed under conditions of greatest solubility, that is, in hot acid solution. Slow neutralization and efficient stirring are essential to give large crystals with minimum coprecipitation.

3. Prolonged standing is inadvisable since magnesium oxalate may in time precipitate.

4. If calcium is to be determined volumetrically, use a filtering crucible.

5. The solution contains 1–2 g per liter.

6. The sample of wash water should be acidified with HNO_3 before addition of $AgNO_3$ to test for chloride; otherwise a precipitate of $Ag_2C_2O_4$ may be obtained.

Errors. 1. Possible loss of calcium in preceding determinations.

2. Incomplete ignition. Unless the precipitate is broken up, it is difficult to obtain complete conversion to the oxide.

3. Readsorption of carbon dioxide and water during cooling and weighing. Keep the crucible covered, and make the weighings rapidly. The proper weights should be placed on the balance pan before the crucible is removed from the desiccator.

4. Coprecipitation of sodium or magnesium oxalates. Without a double precipitation this may lead to quite a large error.

Magnesium

Procedure. Add 50 ml nitric acid to the combined filtrate and washings from the calcium determination, and evaporate to dryness (note 1) on the steam bath. If a large residue is obtained, add nitric and hydrochloric acids, and again evaporate. Add to the residue 2 ml hydrochloric acid and 50 ml of water, and digest until all salts are in solution. A slight insoluble portion is probably silica. This will be removed in the first filtration. Dilute the solution to 150 ml, add 10 ml ammonium phosphate reagent (note 2) and a few drops of methyl red indicator, cool in a bath of ice water, and precipitate by slowly neutralizing with concentrated filtered ammonia. After the solution is neutral, continue to stir vigorously for a few minutes; then add 5 ml ammonia in excess, and allow the solution to stand 4 hours in ice water or overnight at room temperature. Filter the solution, retaining as much of the precipitate as possible in the beaker; wash twice by decantation with cold ammonia (1 : 20). Reject the filtrate and washings. Place the beaker with precipitate beneath the funnel, dissolve the precipitate by pouring through the filter paper 50 ml warm hydrochloric acid (1 : 9), and wash the paper with several small portions of hot water. Discard the paper. Dilute the solution to 150 ml, add 0.5 ml ammonium phosphate reagent, and again precipitate as before. Follow one of the methods on page 372 for the final filtration and weighing. Calculate the weight of magnesium oxide in the precipitate and the percentage in the sample.

Notes. 1. For destruction of ammonium salts and oxalic acid.

2. Prepare the reagent by dissolving 25 g diammonium hydrogen phosphate in 100 ml of water. Filter if the solution is not clear.

Errors. See discussion in analysis of phosphorus. The errors in the determination of magnesium are for the most part the same as those in the determination of phosphorus, except for loss of magnesium which may occur during preceding stages of the analysis. This loss is most likely to occur in the precipitation of calcium. Solubility losses may be important when dealing with such a soluble precipitate as that of magnesium ammonium phosphate, but, if precipitation and washing are properly done, the loss is negligible. It is imperative to use only a small quantity of wash water and to keep the temperature low. Some mechanical loss may occur through failure to remove particles of precipitate which cling to the walls of the beaker. Scratching of the wall by the stirring rod should be carefully avoided. Any calcium and manganese which may remain in solution will be precipitated at this point, the former as the phosphate, the latter as manganese ammonium phosphate.

Questions

1. Construct an outline for a scheme of qualitative analysis of a dolomitic limestone, considering only the major constituents.

2. If a limestone rock which contains SiO_2, CO_2, and metallic oxides of Fe, Al, Ca, Mg, and K is fused with sodium carbonate and the melt is extracted with water, what is the composition of the residue and of the solution?

3. What modification of the procedure given for limestone analysis would be needed for a silicate rock?

4. Suggest a method for the determination of ferrous iron, as FeO, in a rock analysis.

5. What does "loss on ignition" determine? Is it higher or lower than the CO_2 content of a limestone? Why does ignition aid in the subsequent acid decomposition of the sample?

6. Under what circumstances might an analyst report a much higher percentage of silica than is actually present in a silicate rock?

7. Should the constituents of apatite, $Ca_5F(PO_4)_3$, be reported as the oxides? Why?

8. An analyst made the following errors in the procedure for mixed oxides. In each case tell what might happen because of the faulty procedure, and what effect it would have on the results.

(a) Use of ammonia which contained $(NH_4)_2CO_3$.

(b) The presence of insufficient HCl in the solution before ammonia was added.

(c) Ignition of the precipitate under such conditions that there was an insufficient access to air.

9. If a mineral contains an amount of P_2O_5 considerably in excess of R_2O_3, could the separation of calcium and mixed oxides be accomplished by the methods of this chapter?

10. Why must a large excess of ammonia be avoided in precipitation of the mixed oxides?

11. An R_2O_3 precipitate is fused with $KHSO_4$ and the resulting mixture treated with water. If the precipitate is assumed to consist of Fe_2O_3, TiO_2, Al_2O_3, and SiO_2, what is the composition of the aqueous solution after fusion?

12. What are the difficulties involved in weighing calcium as the carbonate? As the sulfate? If platinum crucibles are not available, what method should be used for the determination of calcium?

13. Why must ammoniacal solutions never be allowed to stand in glass vessels for a long time during a mineral analysis?

14. Why is the precipitation of calcium oxalate started in acid solution? Why is the process of reprecipitation used? Why is a period of digestion longer than several hours harmful? Discuss three ways in which the precipitate may be treated to determine the percentage of CaO.

15. What would be the effect on the percentage of CaO reported of each of the following errors?

(a) Incomplete decomposition of a complex calcium silicate.

(b) Ignition of precipitated CaC_2O_4 at too low a temperature.

(c) Postprecipitation of MgC_2O_4.

(d) Coprecipitation of Ca^{+2} in the mixed oxides determination.

(e) Presence of CO_2 in the NH_3 used for mixed oxide determination.

(f) The R_2O_3 precipitate is allowed to stand overnight in equilibrium with its supernatant solution and the atmosphere.

16. What would be the effect on the reported percentage of MgO of each of the following errors?

(a) Too low a concentration of NH_4^+ during the R_2O_3 determination.

(b) Use of a hot ammonia solution for washing $MgNH_4PO_4$.

(c) Partial precipitation as $MgHPO_4$ instead of $MgNH_4PO_4 \cdot 6H_2O$, if the precipitate is weighed as the hydrate; if the precipitate is weighed after ignition to the pyrophosphate.

17. What should magnesium be precipitated as? Weighed as? Why are the precipitations carried out in cold solution? Why is reprecipitation used? What is the purpose of evaporating the solution to dryness several times with HNO_3 before starting the magnesium precipitations?

18. List three places in limestone analysis other than in the actual determination of magnesium where we might account for a loss of magnesium.

19. A dolomite contains Ti, Zr, Cr, Ce, P, V, and Mn. State the formula for the precipitate given by each of these. Where, in the proximate analysis, will the precipitate be formed?

Problems

1. A 1.000-g sample of limestone gave a precipitate of Fe_2O_3 and Al_2O_3, weighing 0.0800 g. The precipitate of mixed oxides was fused with $KHSO_4$, dissolved in sulfuric acid, and passed through a Jones reductor. The iron was titrated with 20.00 ml 0.0100 N permanganate solution. Calculate the percentage of FeO and of Al_2O_3 in the limestone (assume all iron present as FeO in the original sample).   *Ans.* 6.40% Al_2O_3; 1.44% FeO.

2. Calculate the percentage of CaO in a limestone from the following data:

Weight of sample	0.7500 g
Volume $KMnO_4$ to titrate acid solution of CaC_2O_4	15.00 ml

40.00 ml $KMnO_4$ is equivalent to 35.00 ml $Na_2S_2O_3$.
35.00 ml $Na_2S_2O_3$ is equivalent to 0.2300 g As_2O_3.

Ans. 6.521%.

3. From a 0.8125-g sample of limestone there is obtained an impure ignited SiO_2 which weighed 0.0837 g. The residue after treatment of the SiO_2 with HF and H_2SO_4 weighed 0.0045 g.

(a) What is the reported percentage of SiO_2 without purification?

(b) What is the reported percentage of SiO_2 after purification?

(c) On the basis that the total impurity in the SiO_2 was calcium, which was weighed as CaO with the SiO_2 and as $CaSO_4$ after the purification, what error is made by not taking into account the differences in weight between the CaO and $CaSO_4$?   *Ans.* (a) 10.30%.

4. Because of insufficient ignition a supposed CaO precipitate from a 1.000-g sample of limestone weighs 0.4125 and contains 5.30 per cent of its own weight

of $CaCO_3$. What is the reported percentage of CaO, and what is the true percentage of CaO?

5. 0.2000 g of an ignited precipitate of $Mg_2P_2O_7$ contains 10.00 per cent of its own weight of $Mg_3(PO_4)_2$. What should the ignited product have weighed had all the magnesium been properly precipitated as $MgNH_4PO_4 \cdot 6H_2O$? If it is assumed that all the magnesium is present in the mineral as $MgCO_3$, what is the correct percentage of this constituent in the sample which weighs 0.5000 g? *Ans.* 205.4 mg; 31.12%.

6. A 0.2564-g sample of a rock which contains only calcium and magnesium carbonates gives 0.1428 g of mixed oxides after ignition. What weight of rock will be needed to furnish 10.00 liters of carbon dioxide, measured at standard conditions?

7. If it is assumed that in a limestone all the calcium and magnesium are present as carbonates and that there are no other carbonates in the mineral, what is the percentage of CO_2 in a rock which analyzes 40.07 per cent CaO and 11.67 per cent MgO? *Ans.* 44.18%.

8. A sample is 20.00 per cent CaO and 25.00 per cent MgO. Could this sample be a mixture containing only $CaCO_3$ and $MgCO_3$?

9. A limestone analyzes as follows:

$CaCO_3$	91.7%
$MgCO_3$	3.15%
SiO_2	1.75%
R_2O_3	1.32%
Volatile (drying)	1.84%

What is the composition of the resultant lime produced by strong ignition? Assume only the loss of CO_2 and volatile matter on heating.

10. A rock sample which weighed 0.5200 g gave a 0.0416-g residue of sodium and potassium chlorides. From this residue a 0.0648-g precipitate of potassium chloroplatinate was obtained. Calculate the percentage of sodium oxide and potassium oxide in the sample. *Ans.* 2.21% Na_2O; 2.41% K_2O.

PART FIVE — LIGHT AND

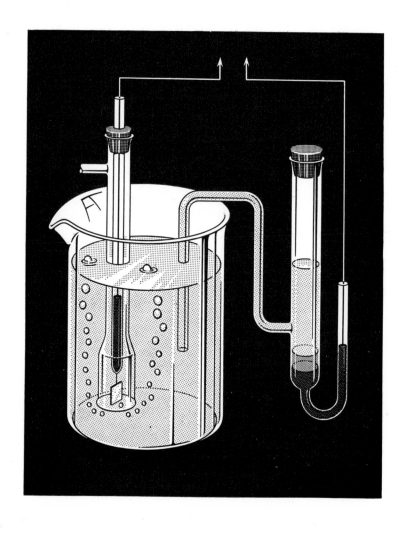

ELECTRICAL METHODS

OF ANALYSIS

CHAPTER 21

Colorimetric Analysis

The basic principle in colorimetry is that the amount of light absorbed by a given species in solution is proportional to the power[1] of the incident light and to the concentration or number of the absorbing species in the path of the beam. This relation leads to a simple mathematical expression that is widely used.

Consider a beam of monochromatic light of power P, incident on a thin layer of solution which contains an absorbing species at concentration c. The layer is of infinitesimal thickness, dx, so that the light is essentially of the same power throughout the layer. As the light passes through the layer, the decrease in power, $-dP$, is proportional to the incident power, to the thickness of the layer, and to the concentration, or

$$-dP \propto Pc\, dx$$

Introducing a proportionality factor gives the equation

$$(1) \qquad\qquad -dP = kPc\, dx$$

For a layer of finite thickness b, the power of the emergent beam is given by integration of equation (1) between the limits $x = 0$ and $x = b$.

---

[1] The "power" of a light beam is frequently called the "intensity." It is the energy per unit time. When a beam of light strikes the electrode of a phototube, the current produced is proportional to the power of the beam.

At $x = 0$, P is P_0, the power of the incident beam.   Then

$$-\int_{P_0}^{P} \frac{dP}{P} = kc \int_{x=0}^{x=b} dx$$

Integration gives

(2) $$\ln \frac{P}{P_0} = -kcb$$

or,

$$\ln \frac{P_0}{P} = kcb$$

Converting to $\log_{10}$, we have

$$\log_{10} \frac{P_0}{P} = 0.4343kcb$$

Combining the constants gives

(3) $$\log \frac{P_0}{P} = abc = A$$

where A is known as the *absorbance* of the sample, as measured.   Obviously A is a function of the wavelength, the concentration, and the length of the light path through the sample.

The constant a is specific for each substance, at each wavelength of light.   Its units depend on the concentration units employed.   When the concentration is expressed in moles per liter, the constant is known as the molar extinction coefficient and is usually denoted by the symbol ϵ (epsilon).

The relation of (3) is usually known as the Lambert-Beer law equation.   It is also employed in the logarithmic form

(4) $$P = P_0 10^{-abc}$$

In most colorimetric measurements we do not evaluate the constant a, but rather compare solutions of unknown concentration with solutions of known concentration, known as standards.   The comparisons may be made in various ways—by measuring the relative powers of emergent beams after they have passed through the solutions, by the concentrations used to give the same transmissions of the two solutions, or by the depths of layer required to give equal transmission.   Applications of all three methods are shown in the following discussions.

Visual Colorimetry

The power of a light beam cannot be measured visually, but we can match beams of the same power and the same color distribution, by using either Nessler tubes or a colorimeter.

Nessler Tubes. Nessler tubes are the simplest and oldest visual-colorimetry instruments. They are long glass tubes with flat ends, so arranged that light passes longitudinally. By comparing the unknown solution with standards of varying concentration, we find that concentration whose transmission appears identical or nearly so to that of the sample. Illumination and the depth of liquid for the tubes compared must be the same. The method is laborious, because of the number of standard solutions required in a comparison, and high precision cannot be achieved. Nevertheless, it has many useful applications.

Visual Colorimeter. This instrument (usually a Duboscq type, Fig. 21–1) measures the depth or thickness of sample and standard solutions to give the same transmission. Tubes containing the solutions to be compared are mounted side by side. White light is reflected from a mirror so as to pass through both tubes. Plungers of fixed length dip into the two solutions and the cups are raised or lowered until the light passing through the two solutions appears to be the same. The eyepiece is so arranged that we observe two half circles, one for each of the solutions. When the two halves appear identical, the depth of layer in the two tubes is read from graduated scales. The concentration of the unknown solution is computed from the known concentration of the standard and the lengths of the light paths to give equal transmission, as follows.

At equal transmission we have the relation

$$P_{\text{standard}} = P_{\text{unknown}}$$

From equation (4) we have

$$P_0 10^{-ab_1c_1} = P_0 10^{-ab_2c_2}$$

This gives the simple relation

$$b_1c_1 = b_2c_2$$

Since c_2 is known, that of the standard solution, we have

$$c_1 = \frac{b_2}{b_1} \cdot c_2$$

Substituting the measured values of c_2, b_1, and b_2 gives a value for the unknown concentration, c_1. Since the constant a is the same for the two solutions, it is not necessary to determine its value. Also we do not need to know the value of P_0.

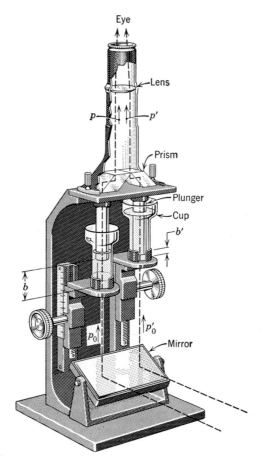

FIGURE 21–1. Light path in a Duboscq colorimeter.

Photoelectric Colorimetry and Spectrophotometry

The visual colorimeter is rapidly being displaced by photoelectric instruments which eliminate the need for subjective measurements by the human eye. There are two types of photoelectric instruments:

(1) filter photometers (Fig. 21–2) which use optical filters to obtain selected wavelengths of light for the measurements and (2) spectrophotometers which use prisms or gratings to give monochromatic light. Both types employ photoelectric cells to measure the light power in terms of a meter reading. Many of the spectrophotometers cover the ultraviolet region of the spectrum as well as the visible. This greatly extends their usefulness since many compounds that are colorless in

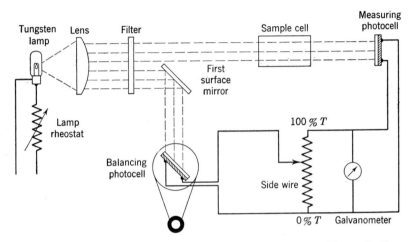

FIGURE 21–2. Photovolt Lumetron filter photometer. (Courtesy Photovolt Corporation.)

the visible region absorb strongly in the ultraviolet. Recently there have been developed infrared spectrophotometers for measurement of absorption of invisible heat rays. Both the ultraviolet and the infrared spectrophotometer are today indispensable in a modern chemistry laboratory, for analyses, for identification, and for proofs of structure of compounds.

All the photoelectric instruments measure the power of the light reaching the detecting cell. This measurement is not in absolute units but in the relative values of a meter reading. A measurement at a given wavelength requires three readings: sample cell filled (1) with the solvent, which gives a value for P_0, (2) with standard sample in the cell, and (3) with unknown in the cell. Since the meter readings are proportional to the light power reaching the detecting cell, the ratio of meter readings for the cell filled with pure solvent and for sample in the cell is the P_0/P ratio. The log of this ratio is the absorbance of the sample. We see from (3) that the absorbance is a linear function of concentration, for a given substance and a given cell. Also used in photometry is the term transmittance, defined as the ratio P/P_0 and measured by the ratio of the meter reading for sample to that for the cell filled with pure solvent. When this ratio is multiplied by 100, it is known as the per cent transmittance. If the meter is adjusted to give a reading of 100 for P_0, the reading

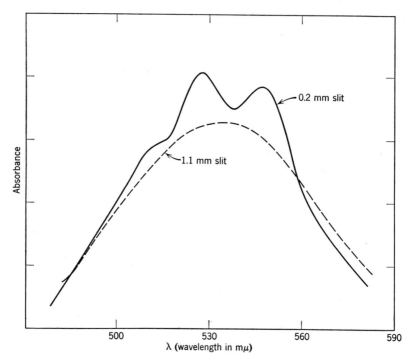

FIGURE 21-3. Absorption spectrum of $KMnO_4$ solution. Student data taken with Beckmann model-DU spectrophotometer. Shows effect of slit width on resolution.

obtained for a sample is the per cent transmittance. Prior to the analysis of samples, we construct a working curve, a plot of absorbance versus concentration or of per cent transmittance versus concentration. The former has the advantage that the relation between absorbance and concentration is linear. The curve is constructed from measurements witth a graded series of standard solutions which cover the desired concentration range.

Spectrophotometers may be used analytically, by measuring absorbance at a selected wavelength; in addition, they may be used to determine the absorption spectrum of liquids and solutions. A typical spectrum is shown in Fig. 21-3, which gives a plot of absorbance versus wavelength for a potassium permanganate solution. The upper curve, showing two distinct peaks, is made with a narrow slit. When a wider slit is used, we obtain the lower curve, which does not resolve the peaks. This is due to the fact that the light transmitted by the wide slit covers a rather wide wavelength region and permits overlapping of the wavelengths that are most highly absorbed by the solution. In determining absorption as a function of wavelength, it is obviously desirable to use as narrow a slit as the sensitivity of the instrument permits.

Validity of Beer's Law. Although the Beer law is quite accurate for a colored substance such as permanganate ion, deviations may occur when colored ions are involved in reactions such as dissociation, polymerization, and formation of complex ions. For example, the constant a for absorption of light by CrO_4^{-2} ion is found to vary with dilution because this ion reacts with water to form the $HCrO_4^-$ ion which has a different absorption spectrum from that of the chromate ion. As the solution is diluted, the ratio chromate–dichromate ion varies with the concentration. The presence of colloidal particles which cause turbidity of the solution also causes deviations from Beer's law.

Nephelometry. When light traverses a suspension of finely divided particles, the intensity is decreased because the individual particles scatter the light. This effect can be used in analyses, but not with as high precision as colorimetry may give. If observations are made of transmitted light, as in colorimetry, the process is known as turbidimetry. The observation of light may also be made at right angles to the beam, a process known as nephelometry. In both these methods it is necessary to prepare the suspension of the insoluble susbtance for the standard and the sample in exactly the same manner, to insure that the particles have the same size distribution. Protective colloids are often employed, to favor formation of a stable suspension. Small amounts of chloride or of sulfate ions can be determined quite satisfactorily by these methods.

Determination of Manganese in Steel

Small amounts of manganese can be quickly and accurately determined colorimetrically, after oxidation to permanganate, using either visual colorimeters or photometers. The analysis of steel samples is an excellent exercise to illustrate the method, which is applicable to alloys, minerals, biological samples, and in general to any material containing small percentages of manganese. The amount of manganese in the final solution should not exceed 20 mg per liter. The general procedure is to dissolve the sample in a suitable acid, remove chloride ion by evaporation if this is present, and oxidize the manganese to permanganate by treatment with potassium periodate. Steel samples are treated with ammonium persulfate prior to the periodate oxidation, to destroy undissolved particles of carbon which would interfere with the color measurement. Interference by ferric ion is prevented by adding phosphoric acid, which forms a colorless ferric phosphate complex.

Procedure.   Preparation of Steel Sample. Dissolve weighed portions of about 0.5 g each in 50 ml approximately 4 M nitric acid and boil to expel oxides of nitrogen. Add about 1 g ammonium persulfate, and continue to boil the solution for about 10 minutes. Add the persulfate in small portions. If there is any precipitate or color in the solution after this treatment, add a small amount of sodium sulfite dissolved in water and boil for another 5 minutes. Cool, add 10 ml sirupy phosphoric acid, and dilute to about 100 ml. Add 0.5 g potassium periodate in small portions and stir until dissolved. Boil the solution for 5 minutes to oxidize the manganese. Cool, transfer the solution quantitatively to a 250-ml volumetric flask, and dilute to volume. Use this solution for the colorimetric comparison with standards.

Colorimetric Standards. For best results the sample solution should be compared with solutions containing the same amount of iron as in the sample, and with known amounts of manganese which has been oxidized by the same procedure as used for the sample. The best standards are prepared from analyzed steel samples if these are available in the proper range of manganese content, but ordinarily such a range of samples is not at hand. Lacking these, we can either prepare the standard by adding a known amount of Mn or $KMnO_4$ to a 0.5-g sample of pure iron wire and then proceeding just as in the preparation of the sample, that is, first reducing the permanganate to Mn^{+2} and then reoxidizing with periodate. If this is done, it is necessary also to analyze the iron wire used and to make a correction for the amount of manganese it contributes to each portion of standard. In most student work it is adequate to prepare the colorimetric standards directly, by dilution of a standardized solution of potassium permanganate, and to use these diluted solutions without further treatment. When this is done, the standard solutions should be used the same day they are prepared, since reducing matter in the water used may cause some reduction of the dilute solutions.

Procedure.   Preparation of Standards. Obtain a 40–50 ml portion of standardized $KMnO_4$ (approximately 0.1 N) from the instructor, in a dry flask. Record the exact normality of the solution. Pipet 25 ml of the solution into a 250-ml volumetric flask and dilute to volume. Calculate the exact weight of manganese per milliliter of solution. Measure from a buret the exact volumes of the diluted solution to give respectively 0.50, 1.00, 2.50, and 5.00 mg Mn, delivering each into a 250-ml volumetric flask. Dilute to volume. Compute the weight of manganese in milligrams per milliliter for each of the solutions.

Analysis with Filter Photometer. Fill the cuvette with water and adjust the zero reading. Remove the water, rinse the cuvette with the most dilute of the standard solutions, fill, and determine the absorbance. Proceed in the same way for each of the standard solutions. Plot the absorbance as a function of milligrams of manganese per milliliter of solution. A straight line should be obtained.

Pour a little of the sample solution into a dry flask and add hydrogen peroxide, a drop at a time, until the color of permanganate ion disappears. The amount of H_2O_2 required should not cause enough dilution of the solution to affect the measurements. Swirl the solution until the bubbles of oxygen escape. Rinse the cuvette with this solution and fill with it. Use this to adjust the photometer zero reading. This procedure corrects for any color not due to permanganate ion in the sample solution. Now rinse and fill the cuvette with sample solution and determine the absorbance. Use this value to determine from the working curve the weight of manganese per milliliter of the solution. Compute the total weight of manganese in the sample and its percentage in the steel.

Analysis with Visual Colorimeter. Compare the flask containing the sample solution with the series of standards, and select the standard whose color is nearest that of the sample.

Place the colorimeter cups into position, and carefully raise each cup until it just touches the bottom of the plunger. If the scale reading is not zero at this point, consult the instructor.

Clean the colorimeter cups and prisms with water and then rinse with the standard solution chosen for the comparison by filling the cup about two-thirds full, raising it into position, and discarding the solution. Refill the cups about two-thirds full with the standard solution and place in position. Raise the left-hand cup until the plunger extends about halfway into the solution and then adjust the right-hand cup until the eyepiece shows equality in the two half circles. Read the scale position of the right-hand cup (leave the left-hand cup fixed—its position does not enter into the determination). Remove the right-hand cup, empty, rinse, and refill with sample solution. Replace in position and adjust the plunger depth to give equality of the two half circles in the eyepiece. Read the scale position. The two readings for the right-hand cup are the depths of liquid to give the same transmission for standard and sample solutions. From these depths and the concentration of the standard solution, compute the weight of manganese per milliliter of sample solution, the total weight of manganese in the sample, and the percentage.

Determination of Ammonia by Nessler Reagent

When an alkaline potassium mercuric iodide solution is added to a solution containing ammonium salts, the ammonia liberated by the hydroxide ion reacts with the reagent to give a colored compound

$$2HgI_4^{-2} + 2NH_3 = NH_2Hg_2I_3 + NH_4^+ + 5I^-$$

The colored substance is colloidal and flocculates on long standing. When freshly prepared it is highly dispersed and may be used for colorimetric determination of ammonium ions. The method, one of the earliest colorimetric procedures to be used in quantitative work, is known by the name of its discoverer, Nessler, and the alkaline solution of mercuric potassium iodide is known as Nessler's reagent. This reagent may be used for colorimetric determination of small amounts of ammonium salts and of nitrogenous compounds that can be converted into ammonium salts. The method we describe illustrates the use of Nessler tubes in a colorimetric analysis. If preferred, the color comparison can be made by visual or photoelectric colorimeters instead of Nessler tubes.

Procedure.   Determination of Ammonia in Water. Prepare Nessler reagent as follows. Dissolve 13 g potassium iodide in 25 ml of ammonia-free water (note 1) and add, with constant stirring, a cold saturated solution of mercuric chloride until the precipitate at first formed no longer redissolves (note 2). Filter and add a solution of potassium hydroxide prepared by dissolving 55 g of the solid in 150 ml of water, allowing the precipitate of carbonate to settle, and decanting the clear solution. After addition of base, dilute to approximately 250 ml, mix, and add a saturated solution of mercuric chloride drop by drop, keeping the solution well mixed, until a slight permanent precipitate is formed. Test the reactivity of the reagent by adding 0.5–10 ml of distilled water containing 0.5 ml standard ammonium chloride solution. A brown tint should be apparent in 1 minute.

The comparison solutions are made from an ammonium chloride solution of such strength that 1 ml contains 0.01 mg nitrogen. Prepare by dissolving 3.82 g ammonium chloride in ammonia-free water, diluting to 1 liter, mixing well, and diluting a 10-ml portion to 1 liter.

Place 50 ml distilled water in a clean distilling flask, add 2 ml saturated sodium carbonate solution, connect with a condenser, and distill off 30 ml. Catch the next 10 ml of distillate in a Nessler tube. Cool and add 0.5 ml Nessler reagent. If no color develops within 5 minutes, the apparatus is free of ammonia; if the apparatus is not free of ammonia, add more water and continue the test until a 10-ml portion of distillate is free of ammonia. Add to the flask a 100-ml sample of the water to be tested, and distil at a rate of 10 ml per minute, collecting five 10-ml samples in Nessler tubes; then collect a sixth sample and immediately test it for ammonia. If none is found, discontinue the distillation; otherwise catch eight samples (note 3). Add 0.5 ml Nessler reagent to the last fraction and compare (note 4) the depth of color with that of each tube in a series of 10-ml samples containing respectively 0.5-, 1-, 2-, 3-, 4-, and 5-ml portions of standard ammonium chloride solution and 0.5 ml of Nessler reagent, each being diluted to exactly 10 ml before addition of Nessler reagent. The comparison should not be made until 5 minutes after the time of mixing the reagents. If the amount of ammonia in the last fraction falls within the limits of range of the standards, compute the amount present by matching the intensity of color in test and in standard. The nearest 0.5 ml of standard ammonium chloride solution may be computed by interpolation; if, for example, the sample tube shows a color considerably lighter than the 4-ml standard but darker than the 3-ml standard, it may be assumed to correspond to a 3.5-ml standard. If the tint in any fraction is less than the lightest standard tube, the amount of ammonia present is considered negligible. If it is greater than the amount in the 5-ml standard, transfer the solution to a 50-ml Nessler tube, dilute to the mark, and withdraw a 10-ml portion for the test. Continue in this manner with the other fractions, taking $\frac{1}{5}$ aliquot for the comparison whenever the intensity of color in a 10-ml fraction becomes greater than that of any standard tube. Determine in this manner the amount of ammonia in each tube and the total amount in the sample.

Notes. 1. These experiments must be carried out in a room free of ammonia fumes and with ammonia-free water. Since the laboratory-distilled water often contains appreciable amounts of ammonia, it is advisable to redistill all the water from an alkaline solution containing a little $KMnO_4$ and keep in stoppered bottles until used.

2. The reactions are:

$$Hg^{+2} + 2I^- = HgI_2$$
$$HgI_2 + 2I^- = HgI_4{}^{-2}$$

No precipitate forms as long as iodide ion is present in excess. It is essential that the solutions be well mixed in order to prevent local deficiency of iodide ion and precipitation of mercuric iodide.

3. Distillation is continued until all the ammonia is removed from the sample. This method eliminates other ions in the water which may interfere with the Nessler test. Calcium and magnesium, in particular, cause trouble, giving a precipitate with the reagent.

4. The comparison is conveniently made by boring holes in a cardboard box so as to permit the column of liquid to be enclosed within the box while allowing the bottom of the tube to project beneath the bottom. A white tile provides an excellent background for observation of the tubes.

Problems

1. Manganese in steel is determined by use of a Duboscq colorimeter. The standard solution contained 3.000 mg Mn per 250.0 ml. The steel sample which weighed 0.5075 g was dissolved, the manganese oxidized, and the solution diluted to a total volume of 250 ml. Colorimeter readings were 60.0 mm for the standard and 66.8 mm for the sample. What is the per cent of manganese in the steel?   *Ans.* 0.531%.

2. Copper is determined colorimetrically by use of the ammonia complex with a Duboscq colorimeter. A 3.000-g sample was dissolved, properly treated, and diluted to 100 ml. Colorimeter readings were 35.0 mm for the cup with unknown solution and 42.7 mm for a cup containing a standard solution whose concentration is 0.0800 mg Cu per milliliter. Find the per cent of copper in the sample.

3. The following data were obtained with a filter photometer and standard manganese solutions:

Concentration of Standard Solution, mg Mn/250 ml	Photometer Reading
0.50	56
1.00	105
3.00	286
5.00	460

A 0.5264-g sample of steel was dissolved, properly treated to oxidize the manganese, and the solution diluted to 250 ml. Photometer reading with unknown solution in the cuvette was 261. What is the per cent of manganese in the steel?

4. In a certain experiment the per cent transmission was determined to be 50.0 per cent. What is the absorbance under these conditions.   *Ans.* 0.301.

CHAPTER 22

Potentiometric Titrations

It was shown in Chapter 11 that oxidation and reduction reactions are due to a transfer of electrons from a reducing agent to an oxidizing agent and that by separating the two processes the chemical reaction can be made to produce an electric current. When this is done, there is an electromotive force between two electrodes of the system that is related to the concentrations of the reacting ions in the two solutions. The measurement of these oxidation-reduction emf's will be discussed in this chapter, and we shall show how they can be used to follow the course of titrations.

Measurement of Electromotive Force

The potential difference between two electrodes of a cell can be measured by connecting the two terminals to a voltmeter, so that a current flows through the coils of the meter. Deflection of the meter needle is proportional to the current, and thereby proportional to the potential difference. This method is not generally used, however, because if current is drawn from an oxidation-reduction system a reaction occurs at the electrodes and this changes the concentrations of the reacting ions. Thus, the act of measuring the emf changes the very quantities we wish to determine. By proper use of a potentiometer, the emf can be measured while drawing a negligible amount of current from the system. This is the method we generally use.

A simple form of a potentiometer is shown in Fig. 22–1. Current from a working battery, W, flows through a slide wire resistance R. The working battery may be a dry cell or a small storage battery. The resistance of R is large enough to limit the current to a few milli-

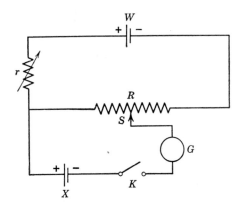

FIGURE 22–1. Simple potentiometer.

amperes, so that W will maintain a reasonably constant voltage across the system. An external resistance, r, adjusts the voltage drop across R to a predetermined value. If, for example, R is 1000 ohms, E_W (voltage of W) is 1.5 volts, and r is 500 ohms, it follows from Ohm's law that the voltage drop across R is 1.0 volt

$$E_R = \frac{1000 \text{ ohms}}{(1000 + 500) \text{ ohms}} \times 1.5 \text{ volts} = 1.0 \text{ volt}$$

The cell X, whose voltage we wish to measure, is connected in opposition to W. That is, its positive pole is connected to the same terminal of R as the positive pole of W. The emf of X must be less than the potential drop across R. If now the slide S is moved until no current flows through X, as shown by the reading of galvanometer G, the potential drop in X is numerically the same as that in the portion of R that is in series with X. If the resistance R is linear, so that the potential fall is uniform with distance, the voltage of X is read from the balance position of the slide S. For example, if R is 100 cm in length, the potential drop is 1.0 volt, and S is at 70 cm, the emf is $70/100 \times 1.0$ volt, or 0.70 volt. This is then voltage of X.

Many commercial pH meters operate on essentially the same principle as the potentiometer of Fig. 22–1. The instrument is first adjusted, by use of a standard cell incorporated into the assembly, to make the dial readings of the correct value. A reference half-cell of the instrument is then connected with the half-cell solution of interest, the dial is turned until a balance is obtained, and the voltage of the whole cell is read directly from the dial position. The dial is also graduated to give directly a reading of the pH in the measured half-cell. The relation between pH and the emf of the cell is discussed later.

Applications of Potentiometric Measurements

Half-cells can be set up in various ways to relate the measured voltage of the assembly to concentrations of ions in the solution. We shall discuss applications in three types of titrations: (1) neutralization, (2) oxidation-reduction, and (3) precipitation.

Relation between Cell Voltage and pH

We have shown in Chapter 11 that cells may be devised in such a way as to relate the pH of one of the half-cells to the measured voltage of the whole cell. An equation was developed to relate the emf of the cell

$$Pt, H_2, H^+ (xM) \parallel KCl(1\,M), Hg_2Cl_2, Hg$$

to the hydrogen ion concentration of the left-hand half-cell. The first electrometric measurements of hydrogen ion concentration were made by just such cells, the experimental arrangement for which is shown later in this chapter.

One half-cell consists of a hydrogen electrode having hydrogen gas in contact with the solution whose pH we wish to measure. The electrode is a platinum wire coated with finely deposited Pt, which serves as catalyst for the reaction

$$H_2 = 2H^+ + 2e^-$$

which is reversible at the electrode. The other half-cell is a standard reference electrode of known emf. Usually we use one of those given on page 216, since these are easy to construct and are reproducible.

Consider the cell

$$Pt, H_2 \,(1\,\text{atm}), H^+ \,(xM) \parallel \text{reference half-cell}$$

The measured potential for the cell is

$$E_{\text{cell}} = E_{\text{hydrogen}} - E_{\text{reference}}$$

By (1) we have

$$E_{cell} = -E_{ferrous-ferric} = -(-0.73 \text{ volt}) = 0.73 \text{ volt}$$

(b) When 2.5 mM Ce^{+4} is added, the solution contains 2.5 mM Fe^{+3}, 2.5 mM Ce^{+3}, 2.5 mM Fe^{+2}, and a small but unknown amount of Ce^{+4}. Here again we know the Fe^{+3}/Fe^{+2} ratio and use equation (2)

$$E_{ferrous-ferric} = -0.77 - 0.059 \log 2.5/2.5 = -0.77 \text{ volt}$$

$$E_{cell} = -(-0.77 \text{ volt}) = 0.77 \text{ volt}$$

This is a generally useful relation that at one-half oxidation the titration half-cell emf is the $E°$ value of the reductant.

(c) When 4 mM Ce^{+4} is added, the solution contains 4 mM Fe^{+3} and 1 mM Fe^{+2}. Using this ratio with equation (2) gives

$$E_{ferrous-ferric} = -0.77 - 0.059 \log 4 = -0.81 \text{ volt}$$

$$E_{cell} = -(-0.81 \text{ volt}) = 0.81 \text{ volt}$$

(d) At the equivalence point, when 5 mM Ce^{+4} is added, the reaction

$$\overset{x}{Ce^{+4}} + \overset{x}{Fe^{+2}} = \overset{5-x}{Ce^{+4}} + \overset{5-x}{Fe^{+3}}$$

is at equilibrium. The numerical values above each of the symbols in the chemical equation are the millimoles of each species present. Since the titration is made starting with 5 mM Fe^{+2}, we have x mM Fe^{+2} remaining unoxidized and $(5 - x)$ mM Fe^{+3} formed. In order to use either equation (2) or (3), we need to evaluate the ratio of oxidized form to reduced form of one of the reactants. It is more convenient to use instead the relation developed in the following paragraph.

The half-cell emf at the equivalence point is half the sum of the half-cell emf's for oxidant and reductant. That is

(4)
$$E_{\text{titration half-cell}} = \frac{E°_{ferrous-ferric} + E°_{cerous-ceric}}{2}$$

$$= \frac{-0.77 - 1.61}{2} = -1.19 \text{ volts}$$

$$E_{cell} = -(-1.19 \text{ volts} = 1.19 \text{ volts}$$

Equation (4) is obtained by adding (2) and (3). This gives

(2) $E_{\text{half-cell}} = E°_{ferrous-ferric} - 0.059 \log Fe^{+3}/Fe^{+2}$

(3) $E_{\text{half-cell}} = E°_{cerous-ceric} - 0.059 \log Ce^{+4}/Ce^{+3}$

$$2E_{\text{half-cell}} = E°_{ferrous-ferric} + E°_{cerous-ceric} - 0.059 \log \left(\frac{Fe^{+3} \times Ce^{+4}}{Fe^{+2} \times Ce^{+3}} \right)$$

Since $Fe^{+3} = Ce^{+3}$ and $Fe^{+2} = Ce^{+4}$, the value of the quantity in parenthesis is 1 and the log term becomes zero. This gives

$$2E_{\text{half-cell}} = E°_{ferrous-ferric} + E°_{cerous-ceric}$$

$$E_{\text{half-cell}} = \frac{-1.61 - 0.77}{2} = -1.19 \text{ volts}$$

$$E_{cell} = 1.19 \text{ volts}$$

An alternative method for computing the emf, by using the equilibrium constant, is given in the footnote below.[1]

(e) When 6 mM Ce^{+4} is added, the solution contains 5 mM Ce^{+3} and 1 mM Ce^{+4}. We use these values in (3) to compute the half-cell emf.

$$E_{\text{cerous-ceric}} = -1.61 - 0.059 \log 1/5 = -1.57 \text{ volts}$$

$$E_{\text{cell}} = 1.57 \text{ volts}$$

A plot of the cell emf values computed in this problem is shown in Fig. 22–2. This figure also shows several experimental titration curves for the ferrous-ceric system, in solutions of different acids. It will be noted that no two of the curves are identical and that the theoretical curve does not exactly agree with any one of the experimental ones. (To prevent overlapping, the curves are shifted on the x-axis. If plotted with the same origin for all, the steep portions of all the curves fall at the same volume of ceric solution.)

The differences in the experimental curves of Fig. 22–2 illustrate an important principle. The calculations are based on $E°$ values which are for unit activity of all reactants. In solutions containing anions of acids, the cations of the system form various complex ions whose formulas have not been established. When this occurs, the con-

---

[1] In the reaction we have the half-cells

$$Fe^{+2} = Fe^{+3} + e^- \qquad E° = -0.77 \text{ volt}$$

$$Ce^{+3} = Ce^{+4} + e^- \qquad E° = -1.61 \text{ volts}$$

Combining these gives

$$Fe^{+2} + Ce^{+4} = Fe^{+3} + Ce^{+4} \qquad E°_{l-r} = 0.84 \text{ volt}$$

At equilibrium,

$$\log K_e = \frac{E°_{l-r}}{0.059} = \frac{0.84}{0.059} = 14.24$$

$$K_e = \frac{[Fe^{+3}][Ce^{+3}]}{[Fe^{+2}][Ce^{+4}]} = 10^{14.24}$$

Let

$$x = \text{millimoles } Fe^{+2} = \text{millimoles } Ce^{+4}$$

$$5 - x = \text{millimoles } Fe^{+3} = \text{millimoles } Ce^{+3}$$

$$\left(\frac{5-x}{x}\right)^2 = 10^{14.24}$$

$$\frac{5-x}{x} = \frac{[Fe^{+3}]}{[Fe^{+2}]} = 10^{7.12}$$

From equation (2)

$$E_{\text{half-cell}} = -0.77 - 0.059 \log (10^{7.12})$$

$$= -0.77 - 0.42 = -1.19 \text{ volts}$$

centration of the free cation is not its formality or molarity. For example, if say 50 per cent of the ferric ion is tied up as a chloride complex in the titration made in HCl, the actual concentration of Fe^{+3} is not the 1 mole per liter that is present in the solution but only 0.5 mole per liter. The fact that the cell emf has its highest values

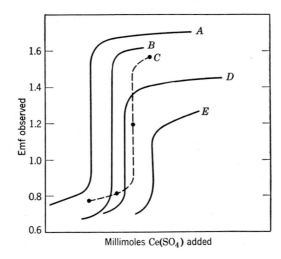

Millimoles Ce(SO_4) added

FIGURE 22–2.   Electrometric titration of ferrous ion by ceric ion in 1 N solutions of various acids. The voltages are for a platinum electrode with reference to a standard hydrogen electrode. A, perchloric acid; B, nitric acid; C, calculated values; D, sulfuric acid; E, hydrochloric acid. (Experimental curves reproduced by permission of G. F. Smith and C. A. Getz, *Ind. Eng. Chem., Anal. Ed.*, **10**, 191.)

in $HClO_4$ solution is indication that the amount of complex formation is least in this medium and that perhaps the $E°$ value of Table A–5 is too low for the cerous-ceric couple. The computed curve shows best agreement with the experimental one made in HNO_3 solution.

As noted previously, we have assumed use of a standard hydrogen reference electrode in these calculations of the cell emf of a titration. In practice we use one of the metal–metal ion reference electrodes described on page 216. The emf values obtained are corrected to a hydrogen electrode standard by subtracting the emf of the reference cell from the measured voltage. Thus, if using a normal calomel electrode gives a cell voltage of 0.30 volt, we correct to the hydrogen electrode standard by subtracting −0.28 volt. This gives a value of 0.58 volt.

We have selected a simple redox system involving a one-electron change to illustrate the computations of a redox titration. When the oxidant or reductant requires two or more electrons, the computations are a little more complicated, but the principles are the same as in the ones we have made.

When the reactants are substances for which the numbers of moles are the same for the oxidized and the reduced forms, the following relations may be used to construct a titration curve, without going through the detailed calculations given above.

1. At half oxidation

$$E_{\text{titration half-cell}} = E^\circ_{\text{reductant couple}}$$

2. At the equivalence point

$$E_{\text{titration half-cell}} = \frac{E^\circ_{\text{oxidant couple}} + E^\circ_{\text{reductant couple}}}{2}$$

when the reaction requires only one electron. If the reaction requires n_1 electrons for the reductant and n_2 electrons for the oxidant, we have the relation

$$E_{\text{half-cell}} \text{ (at E.P.)} = \frac{n_1 E^\circ_{\text{reductant}} + n_2 E^\circ_{\text{oxidant}}}{n_1 + n_2}$$

Thus, if Fe^{+2} is titrated by MnO_4^- in a solution whose hydronium ion concentration is $1\,M$, it can be shown by the method used on page 421 that

$$E'_{\text{half-cell}} \text{ (at E.P.)} = \frac{E^\circ_{Fe^{+2},Fe^{+3}} + 5E^\circ_{Mn^{+2},MnO_4^-}}{6}$$

3. When the amount of oxidant added is twice the amount used at the equivalence point, the ratio of the oxidized and reduced forms present is 1 and we have the relation

$$E_{\text{half-cell}} = E^\circ_{\text{oxidant couple}}$$

Selection of Indicator for Redox Titration.   One method of selecting an indicator for a redox titration was described in Chapter 11. The principle of this method is that the half-cell transition emf of the indicator should be of the proper value that a color change will occur when essentially all the reducing agent is oxidized. Two other methods for selecting an indicator can now be considered.

1. When an experimental titration curve is prepared, this is used to select the indicator. If the steep portion of the curve is nearly vertical, as in the curves of Fig. 22–2, we may use any indicator whose transition emf has a numerical value within the emf range of the

steep portion.   It is, of course, necessary to change the sign of the cell emf to obtain the half-cell indicator emf, since $E_{cell} = -E_{titration\ half-cell}$.

An experimental titration curve is the best method to use for selection of an indicator, because of the deviations that may be found between computed cell emf's and the experimental values, when there is complex formation involving ions of the redox system.   We note in Fig. 22–2 that if the computed titration curve were used to select an indicator for titration of ferrous ion by ceric ion in HCl solution, there might be a serious error.

2. If an experimental titration curve is not available, the $E°$ values of oxidant and reductant   may be used to compute the titration half-cell emf at the equivalence point.   The transition emf of the indicator should be the same as the half-cell emf at this point.   In using this method we should bear in mind the possibility of error due to complex formation of some ions in the system.

Precipitation Titrations

When an ion of a precipitate is the oxidized form of a redox system that is stable in aqueous solution, the course of a precipitation titration can be followed potentiometrically.   For example, if Ag^+ ion is involved in a precipitation, we can set up a cell whose emf is a function of the Ag^+ concentration.

A potentiometric titration of chloride ion by silver ion may be made by use of the cell

$$Hg,\ Hg_2Cl_2,\ KCl\ (1\ M)\,\|\,Ag^+\ (xM),\ Ag$$
(normal calomel cell)

The right-hand half-cell is a silver electrode dipping into the chloride solution, which contains some silver ion as soon as $AgNO_3$ is added and a precipitate of AgCl is present

$$E_{cell} = E_{calomel} - E_{Ag}$$

$E_{calomel}$ is -0.28 volt. E_{Ag} is for the reaction

$$Ag = Ag^+ + e^- \qquad E° = -0.80\ volt$$

By the Nernst equation,

$$E_{Ag} = E°_{Ag} - 0.059\ log\ [Ag^+]$$
$$= -0.80 - 0.059\ log\ [Ag^+]$$

Substituting these values gives

$$E_{cell} = -0.28 - (-0.80 - 0.059 \log [Ag^+]$$

$$= 0.52 + 0.059 \log [Ag^+]$$

Before the equivalence point is reached, the concentration of Ag^+ is computed from the K_{sp} of AgCl (1×10^{-10}) and the concentration of Cl^- in the solution.

Problem. What is the emf of the cell just described when 49 ml 0.1 M AgNO₃ is added to 50 ml 0.1 M NaCl solution?

SOLUTION: The solution contains 4.9 mM AgCl and 0.1 mM Cl^- in 99 ml.

$$[Cl^-] = \frac{0.1}{99} = 10^{-3} M$$

$$[Ag^+] = \frac{1 \times 10^{-10}}{10^{-3}} = 10^{-7}$$

$$E_{cell} = 0.52 + 0.059 \log (10^{-7})$$

$$= 0.52 - 0.41 = 0.11 \text{ volt}$$

At the equivalence point the concentration of Ag^+ is $\sqrt{K_{sp}}$ or $10^{-5}M$. Past the equivalence point the concentration of Ag^+ is computed from the millimoles of silver nitrate in excess and the volume of the solution.

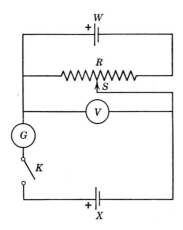

FIGURE 22–3. Simple slide wire potentiometer for electrometric titrations.

Experimental

If a pH meter is not available, a simple potentiometer can be constructed from readily available equipment, as shown in Fig. 22–3. The working battery W is two dry cells in series. The slide wire rheostat R should have a resistance of 500–1000 ohms. Voltmeter V is a small box type, with a range of 2–3 volts. Galvanometer G is a portable student type of low sensitivity. K is a tapping key. The titration cell X is connected so that its polarity is in opposition to that of W.

To measure the voltage of X, move slide S until the point is reached at which the galvanometer shows only a small deflection when the tapping key is closed. Do not hold K down, since this closes the circuit and draws current from X. When S is adjusted to the balance point, read the voltage of the cell from the voltmeter.

Electrodes. Hydrogen and calomel electrodes are shown in Fig. 22–4. To prepare the hydrogen electrode for use, clean it in nitric acid, rinse well, immerse in a 1–5 per cent solution of platinic chloride, and electrolyze with 2–4 volts potential until a coat of platinum black

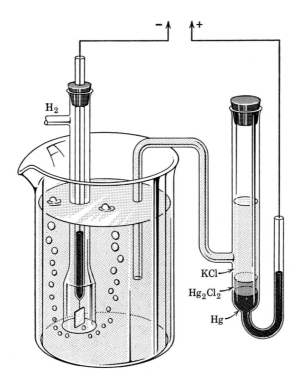

FIGURE 22–4. Hydrogen and calomel electrode assembly for electrometric titration.

is deposited. The electrode should be the cathode in this electrolysis, and a similar platinum electrode should be used as the anode. Rinse the plated electrode in distilled water, electrolyze for 1 minute in dilute sulfuric acid solution (using the plated electrode as cathode), and keep in distilled water until needed. To prepare the calomel electrode, triturate calomel with mercury in a mortar, to reduce any mercuric chloride present. Put the paste of mercury and calomel into the electrode vessel, and fill the vessel with a solution of 1 M potassium chloride which has been saturated with calomel.

Acid-Base Titrations. Assemble the apparatus as shown in Fig. 22–4. Place in the beaker a measured volume of the acid to be titrated and add water until the solution covers the mouth of the hydrogen bell. Pass a stream of hydrogen[2] through the bell at such a rate that there is a steady stream of bubbles of gas escaping through the solution. Run a little potassium chloride from the calomel electrode vessel, so as to obtain a junction of fresh solution. Connect the two electrodes to a potentiometer. (The mercury is the + electrode.) Measure the voltage. Titrate with standard sodium hydroxide solution, measuring the voltage after each addition of base. Only a few measurements need be made until the equivalence point is approached, but in the vicinity of this point the voltage should be measured after the addition of each drop of reagent. Continue the titration 5–10 ml beyond the equivalence point or until the addition of base causes little change in voltage. Calculate the pH for each reading. Plot the results, as shown in Chapter 7.

Titration with Quinhydrone Electrode. Arrange the apparatus as shown in Fig. 22–4, except that the hydrogen electrode is replaced by a polished platinum electrode and no hydrogen gas is passed through the solution. Connect this electrode to the positive terminal of the potentiometer and the calomel half-cell to the negative terminal. Place in the beaker a measured volume of acid, add solid quinhydrone, and proceed with the titration as in the previous exercise. In the quinhydrone titration the voltage decreases. When it reaches zero, reverse the polarity of the electrodes and continue the titration, recording all voltage readings after reversal as negative values. Calculate pH values from the voltage values obtained. Plot the titration curve.

Oxidation-Reduction Titration. Arrange the apparatus just as in the quinhydrone titration. Place in the beaker a measured volume of ferrous sulfate solution and add 10–15 ml sulfuric acid. Titrate with permanganate or ceric sulfate solution. Plot voltage as ordinate versus volume of reagent added as abscissa. The end point of the titration is recognized by the abrupt break in the curve.

Differential Titration. The sharpness of the end point in a potentiometric titration can be increased if we measure the change in

---

[2] Use compressed hydrogen, from a tank, if it is available. The gas should be purified by passage through a wash bottle filled with alkaline pyrogallol, which removes oxygen. If compressed hydrogen is not available, prepare the gas from arsenic-free zinc and hydrochloric acid, in a Kipp generator. Gas prepared in this manner should be purified by passage through successive wash bottles containing solutions of potassium permanganate and sodium hydroxide.

voltage for each addition of reagent, rather than the total voltage. This can be done experimentally by measuring the potential difference between two identical electrodes that dip into the same solution, but with one of the electrodes in equilibrium with the concentrations prevailing before a portion of reagent is added and the other in equilibrium after the addition. One electrode is inside a glass sleeve that partially isolates it from the bulk of the solution. When the sleeve is lifted and the solution stirred, E is zero, since concentrations inside and outside the sleeve are the same. When reagent is added, the concentration changes for the region outside the sleeve but remains the same inside the sleeve. Consequently there is a potential difference between the two electrodes. This is measured; then the sleeve is lifted and the solution stirred before the next portion is added.

The cell voltage is divided by the volume of reagent added in each portion, to obtain the ratio $\Delta E / \Delta V$. This ratio is plotted versus the total volume of reagent at each point. Just before the equivalence point is reached the $\Delta E / \Delta V$ curve rises sharply, and just beyond the equivalence point the curve falls. The intersection of the two branches marks the equivalence point.

The sensitivity of a differential titration is so great that this method can be used to obtain end points for titrations that are otherwise not feasible. For example, this method can be used for titration of $0.1\,M$ acetic acid by $0.1\,M$ ammonia solution.

Procedure. Mount two plain platinum electrodes so that they dip into the same beaker and connect to the terminals of a potentiometer. Make a short glass tube that fits over one electrode, so that when the tube is lowered it reaches the bottom of the beaker. Take care that it does not make so tight a seal at the bottom that conduction of ions in the solution is prevented.

Place a measured portion of acid in the beaker and add quinhydrone to saturate the solution. Titrate with standard base solution, measuring the voltage after each addition, then lifting the tube and stirring before the next portion is added. As the equivalence point is approached, add base in small portions. Continue the titration to a point well beyond the equivalence point. Plot the data as described.

Titrations with a pH Meter

A pH meter is a self-contained potentiometer system so arranged that when used with a glass electrode the readings are in pH units, or when used with a plain platinum electrode the readings are in millivolts. Since there are a number of excellent instruments available today, we shall not give detailed instructions for any particular make or model. These directions will be given by the instructor before the experiment is started.

Certain general instructions are applicable to all instruments.

1. The glass and reference electrodes are fragile and must be protected from damage. Typical commercial electrodes are shown in Fig. 22–5. Always keep the electrodes in a buffer solution or in distilled water when not in use. Avoid damaging the electrodes in moving the instrument.

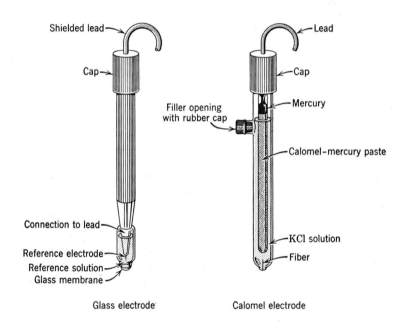

Shielded lead

Cap

Filler opening with rubber cap

Connection to lead

Reference electrode
Reference solution
Glass membrane

Glass electrode

Lead

Cap

Mercury

Calomel-mercury paste

KCl solution
Fiber

Calomel electrode

FIGURE 22–5.   Commercial electrodes.

2. The glass electrode must not be used in a strongly basic solution. Always titrate acid by base, never the reverse. If readings of high pH are required, use special electrodes that will function up to pH 14 (0.1 M NaOH solutions).

3. Time must be allowed to warm up the instrument after it is turned on. Initially there is a rapid drift, as tube filaments and resistors come to operating conditions. Do not attempt to make readings or adjust the instrument until after the recommended warm-up time.

4. Set the temperature adjustment according to instructions, before adjusting the instrument. This is a compensating potentiometer to

care for the effects of temperature changes on the emf's of the glass and reference electrodes.

5. Adjust the instrument to give proper scale readings before starting a measurement. Ask instructions for this adjustment. For pH measurements adjust by using a buffer solution of known pH, preferably a pH near that of the solution to be measured. Insert the electrodes in a beaker of the reference solution and adjust to make the needle reading correspond to the proper value. Before making emf measurements, ask instructions for adjusting to zero reading.

After adjusting the instrument, set the marker provided for this purpose to the proper position and readjust to bring needle readings into coincidence with the marker at frequent intervals, as required to compensate for drift. Some types of instruments show much greater drift than others.

6. When changing electrodes from one solution to another, raise them from the beaker and rinse with a stream of water from a wash bottle. After rinsing, put the electrodes into the fresh solution immediately. Never allow the glass electrode, in particular, to become dry.

7. Use mechanical stirring for all titrations if this is provided. Before starting the stirrer, make certain that it does not strike either electrode.

8. Take time to arrange a neat and workman-like setup before starting a titration. The electrodes are held in clamps on a ringstand or support. The meter is placed adjacent to the titration beaker, taking care that there is no danger of spilling any liquid on it. Clear off other equipment in the vicinity, to allow sufficient work space. The meter and electrodes are expensive and precise instruments which should be handled with care at all times.

Acid-Base Titrations. Instructions are given for a typical acid-base titration. Phosphoric acid is selected, because the curve shows two breaks, corresponding to neutralization of the first and second hydrogen ions. The HPO_4^{-2} ion is such a weak acid that it is not neutralized in the titration.

If it is desired to use potentiometric titration for standardization of the NaOH solution, weigh out a portion of dried potassium acid phthalate, dissolve in water, and titrate by the NaOH solution according to the instructions below for the titration of phosphoric acid.

Procedure. Prepare and standardize a supply of approximately 0.2 N NaOH, carbonate-free. An approximately 0.1 M H_3PO_4 solution is provided. The NaOH titration is used to compute the molarity of the acid.

Pipet out a 25-ml portion of the H_3PO_4 solution. Put this in the titration beaker. Adjust the meter by use of the standard buffer solution provided. Rinse off the electrodes and clamp in place in the titration beaker. Add water to bring the liquid level to proper height. Set up the stirrer and test to see that it does not strike either of the electrodes. Rinse a buret and fill with the standard NaOH solution. Clamp in place to deliver into the titration beaker. Take care to leave space for turning the stopcock without striking the electrodes. Add two drops phenolphthalein indicator to the acid. Read the pH of the solution and record. Record the buret reading. Titrate the acid by NaOH, taking pH readings after each addition of base. Record the pH and the buret reading for each addition. Initial additions may be in 3–5 ml portions, but as the end point is approached add only a drop at a time. Note that there are two end points, corresponding to the two reactions

$$H_3PO_4 + NaOH = NaH_2PO_4 + H_2O$$

$$NaH_2PO_4 + NaOH = Na_2HPO_4 + H_2O$$

Add base a drop at a time as the phenolphthalein point is approached, and determine the pH at the indicator end point as accurately as possible. Continue the titration well past the phenolphthalein end point (5–10 ml past).

Plot the titration curve on large-scale graph paper, so spaced that an 8½ x 11-inch sheet is nearly filled, giving pH values on the y-axis and the total volume of NaOH added on the x-axis. Show the point of indicator change on the graph. Compute the molarity of the acid, from the volumes used for the end points and the molarity of the base. Report the two values obtained.

Use the same data to plot a differential titration curve. For each addition of base compute the pH change. Divide this by the volume added in that portion to obtain the ratio

$$\frac{\Delta\,pH}{\Delta\,V} = \frac{\text{change in pH}}{\text{volume NaOH added}}$$

Plot $\Delta pH/\Delta V$ on the y-axis against the total volume NaOH on the x-axis, and draw the best smooth curve through the points. This plot should show sharp peaks at the two end points of the titration.

Oxidation-Reduction Titration.   The sample is an analyzed mixture of a soluble iron salt. It is dissolved in HCl, and $SnCl_2$ is added to reduce any ferric ion present. The solution is then directly titrated with a standard solution of ceric sulfate. It is not necessary to remove the excess stannous chloride by mercuric chloride, as was done for titrations employing redox indicators, because the potentiometric titration curve shows two breaks which correspond respectively to oxidation of the stannous ion and the ferrous ion. The volume of solution used in going from the first break to the second is the net amount consumed in oxidation of ferrous ion.

Procedure. Weigh out a sample of proper size to furnish about 3 meq iron. Dissolve in 10 ml hydrochloric acid. Heat to near boiling and add stannous chloride solution (see note 2, page 277), a drop at a time, until the yellow color of ferric ion has changed to the green of ferrous ion; then add five or six drops in excess.

Attach the plain platinum electrode to the meter, as instructed. Place the sample beaker, containing the reduced iron sample, in position and insert the two electrodes. Set up the stirrer. Add water if necessary to bring the liquid level to the proper position. Rinse and fill a buret with standardized 0.1 N ceric sulfate solution. Mount the buret above the titration beaker. Adjust the meter to give milli-volt readings.

Titrate the iron solution with ceric sulfate, taking an initial voltage reading and a reading after each addition of reagent. Add the ceric sulfate in small portions in the first sample titrated, to locate the approximate breaks corresponding to oxidation of stannous ion and of ferrous ion. In later portions of sample take just enough points to locate the breaks accurately.

Plot the voltage readings versus volume of ceric sulfate, and determine the two break points. From the volume of solution used in going from the first to the second break and the normality of the solution, compute the weight of iron and the percentage in the sample.

Questions

1. Replot the titration data given in Chapter 7 for HCl against NaOH, with emf values on the y-axis for titration:
   (a) With a hydrogen and normal calomel electrode.
   (b) With a quinhydrone and normal calomel electrode.

2. Why is not the quinhydrone electrode suitable for measuring the pH of a strongly basic solution?

3. Draw in some detail the cell which could be used for the determination of Fe^{+2} by electrometric titration with $K_2Cr_2O_7$.

4. Describe a method for the determination of $E°$ for the following oxidation-reduction systems:

    (a) Ferrous-ferric.

    (b) Quinone-hydroquinone.

    (c) Ceric sulfate-cerous sulfate.

    (d) An oxidation-reduction indicator.

List the sources of error which might affect each of the measurements.

5. How do you explain the fact that the stoichiometric emf found for ceric-ferrous titrations varies with the acid medium employed?

6. Ferric iron forms a complex with chloride ion. If HCl is added to an equimolar mixture of ferric and ferrous ions, how will the half-cell emf be changed?

7. Which titration will have the larger emf change at the equivalence point, silver ion with chloride ion or silver ion with iodide ion?

8. From the data cited in question 1 (a), calculate the rate of change of emf with volume ($\Delta E/\Delta V$) for the volumes used in obtaining the various points in the titration curve. Plot $\Delta E/\Delta V$ as ordinate against V as abscissa. What advantage does this curve have in the accurate location of the stoichiometric volume in an electrometric titration?

Problems

1. What is the emf at the equivalence point for the silver chloride titration if it is made by means of the cell,

$$\text{Hg, Hg}_2\text{Cl}_2\text{, KCl (1 }M\text{) } || \text{ KNO}_3 \text{ } || \text{ Ag}^+\text{, Ag}$$

(normal calomel electrode)

Ans. 0.22 volt.

2. Derive equations for pH in terms of emf for the cells:

    (a) Ag, AgCl, KCl (0.1 M) + H$^+$ (xM), Q + H$_2$Q, Pt

    (b) Pt, H$_2$, H$^+$ (xM) $||$ HCl (0.1 M), AgCl, Ag

3. If 50 ml 0.1 M silver nitrate solution is titrated by 0.1 M potassium iodide solution, compute the emf when the following quantities of iodide have been added: 0, 10, 40, 49, 49.9, 50.0, 50.1, 51, 60 ml. Plot the titration curve, with emf on the y-axis. Use the cell of problem 1.

4. What is the emf of each of the following cells:

    (a) Normal calomel electrode $||$ sodium acetate (0.1 M), hydrogen electrode.

    (b) Normal calomel electrode $||$ boric acid (0.03 M), quinhydrone electrode.

    (c) Normal calomel electrode $||$ AgNO$_3$(0.01 M) + NH$_3$(1 M), Ag.

    (d) Pt, H$_2$(740 mm), KOH (0.001 M) $||$ saturated calomel cell.

    (e) Decinormal calomel cell $||$ NH$_4$Cl (0.1 M), Q + H$_2$Q, Pt.

Ans. (a) -0.80 volt; (e) 0.06 volt.

5. In an electrometric titration 50 ml 0.1 M ferrous sulfate solution is titrated with 0.1 M ceric sulfate solution. Compute the emf when 10, 20, 40, 49, 50, 51, 60 ml ceric sulfate has been added (a) with a normal calomel reference electrode, and (b) with a hydrogen reference electrode. Plot the results obtained in (b), and compare with the curve given for this titration in sulfuric acid solution.

6. If pH is determined by a hydrogen electrode combined with a normal calomel electrode, what is the error in the pH value when the voltmeter used is in error by 2 mv?   *Ans.* 0.034 pH unit.

7. What is the pH when the emf of a quinhydrone-saturated calomel cell is zero?   *Ans.* 7.63.

8. The emf of the cell:

Decinormal calomel cell $\|$ CH_3NH_3Cl (0.1 M), Q + H_2Q, Pt

is 0.02 volt. Find the pH of the methyl ammonium chloride solution and the ionization constant of the base.

9. (a) Find the pH of a buffer solution in which the saturated calomel-quinhydrone cell combination shows an emf of 0.150 volt. The quinhydrone electrode is the positive pole.

(b) How many grams of sodium acetate must be added to 250 ml 0.15 M acetic acid to produce the buffer of this problem?

CHAPTER 23

Analysis by Electrodeposition

Theory

Electrolysis. The decomposition of a solution by passage of an electric current is known as *electrolysis*. During electrolysis positive ions migrate to the electrode connected to the negative terminal of the source of current, and negative ions migrate to the positive electrode. At the electrodes positive ions take up electrons and negative ions give up electrons. If there are in a solution several ions which may undergo electrolysis, the positive ion discharged is that with the greatest tendency to take up electrons, whereas the discharged negative ion is the one with the least tendency to hold electrons.

Laws of Electrolysis. The amount of an element set free by the passage of a current is proportional to the current which passes. Faraday's law states that "the same quantity of electricity sets free at the electrode the same number of equivalent weights of any substance." The practical unit of electricity is based on this law. One faraday of electricity, the unit quantity, will liberate one equivalent weight of any substance. The faraday is 96,500 *coulombs*.

The current passing through a solution is measured in *amperes*, that is, coulombs per second. The current is related to the applied potential and the resistance by Ohm's law,

$$\text{Effective emf} = \text{current} \times \text{resistance}$$

The effective emf is the applied emf minus the decomposition voltage (see page 438), since the decomposition voltage is in effect an opposing emf. That is,

$$E_{applied} - E_{decomp.} = IR$$

The resistance R is measured in ohms. It is proportional to the distance between the electrodes and inversely proportional to the area of the electrodes.

Electrode Reactions. The reaction which occurs at an electrode depends on the ions present in solution and the nature of the electrode. In aqueous solution, with inert electrodes of platinum, hydrogen is liberated at the cathode (negative electrode) when the other positive ions present are substances above hydrogen in Table A–5. Substances below hydrogen in Table A–5 are reduced. Likewise, oxygen (from (water) is liberated at the anode if other ions present are more difficult to oxidize than the OH^- ion. If the charged electrode is readily oxidized or reduced, no element is liberated, but instead the electrode is attacked. A few examples will illustrate these statements.

Example 1. Silver nitrate is electrolyzed between silver electrodes.

At cathode $Ag^+ + e^- = Ag$

At anode $Ag - e^- = Ag^+$

Example 2. Silver nitrate is electrolyzed between platinum electrodes.

At cathode $Ag^+ + e^- = Ag$

At anode $2H_2O - 4e^- = O_2 + 4H^+$

Example 3. Sodium chloride is electrolyzed between a silver chloride cathode and a silver anode.

At cathode $AgCl + e^- = Ag + Cl^-$

At anode $Cl^- + Ag - e^- = AgCl$

There may also occur at the electrodes a change in valence without liberation of an elementary substance. For example, Fe^{+2} may be oxidized at an anode.

$$Fe^{+2} - e^- = Fe^{+3}$$

and Fe^{+3} may be reduced at a cathode,

$$Fe^{+3} + e^- = Fe^{+2}$$

Decomposition Potential. If current is passed through a solution, no appreciable decomposition occurs unless the voltage applied to the electrodes exceeds a certain minimum value, which is known as the decomposition potential of the ions involved. The decomposition potential is the emf set up by the cell resulting from the electrolysis. Suppose, for example, that 1 M solution of zinc iodide is electrolyzed between platinum electrodes. Zinc and iodine are liberated:

$$Zn^{+2} + 2e^- = Zn \quad \text{at} - \text{electrode}$$

$$2I^- - 2e^- = I_2 \quad \text{at} + \text{electrode}$$

As soon as this reaction occurs, there is built up the cell,

$$Zn, Zn^{+2} + I^-, I_2, Pt$$

The emf of this cell is readily computed by the methods of Chapter **11**

$$
\begin{array}{ll}
Zn = Zn^{+2} + 2e^- & E^\circ = +0.76 \\
2I^- = I_2 + 2e^- & E^\circ = -0.54 \\
\hline
Zn + I_2 = Zn^{+2} + 2I^- & E^\circ_{l-r} = +1.30
\end{array}
$$

In a 1 M solution of ZnI_2 the concentration of Zn^{+2} is 1 and the concentration of I^- is 2 M. Substitution of these values for the activities of the ions gives

$$E_{l-r} = 1.30 - \frac{0.059}{2} \log (1)(2)^2$$

$$= 1.28 \text{ volts}$$

It is noted that in this cell, the I_2 is the positive electrode and the Zn the negative. Thus the cell emf is in opposition to the applied voltage, and in order to pass any current through this cell the applied voltage must exceed the cell emf of 1.28 volts. The cell emf is therefore known as the decomposition voltage.

The student should at this point construct an electrical diagram for the electrolysis circuit and mark the voltage and polarity at each electrode, to understand that the cell emf opposes the applied voltage.

Overvoltage. When the decomposition potential of a cell is computed, as previously, it is often found that application of the equilibrium voltage will not cause electrolysis. In some cases it may be necessary to exceed the theoretical value by as much as 1 volt or more before electrolysis will begin. The excess potential, above the decomposition potential, is known as *overvoltage*.

Overvoltage is chiefly of importance at electrodes where a gas may be liberated; usually little or no overvoltage is found for the liberation of metals. The liberation of a diatomic gas (practically all those encountered in analytical depositions are diatomic) must take place in two steps:

$$X^+ + e^- = X \text{ (atoms)}$$

$$2X = X_2 \text{ (molecules)}$$

One of the explanations proposed for overvoltage is that the liberated atoms accumulate on the electrode surface to a varying degree, depending on the speed of liberation (governed by the current density[1]) and the nature of the surface. This accumulation of X atoms on the surface causes a back emf to develop, to a degree proportional to the surface concentration of the atoms. The freshly deposited surface of a platinized Pt electrode[2] is a good catalyst for the reaction

$$H_2 = 2H$$

and this surface, which is used in the hydrogen electrode, has a very low hydrogen overvoltage. Other metal surfaces, such as Hg, have high hydrogen overvoltages. Illustrations of the effects of current density and the nature of the metal surface on overvoltage are shown in the typical values of Table 23–1.

TABLE 23–1.   Approximate Hydrogen Overvoltages on Different Metals*

Current Density, amp per cm²	Platinized Pt	Smooth Pt	Cu	Fe	Hg	Ni	Al
10^{-4}	0.003	—	0.35	0.22	0.28	—	0.50
10^{-3}	0.015	0.024	0.48	0.40	0.9	0.56	0.57
10^{-1}	0.038	0.19	0.80	0.82	1.06	1.05	1.0
1	0.048	0.68	1.25	1.29	1.13	1.24	1.29

* *International Critical Tables,* VI, 339–340.

From the analytical chemist's point of view the overvoltages of importance are those of hydrogen and oxygen, particularly the former. In nearly all electrolyses there is some evolution of hydrogen at the cathode, from local exhaustion of metal ions in the vicinity. If it were not for hydrogen overvoltage, this effect would predominate to

[1] The current density at an electrode is the current passing per unit area of electrode surface. Usually the density is stated in amperes per square centimeter, N.D., or in amperes per square decimter, N.D.$_{100}$.

[2] A platinized platinum electrode is made of platinum covered by a coat of electrolytically deposited platinum. The term is used to differentiate the electrolytically deposited surface from the surface of a massive electrode. Platinized platinum electrodes are usually employed when equilibrium is sought between a gas and its ions.

such an extent that complete deposition of certain metals would be difficult. Often the analyst may, by increasing the overvoltage, make depositions that could not otherwise be made. For example, a mercury cathode (a pool of mercury, connected to the negative terminal) is often employed to take advantage of the high hydrogen overvoltage on mercury. By means of the mercury cathode, iron and zinc can be completely separated from aluminum. The amalgamation of zinc for use in a Jones reductor prevents evolution of hydrogen because of the high overvoltage of hydrogen on mercury.

A striking illustration of the effect of overvoltage is provided in the familiar experiment of placing a strip of amalgamated zinc in dilute sulfuric acid and touching the zinc with a platinum wire. When the wire is removed, there is little evolution of hydrogen. When the wire is in contact with the zinc, there is copious evolution, but all the gas appears to come from the platinum wire. Because of the high overvoltage on the zinc surface the reaction,

$$H^+ + e^- = H$$

proceeds very slowly. When the platinum wire is touched to the zinc surface, the electrons released by the reaction,

$$Zn - 2e^- = Zn^{+2}$$

may flow into the platinum wire; there, because of the low overvoltage, the deposition of hydrogen may proceed rapidly.

$$2H^+ + 2e^- = H_2$$

Separation by Electrolysis. If the decomposition voltages for two compounds in the same solution lie far apart, one of the substances may be almost completely deposited by application of a potential which is between the two decomposition voltages. For example, the decomposition voltage of a zinc sulfate solution is 2.55 volts, of a sulfuric acid solution 1.67 volts, and of a copper sulfate solution 1.49 volts. If a potential of 2.0 volts is applied to a sulfuric acid solution of zinc and copper sulfates, the copper is almost completely deposited, some hydrogen is liberated, but no zinc ion is deposited.

Because of overvoltage effects and the effect of pH on electrolysis (see below), the exact computation of decomposition voltages from aqueous solutions is complicated. The completeness of separation of two metals is, however, readily computed by a consideration of the equilibrium relations. For example, we may compute the concentration of cupric ion at which there will be simultaneous deposition of

methods are largely based on the procedures of the American Society for Testing Materials. In the ASTM methods separate samples are taken for the tin determination and for the copper-lead determination, because of the widely different amounts of these constituents; in student analyses all the constituents listed above may be determined from a single sample. If increased accuracy is desired for the tin analysis, a large portion may be taken to start the analysis, and after removal of tin $\frac{1}{5}$ aliquot portions may be used for the other constituents.

Tin. The sample is dissolved in nitric acid. Tin forms hydrous stannic oxide $SnO_2 \cdot xH_2O$, which remains as a precipitate when the other constituents go into solution. It is separated by filtration, ignited, and weighed as the anhydrous oxide. The precipitate is always contaminated with some copper, lead, and iron. In exact analyses it is necessary to purify by fusing the precipitate with a mixture of sodium carbonate and sulfur, which converts the tin into soluble sodium sulfostannate, while leaving the impurities as sulfides. The fused sample is leached with water, the soluble sodium sulfostannate is brought into solution, and the sulfides are separated by filtration. They are ignited to oxides, and the weight is subtracted from the weight of impure stannic oxide. After ignition the recovered impurities are dissolved in nitric acid and added to the bulk of solution from which tin has been separated. An alternative method of purification is to mix the impure stannic oxide with ammonium iodide and heat. Stannic iodide is volatilized quantitatively, leaving the iodides of copper, lead, and iron. These are converted to the oxides by treatment with nitric acid, evaporation, and ignition.

Lead. In industrial laboratories lead is commonly determined by deposition as PbO_2 simultaneously with the deposition of Cu. This method is not recommended for student use; it is preferable to remove lead as the sulfate prior to electrolysis of copper. The sample solution, after removal of tin, is acidified with sulfuric acid and evaporated until all nitrates are removed. On dilution lead sulfate precipitates. It is separated by filtration with a porous-bottom crucible, since a filter paper cannot be used, and is dried and weighed as the sulfate.

Copper. Usually copper is determined electrolytically, after removal of lead, as described in the preceding section. When electrolytic equipment is not available, a gravimetric method based on precipitation as cuprous thiocyanate may be used.

Iron. After removal of copper, the solution is treated with bromine or nitric acid to oxidize the iron to the ferric state. Hydrous ferric oxide is precipitated by the addition of ammonia. The precipitate may be ignited to the oxide and weighed, or dissolved and the iron determined volumetrically. Student samples frequently contain so little iron that this determination is often omitted. Consult the instructor before starting it.

Zinc. In many commercial analyses zinc is determined by difference. When an independent determination is desired, zinc may be precipitated as sulfide, after removal of the other constituents, converted to the sulfate, and weighed as such. Alternatively, the zinc may be precipitated as zinc ammonium phosphate and ignited to the pyrophosphate, as in the determination of magnesium or the zinc may be determined by electrodeposition onto a copper electrode from hot, alkaline solution.

Procedure

Tin. Weigh out 1-g portions of sample into 150-ml beakers, and slowly add to each 25 ml nitric acid (1 : 1), taking care that the reaction (note 1) does not become too violent. Cover the beaker with a watch glass supported on glass hooks, and evaporate on a steam bath to a volume of 10 ml. Dilute to 50 ml, digest (do not boil) at 80–90°C for 10 minutes, and filter while hot. Addition of macerated filter paper may prove advantageous if the amount of precipitate is large. Wash the precipitate 10–20 times with hot nitric acid (1 : 20). Reserve the filtrate and washings. Transfer the precipitate to a porcelain crucible, char the paper at a low temperature, and ignite. Complete the ignition to constant weight by use of a blast lamp or Meker burner for half-hour periods. From the weight of stannic oxide, SnO_2, calculate the percentage of tin in the sample (note 2).

Notes. 1. The reaction is

$$3Sn + 4NO_3^- + 4H^+ + yH_2O = 3SnO_2 \cdot xH_2O + 4NO$$

2. If purification of the stannic oxide is desired (consult instructor), proceed as follows. Mix the impure stannic oxide in the crucible with eight times its weight of a mixture of equal parts of anhydrous sodium carbonate and sulfur. Cover the crucible, and heat, with a small flame, until sulfur ceases to burn at the edge of the cover. Cool, place the crucible and cover in a 250-ml beaker, and add sufficient water to cover both. Warm until the fused mass is completely dissolved, remove the crucible and lid, and wash in a stream of water. Add 0.2–0.5 g sodium sulfite, and boil for a few minutes to decompose the polysulfides present. Filter through a small paper, and wash thoroughly with dilute ammonium sulfide solution which should contain no polysulfide. Transfer the paper and precipitate to a porcelain crucible, and ignite to constant weight. Deduct the weight of

this precipitate from the total weight of impure stannic oxide. Dissolve the residue in dilute nitric acid, and add the solution to the main filtrate, which contains the other constituents.

Lead. Add 4 ml sulfuric acid to the filtrate from the tin precipitate, and expel nitric acid by evaporating (note 1) until fumes of sulfur trioxide are evolved. Cool, wash down the sides of the beaker with a little water, and again evaporate until fumes are evolved. Cool, add 25 ml water, and digest on the steam bath for 30 minutes (note 2); then add 50 ml water, and allow the solution to stand until it is at room temperature. Filter through an ignited and weighed filtering crucible (note 3), and wash with small portions of sulfuric acid (1 : 20). Reserve the filtrate and washings for the determination of copper. Ignite for 30 minutes at 500–600°C. From the weight of lead sulfate, calculate the percentage of lead in the sample.

Notes. 1. The solution must be evaporated in order to remove nitric acid, which markedly increases the solubility of lead sulfate. Use a hot plate (Hood).

2. Prolonged digestion is necessary, after evaporation and dilution, in order to bring all sulfates other than lead into solution.

3. A filtering crucible is necessary for ignition of the precipitate of lead sulfate, since this precipitate is easily reduced by carbon if a filter paper is used. The ignition temperature must be sufficiently high to volatilize sulfuric acid and water, but must not exceed 700°C since lead sulfate is decomposed above this temperature.

Copper. Evaporate the combined filtrate and washings from the lead determination to less than 100 ml, and transfer quantitatively to an electrolytic beaker. Add 1 ml nitric acid, and electrolyze according to the procedure on page 446, beginning with the star. Report the percentage of copper in the sample. Reserve the solution for the zinc determination.

Iron. If this determination is desired (consult instructor), transfer the solution from the electrolytic beaker to a 400-ml beaker, and precipitate iron by addition of ammonium hydroxide following the procedure on page 363. Ignite, and weigh as Fe_2O_3. Report the percentage of Fe. Reserve the filtrate and washings from both precipitations for the determination of zinc.

Zinc as Pyrophosphate. If iron was not determined (note 1), transfer the solution from the electrolytic beaker to a 400-ml beaker. Dilute to 150 ml, and neutralize with $6 N$ ammonia to the methyl red end point. Heat nearly to boiling, and slowly add 30 ml of a solution prepared by adding 20 g diammonium hydrogen phosphate to 80 ml of water (note 2). Digest one-half hour on the steam bath, with frequent stirring, allow to stand until the solution is at room temperature, and filter through a previously ignited and weighed porcelain filtering crucible. Wash with dilute diammonium phosphate solution (note 3) until a test portion *acidified with HNO₃* gives no precipitate with silver nitrate. Complete the washing by using several small portions of a mixture of equal volumes of alcohol and water, to remove the diammonium hydrogen phosphate from the precipitate. Allow the precipitate to suck dry after each addition of alcohol wash.

Dry the crucible and precipitate, and ignite in a 900°C oven for 30-minute periods until constant weight is obtained. From the weight of $Zn_2P_2O_7$ compute the weight of Zn in the sample.

Notes. 1. If iron was determined, it is necessary to destroy ammonium salts before starting the zinc determination. Add to the combined filtrates from the first and second precipitations of iron 25 ml each of nitric and hydrochloric acids, and evaporate until fumes of sulfur trioxide are evolved. Dilute to 150 ml, and proceed with the zinc determination.

2. The solution should be just neutral to phenolphthalein. Add a drop of indicator, and adjust if necessary by adding very dilute ammonia until a faint pink color is obtained.

3. Prepare the wash liquid by diluting 1 volume of the 20 per cent solution with 20 volumes of water.

Alternative Procedure for Copper

Copper as Thiocyanate. Combine the filtrate and washings from the lead determination. Concentrate to about 200 ml if the volume is greater than this. Transfer quantitatively to a 250-ml volumetric flask, dilute to the mark, and mix. Withdraw two 50-ml aliquot portions into conical flasks.

Neutralize with sodium hydroxide until a precipitate just begins to form, and then add 30–50 meq sulfuric acid of any convenient strength. Pour in with stirring 20 ml approximately 0.5 M sodium or ammonium bisulfate solution, and then add (note 1) slowly from a pipet 10 ml approximately 1 M ammonium or potassium thiocyanate solution. Stopper the flask to precent escape of SO_2, and allow to stand 1–2 hours. At the end of this time the precipitate should be white (note 2) and the liquid above it clear.

Filter through a filtering crucible that has been brought to constant weight at 120°C. Wash 10 times with 10-ml portions of a solution which contains about 1 g per liter ammonium thiocyanate and 0.1 g per liter ammonium bisulfate, and then wash 6 times with 10-ml portions of a solution containing 20 ml of alcohol to 80 ml of water. The washings need not be saved, for, if zinc is to be determined, a larger portion should be used than the $\frac{1}{5}$ aliquot employed for the copper. Dry at 120°C, and weigh as CuSCN. Alternatively, the instructor may direct to dissolve the precipitate in nitric acid and determine the copper iodometrically.

Notes. 1. A more crystalline precipitate may be obtained by heating before the addition of thiocyanate, but precipitation in the cold followed by standing is usually recommended.

2. If the precipitate is dark-colored, test the thiocyanate for presence of sulfide. A small amount of sulfide will darken the precipitate but will not seriously affect the result of the analysis.

Alternative Procedure for Zinc

Because of the amphoteric nature of zinc hydroxide, it is possible to prepare solutions of Zn^{+2} in equilibrium with $Zn(OH)_4^{-2}$ in strongly basic media. In such solutions the E_{H_2,H^+} is greatly increased because of the low concentration of hydronium ion. The use of a copper electrode on which the overvoltage of hydrogen is high along with the concentration effect makes it possible to deposit zinc at the cathode without the evolution of hydrogen.

Zinc by Electrodeposition.   Do not clean the platinum cathode after the copper determination but retain the weighed electrode with its copper deposit in a desiccator until needed.

Quantitatively transfer the solution remaining from the copper electrolysis to a 300-ml tall form beaker. Remove all the nitric acid by evaporating in a hood to thick fumes of SO_3. *Cool* to room temperature and cautiously add 10 ml of water. Again cool to room temperature and slowly neutralize drop by drop with 30 per cent KOH solution. Finally add 50 ml of the KOH solution and dilute to 150 ml.

Arrange the electrode system with the copper-plated cathode and a platinum anode. The beaker with the zinc solution should be supported at the proper height on a hot plate and the solution kept just below its boiling point throughout the electrolysis. Cover the beaker with a split watch glass. Electrolyze with stirring for 45 minutes at 2.2 amp. Stop the electrolysis as directed for copper on page 446. Wash the electrode, rinse with acetone, and dry for 5 minutes in an oven at 110°C. Cool and weigh. From the weight of deposited zinc compute the per cent of zinc in the brass sample.

Questions

1. Define electrolyte, electrolysis, coulomb, ampere, volt, ohm, current density, cathode, cation, overvoltage, decomposition voltage.

2. In electrodepositions why
   (a) Are gauze electrodes used?
   (b) Are solutions stirred?
   (c) Does the current decrease as the electrolysis proceeds unless the applied voltage is raised?

3. How may direct current be obtained from a 110-volt a-c source? What equipment is needed?

4. Explain how the use of a mercury cathode permits deposition of ions which cannot be deposited on platinum cathodes.

5. Why is it desirable in separations by electrolysis, that one substance be above hydrogen and the other below hydrogen in the electrochemical series? May any metals which are above hydrogen in the electrochemical series be deposited by electrolysis? Explain.

6. What are the products of the following electrolyses:

Cathode	Electrolyte	Anode
Pt	$NaNO_3$	Pt
Pt	KI	Pt
Pt	KI	Ag

7. Which of the following ions is likely to interfere in the electrolytic determination of copper: $PtCl_4^{-2}$, K^+, Ag^+, Zn^{+2}?

8. Why is the ignited SnO_2 precipitate sometimes tinged with color?

9. Why is the solution evaporated to SO_3 fumes in the determination of lead as sulfate?

10. What is the reducing agent in the determination of copper as thiocyanate?

Problems

1. A brass is 4.57 per cent Sn, 7.63 per cent Pb, 68.42 per cent Cu, and the remainder is Zn. From a 0.5000-g sample:
   (a) What weight of SnO_2 would be obtained?
   (b) If the lead had been determined electrolytically, what would a Pt anode which weighs 17.6512 g weigh after deposition is complete?
   (c) What weight of cuprous thiocyanate would be obtained?

2. What time will be required for the deposition of 1.1200 g lead as the metal, by a current of 3.000 amperes? What time will be required for the deposition of this amount of lead as the dioxide, using the same current?

3. A 0.7500-g sample of alloy yields by electrolysis 0.1532 g of a mixture of cobalt and nickel. The precipitate is dissolved and the nickel precipitated by addition of dimethylglyoxime, giving 0.3560 g $Ni(C_8H_{14}N_4O_4)$. Use the empirical factor 0.2032 for Ni in this compound. Calculate the percentages of Co and Ni in the sample.   *Ans.* 9.64% Ni; 10.79% Co.

4. A current of 0.5 ampere is passed through a solution of copper sulfate between platinum electrodes 10 cm^2 in area for 1 hour. What is N.D.$_{100}$ (assume that only one side of each electrode is effective)? How many grams of copper are deposited? How many milliliters of oxygen, at 25°C and 750 mm pressure, are evolved at the anode?   *Ans.* N.D.$_{100}$ = 5; 0.593 g Cu.

5. What is the current density if a current of 1 ampere passes from a solution to a platinum foil electrode which is 1×2.5 cm?

6. A solution is originally 0.1 M in Cu^{+2} and has a pH of 1.2. What is the pH of the solution after 99.9 per cent of the copper has deposited and the corresponding amount of oxygen has been liberated at the anode? Assume no H_2 is liberated.   *Ans.* 0.58.

7. In the analysis of a solution which contains Cu^{+2} and Pb^{+2} the deposition of copper is followed by the liberation of H_2 and the deposition of PbO_2 by the liberation of O_2. In the analysis of a 100-ml portion of a solution whose density is 1.02 and which contains 60 per cent Cu and 5 per cent Pb, is H_2 or O_2 liberated first?

8. A solution is 0.1 M in Cu^{+2}, 0.1 M in Pb^{+2}, and 1 M in H^+. What is the concentration of H^+ when the deposition of Cu and PbO_2 is complete? Assume no evolution of hydrogen or oxygen.   *Ans.* 1.4 M.

9. The bivalent metals A and B are to be separated by electrolysis. What difference in their half-cell emf's is necessary to permit the concentration of B to be reduced to 10^{-6} M while the concentration of A is 0.1 M? Compare the answer with the known potential differences for feasible electrolytic separations.   *Ans.* 0.148 volt.

10. A solution contains Cu^{+2}, Ag^+, and H^+, each at a concentration of 0.1 M. Neglect overvoltage, and

(a) Calculate the decomposition potential for the deposition of Ag and the liberation of O_2 at the anode.

(b) Calculate the decomposition potential for the deposition of copper and the liberation of O_2 at the anode.

(c) At what concentration of silver does the copper begin to deposit?

11. Find the decomposition potential for a 0.1 M $CuSO_4$ solution which is also 1 M in H_2SO_4 and is to be electrolyzed between smooth platinum electrodes at a current density such that the overvoltage of oxygen is 0.5 volt.   *Ans.* 1.42 volts.

12. Recalculate the decomposition potential for problem 11 after 50 per cent of the copper has deposited. Include in the calculation the increase in hydrogen ion concentration.

13. If the overvoltage of hydrogen on copper is 0.8 volt, at what concentration of cupric ion will hydrogen begin to deposit at the cathode in the electrolysis described in problem 11. For the purposes of estimating the hydrogen ion concentration, assume that the deposition of copper is essentially complete.

14. What is the minimum concentration of zinc ion that will permit the deposition of zinc from a solution of pH 7? Assume that a zinc cathode is used and that the overvoltage of hydrogen on it is 0.6 volt. *Ans.* 3×10^{-9} M.

15. Calculate the concentration of hydrogen ion necessary to permit deposition of sodium from a solution of 0.01 M NaCl, assuming no hydrogen overvoltage. What overvoltage of hydrogen would be required, in order that sodium be deposited from a solution in which the concentration of hydrogen ion is 10^{-13} M? If a mercury cathode is used, how will it affect the results of the calculations above? At what concentration of hydrogen ion may sodium be deposited from a 0.001 M salt solution if the amalgam contains 0.1 g sodium per 50 g mercury and the overvoltage of hydrogen is 1 volt?

Tables of Constants

The dissociation constants of these tables are in general given to one significant figure and the solubility product constants only to the nearest power of 10. Many of the constants are known to greater precision than is shown in the tables, and others are not even certain to the precision indicated by the values given. Experimental determination of equilibrium constants involves assumptions about the concentrations of the ionic species present in solutions of given formality; as these assumptions are re-examined, changes are often made. Many of the constants of these tables have been revised in the past ten years, and doubtless other revisions will be made as further work progresses. The values of the tables are therefore to be considered as approximations whiceh are sufficiently valid for the usual computations of equilibrium concentrations. If more precise values are needed, original sources should be consulted. Numerous references to original sources are given in *The Oxidation States of the Elements and Their Potentials in Aqueous Solutions,* by W. M. Latimer (Prentice-Hall, 1952).

TABLE A–I.   Ionization Constants of Weak Acids*

Acid	Reaction	Approximate Ionization Constant at Room Temperature
Acetic	$CH_3CO_2H = CH_3CO_2^- + H^+$	1.8×10^{-5}
Arsenic	$H_3AsO_4 = H_2AsO_4^- + H^+$	3×10^{-4}
Arsenious	$H_3AsO_3 = H_2AsO_3^- + H^+$	6×10^{-10}
Benzoic	$C_6H_5CO_2H = C_6H_5CO_2^- + H^+$	7×10^{-5}
Boric	$HBO_2 = BO_2^- + H^+$	6×10^{-10}
Carbonic	$CO_2 + H_2O = HCO_3^- + H^+$	4×10^{-7}
	$HCO_3^- = CO_3^{-2} + H^+$	5×10^{-11}
Formic	$HCO_2H = HCO_2^- + H^+$	2×10^{-4}
Hydrocyanic	$HCN = CN^- + H^+$	4×10^{-10}
Hydrofluoric	$HF = F^- + H^+$	7×10^{-4}
Hydrosulfuric	$H_2S = HS^- + H^+$	1×10^{-7}
	$HS^- = S^{-2} + H^+$	1×10^{-13}
Nitrous	$HNO_2 = NO_2^- + H^+$	5×10^{-4}
Oxalic	$H_2C_2O_4 = HC_2O_4^- + H^+$	4×10^{-2}
	$HC_2O_4^- = C_2O_4^{-2} + H^+$	5×10^{-5}
Phosphoric	$H_3PO_4 = H_2PO_4^- + H^+$	8×10^{-3}
	$H_2PO_4^- = HPO_4^{-2} + H^+$	6×10^{-8}
	$HPO_4^{-2} = PO_4^{-3} + H^+$	1×10^{-12}
Phthalic	$H_2C_8H_4O_2 = HC_8H_4O_2^- + H^+$	1×10^{-3}
	$HC_8H_4O_2 = C_8H_4O_2^{-2} + H^+$	4×10^{-6}
Propionic	$C_2H_5COOH = C_2H_5COO^- + H^+$	1×10^{-5}
Sulfuric	$HSO_4^- = SO_4^{-2} + H^+$	1.2×10^{-2}
Sulfurous	$H_2O + SO_2 = HSO_3^- + H^+$	1×10^{-2}
	$HSO_3^- = SO_3^{-2} + H^+$	6×10^{-8}

* The hydronium ion is indicated by the symbol H^+.

TABLE A–2.   Equilibrium Constants for Bases

Base	Reaction	Approximate Constant at Room Temperature
Ammonia	$NH_3 + H_2O = NH_4^+ + OH^-$	1.8×10^{-5}
Methyl ammonia	$CH_3NH_2 + H_2O = CH_3NH_3^+ + OH^-$	5×10^{-4}
Dimethyl ammonia	$(CH_3)_2NH + H_2O = (CH_3)_2NH_2^+ + OH^-$	7×10^{-4}
Trimethyl Ammonia	$(CH_3)_3N + H_2O = (CH_3)_3NH^+ + OH^-$	7×10^{-5}
Ethyl ammonia	$C_2H_5NH_2 + H_2O = C_2H_5NH_3^+ + OH^-$	6×10^{-4}
Phenyl ammonia	$C_6H_5NH_2 + H_2O = C_6H_5NH_3^+ + OH^-$	5×10^{-10}

TABLE A–3. Solubility Product Constants at Room Temperature

Compound	Equilibrium	Approximate K_{sp}
Acetates		
Silver acetate	$AgOAc = Ag^+ + OAc^-$	10^{-3}
Bromates		
Silver bromate	$AgBrO_3 = Ag^+ + BrO_3^-$	10^{-4}
Bromides		
Mercurous bromide	$Hg_2Br_2 = Hg_2^{+2} + 2Br^-$	10^{-22}
Silver bromide	$AgBr = Ag^+ + Br^-$	10^{-12}
Carbonates		
Barium carbonate	$BaCO_3 = Ba^{+2} + CO_3^{-2}$	10^{-9}
Calcium carbonate	$CaCO_3 = Ca^{+2} + CO_3^{-2}$	10^{-8}
Lead carbonate	$PbCO_3 = Pb^{+2} + CO_3^{-2}$	10^{-13}
Magnesium carbonate	$MgCO_3 = Mg^{+2} + CO_3^{-2}$	10^{-5}
Silver carbonate	$Ag_2CO_3 = 2Ag^+ + CO_3^{-2}$	10^{-11}
Strontium carbonate	$SrCO_3 = Sr^{+2} + CO_3^{-2}$	10^{-9}
Chlorides		
Mercurous chloride	$Hg_2Cl_2 = Hg_2^{+2} + 2Cl^-$	10^{-18}
Lead chloride	$PbCl_2 = Pb^{+2} + 2Cl^-$	10^{-5}
Silver chloride	$AgCl = Ag^+ + Cl^-$	10^{-10}
Chromates		
Barium chromate	$BaCrO_4 = Ba^{+2} + CrO_4^{-2}$	10^{-10}
Lead chromate	$PbCrO_4 = Pb^{+2} + CrO_4^{-2}$	10^{-16}
Silver chromate	$Ag_2CrO_4 = 2Ag^+ + CrO_4^{-2}$	10^{-12}
Fluorides		
Barium fluoride	$BaF_2 = Ba^{+2} + 2F^-$	10^{-5}
Calcium fluoride	$CaF_2 = Ca^{+2} + 2F^-$	10^{-10}
Lead fluoride	$PbF_2 = Pb^{+2} + 2F^-$	10^{-8}
Magnesium fluoride	$MgF_2 = Mg^{+2} + 2F^-$	10^{-7}
Strontium fluoride	$SrF_2 = Sr^{+2} + 2F^-$	10^{-9}
Hydroxides		
Cupric hydroxide	$Cu(OH)_2 = Cu^{+2} + 2OH^-$	10^{-19}
Ferrous hydroxide	$Fe(OH)_2 = Fe^{+2} + 2OH^-$	10^{-15}
Ferric hydroxide	$Fe(OH)_3 = Fe^{+3} + 3OH^-$	10^{-37}
Lead hydroxide	$Pb(OH)_2 = Pb^{+2} + 2OH^-$	10^{-15}
Magnesium hydroxide	$Mg(OH)_2 = Mg^{+2} + 2OH^-$	10^{-11}
Manganous hydroxide	$Mn(OH)_2 = Mn^{+2} + 2OH^-$	10^{-13}
Zinc hydroxide	$Zn(OH)_2 = Zn^{+2} + 2OH^-$	10^{-17}
Iodates		
Barium iodate	$Ba(IO_3)_2 = Ba^{+2} + 2IO_3^-$	10^{-9}
Calcium iodate	$Ca(IO_3)_2 = Ca^{+2} + 2IO_3^-$	10^{-6}
Cupric iodate	$Cu(IO_3)_2 = Cu^{+2} + 2IO_3^-$	10^{-7}
Lead iodate	$Pb(IO_3)_2 = Pb^{+2} + 2IO_3^-$	10^{-13}
Silver iodate	$AgIO_3 = Ag^+ + IO_3^-$	10^{-8}

TABLE A–3. Solubility Product Constants at Room Temperature (*Continued*)

Compound	Equilibrium	Approximate K_{sp}
Iodides		
Cuprous iodide	$CuI = Cu^+ + I^-$	10^{-12}
Lead iodide	$PbI_2 = Pb^{+2} + 2I^-$	10^{-8}
Mercurous iodide	$Hg_2I_2 = Hg_2^{+2} + 2I^-$	10^{-29}
Silver iodide	$AgI = Ag^+ + I^-$	10^{-16}
Oxalates		
Barium oxalate	$BaC_2O_4 = Ba^{+2} + C_2O_4^{-2}$	10^{-8}
Cadmium oxalate	$CdC_2O_4 = Cd^{+2} + C_2O_4^{-2}$	10^{-8}
Calcium oxalate	$CaC_2O_4 = Ca^{+2} + C_2O_4^{-2}$	10^{-9}
Cupric oxalate	$CuC_2O_4 = Cu^{+2} + C_2O_4^{-2}$	10^{-8}
Ferrous oxalate	$FeC_2O_4 = Fe^{+2} + C_2O_4^{-2}$	10^{-7}
Lead oxalate	$PbC_2O_4 = Pb^{+2} + C_2O_4^{-2}$	10^{-11}
Magnesium oxalate	$MgC_2O_4 = Mg^{+2} + C_2O_4^{-2}$	10^{-4}
Strontium oxalate	$SrC_2O_4 = Sr^{+2} + C_2O_4^{-2}$	10^{-7}
Zinc oxalate	$ZnC_2O_4 = Zn^{+2} + C_2O_4^{-2}$	10^{-9}
Sulfates		
Barium sulfate	$BaSO_4 = Ba^{+2} + SO_4^{-2}$	10^{-9}
Calcium sulfate	$CaSO_4 = Ca^{+2} + SO_4^{-2}$	10^{-5}
Lead sulfate	$PbSO_4 = Pb^{+2} + SO_4^{-2}$	10^{-8}
Strontium sulfate	$SrSO_4 = Sr^{+2} + SO_4^{-2}$	10^{-6}
Sulfides		
Cadmium sulfide	$CdS = Cd^{+2} + S^{-2}$	10^{-26}
Cobalt sulfide	$CoS = Co^{+2} + S^{-2}$	10^{-22}
Cupric sulfide	$CuS = Cu^{+2} + S^{-2}$	10^{-36}
Ferrous sulfide	$FeS = Fe^{+2} + S^{-2}$	10^{-17}
Lead sulfide	$PbS = Pb^{+2} + S^{-2}$	10^{-26}
Manganous sulfide	$MnS = Mn^{+2} + S^{-2}$	10^{-13}
Mercuric sulfide	$HgS = Hg^{+2} + S^{-2}$	10^{-50}
Nickelous sulfide	$NiS = Ni^{+2} + S^{-2}$	10^{-22}
Silver sulfide	$Ag_2S = 2Ag^+ + S^{-2}$	10^{-50}
Zinc sulfide	$ZnS = Zn^{+2} + S^{-2}$	10^{-20}
Thiocyanates		
Silver thiocyanate	$AgSCN = Ag^+ + SCN^-$	10^{-12}

TABLE A–4. Dissociation Constants for Selected Complex Ions

Ion	Equilibrium	Approximate Constant at Room Temperature
Ammonia complex		
Cadmium	$Cd(NH_3)_4^{+2} = Cd^{+2} + 4NH_3$	10^{-7}
Copper (II)	$Cu(NH_3)_4^{+2} = Cu^{+2} + 4NH_3$	5×10^{-14}
Silver	$Ag(NH_3)_2^{+} = Ag^{+} + 2NH_3$	5×10^{-8}
Zinc	$Zn(NH_3)_4^{+2} = Zn^{+2} + 4NH_3$	3×10^{-10}
Chloride complex		
Mercury (II)	$HgCl_4^{-2} = Hg^{+2} + 4Cl^-$	10^{-16}
Silver	$AgCl_2^{-} = Ag^{+} + 2Cl^-$	10^{-5}
Cyanide complex		
Cadmium	$Cd(CN)_4^{-2} = Cd^{+2} + 4CN^-$	10^{-19}
Copper (I)	$Cu(CN)_4^{-3} = Cu^{+} + 4CN^-$	10^{-27}
Mercury (II)	$Hg(CN)_4^{-2} = Hg^{+2} + 4CN^-$	10^{-42}
Silver	$Ag(CN)_2^{-} = Ag^{+} + 2CN^-$	10^{-19}
Iodide complex		
Mercury (II)	$HgI_4^{-2} = Hg^{+2} + 4I^-$	10^{-31}
Hydroxide complex		
Aluminum	$Al(OH)_4^{-} = Al(OH)_3(s) + OH^-$	2.5×10^{-2}
Chromium	$Cr(OH)_4^{-} = Cr(OH)_3(s) + OH^-$	10^2
Lead (II)	$Pb(OH)_3^{-} = Pb(OH)_2(s) + OH^-$	50
Zinc	$Zn(OH)_4^{-2} = Zn(OH)_2(s) + 2OH^-$	10

TABLE A–5. Standard Half-Cell Voltages of Oxidation-Reduction Couples

All values have been rounded off to one or two figures after the decimal point, without relation to how accurately they are known. For a more complete table and more precise values see *Oxidation Potentials*.* The symbol H^+ is used to designate the hydronium ion.

Couple	$E°$
$Li = Li^+ + e^-$	3.04
$Cs = Cs^+ + e^-$	2.92
$K = K^+ + e^-$	2.92
$Ba = Ba^{+2} + 2e^-$	2.90
$Sr = Sr^{+2} + 2e^-$	2.89
$Ca = Ca^{+2} + 2e^-$	2.87
$Na = Na^+ + e^-$	2.71
$2OH^- + Mg = Mg(OH)_2 + 2e^-$	2.69
$Mg = Mg^{+2} + 2e^-$	2.37
$H^- = \frac{1}{2}H_2 + e^-$	2.25
$Al = Al^{+3} + 3e^-$	1.66
$Mn = Mn^{+2} + 2e^-$	1.18
$Zn = Zn^2 + 2e^-$	0.76
$Cr = Cr^{+3} + 3e^-$	0.74
$S^{-2} + Hg = HgS + 2e^-$	0.72
$S^{-2} + 2Ag = Ag_2S + 2e^-$	0.69
$H_2C_2O_4 \text{ (aq)} = 2CO_2 \text{ (g)} + 2H^+ + 2e^-$	0.49
$S^{-2} = S + 2e^-$	0.48
$Fe = Fe^{+2} + 2e^-$	0.44
$Cr^{+2} = Cr^{+3} + e^-$	0.41
$Cd = Cd^{+2} + 2e^-$	0.40
$2I^- + Pb = PbI_2 + 2e^-$	0.37
$2Br^- + Pb = PbBr_2 + 2e^-$	0.28
$Co = Co^{+2} + 2e^-$	0.28
$2Cl^- + Pb = PbCl_2 + 2e^-$	0.27
$Ni = Ni^{+2} + 2e^-$	0.25
$V^{+2} = V^{+3} + e^-$	0.25
$I^- + Cu = CuI + e^-$	0.19
$I^- + Ag = AgI + e^-$	0.15
$Sn = Sn^{+2} + 2e^-$	0.14
$Pb = Pb^{+2} + 2e^-$	0.13
$2I^- + 2Hg = Hg_2I_2 + 2e^-$	0.04
$4I^- + Hg = HgI_4^{-2} + 2e^-$	0.04
$Fe = Fe^{+3} + 3e^-$	0.04
$2I^- + Cu = CuI_2^- + e^-$	0.00
$H_2 = 2H^+ + 2e^-$	0.00
$Br^- + Ag = AgBr + e^-$	−0.09
$H_2S = S + 2H^+ + 2e^-$	−0.14
$Sn^{+2} = Sn^{+4} + 2e^-$	−0.15

* W. M. Latimer, *The Oxidation States of the Elements and Their Potentials in Aqueous Solution*, Prentice-Hall, second edition, 1952.

TABLE A–5. Standard Half-Cell Voltages (Continued)

Couple	$E°$
$Cu^+ = Cu^{+2} + e^-$	-0.15
$Cl^- + Ag = AgCl + e^-$	-0.22
$6OH^- + I^- = IO_3^- + 3H_2O + 6e^-$	-0.26
$2Cl^- + 2Hg = Hg_2Cl_2 + 2e^-$	$-0.27*$
$Cu = Cu^{+2} + 2e$	-0.34
$H_2O + V^{+3} = VO^{+2} + 2H^+ + e^-$	-0.36
$Fe(CN)_6^{-4} = Fe(CN)_6^{-3} + e^-$	-0.36
$2NH_3(aq) + Ag = Ag(NH_3)_2^+ + e^-$	-0.37
$Cu = Cu^+ + e^-$	-0.52
$2I^- = I_2 + 2e^-$	-0.54
$2H_2O + HAsO_2 = H_3AsO_4 + 2H^+ + 2e^-$	-0.56
$4OH^- + MnO_2 = MnO_4^- + 2H_2O + 3e^-$	-0.59
$H_2O_2 = O_2 + 2H^+ + 2e^-$	-0.68
$Fe^{+2} = Fe^{+3} + e^-$	-0.77
$2Hg = Hg_2^{+2} + 2e^-$	-0.79
$Ag = Ag^+ + e^-$	-0.80
$Hg = Hg^{+2} + 2e^-$	-0.85
$CuI = Cu^{+2} + I^- + e^-$	-0.86
$Hg_2^{+2} = 2Hg^{+2} + 2e^-$	-0.92
$HNO_2 + H_2O = NO_3^- + 3H^+ + 2e^-$	-0.94
$2H_2O + NO = NO_3^- + 4H^+ + 3e^-$	-0.96
$2Br^- = Br_2(l) + 2e^-$	-1.07
$2Br^- = Br_2(aq) + 2e^-$	-1.09
$\frac{1}{2}I_2 + 3H_2O = IO_3^- + 6H^+ + 5e^-$	-1.20
$2H_2O = O_2 + 4H^+ + 4e^-$	-1.23
$2H_2O + Mn^{+2} = MnO_2 + 4H^+ + 2e^-$	-1.23
$7H_2O + 2Cr^{+3} = Cr_2O_7^{-2} + 14H^+ + 6e^-$	-1.33
$Cl^- = \frac{1}{2}Cl_2(g) + e^-$	-1.36
$3H_2O + Br^- = BrO_3^- + 6H^+ + 6e^-$	-1.44
$3H_2O + Cl^- = ClO_3^- + 6H^+ + 6e^-$	-1.45
$3H_2O + Pb^{+2} = PbO_2 + 4H^+ + 2e^-$	-1.46
$\frac{1}{2}Cl_2(g) + 3H_2O = ClO_3^- + 6H^+ + 5e^-$	-1.47
$Mn^{+2} = Mn^{+3} + e^-$	-1.51
$4H_2O + Mn^{+2} = MnO_4^- + 8H^+ + 5e^-$	-1.52
$\frac{1}{2}Br_2(l) + 3H_2O = BrO_3^- + 6H^+ + 5e^-$	-1.52
$Ce^{+3} = Ce^{+4} + e^-$	-1.61
$2H_2O + MnO_2 = MnO_4^- + 4H^+ + 3e^-$	-1.69
$Pb^{+2} = Pb^{+4} + 2e^-$	-1.69
$O_2 + H_2O = O_3 + 2H^+ + 2e^-$	-2.07
$2F^- = F_2(g) + 2e^-$	-2.85

* When the cell Hg, Hg_2Cl_2, KCl($1\,M$) is used in problems, the "normal calomel electrode" is indicated. Its emf is -0.28 volts. See p. 216.

Preparation of Indicator Solution

Bromcresol green. Dissolve 0.1 g bromcresol green in 7.2 ml 0.02 N NaOH, and dilute to 250 ml.

Bromphenol blue. Dissolve 0.1 g bromphenol blue in 7.5 ml 0.02 N NaOH, and dilute to 250 ml.

Bromthymol blue. Dissolve 0.1 g bromthymol blue in 8.0 ml 0.02 N NaOH, and dilute to 250 ml.

Dichlorofluorescein. Dissolve 0.1 g dichlorofluorescein in 100 ml 70 per cent alcohol, or dissolve 0.1 g sodium dichlorofluoresceinate in 100 ml of water.

Diphenylamine sulfonate. Dissolve 0.32 g barium diphenylamine sulfonate in 100 ml of water, and add 0.5 g sodium sulfate. Filter to remove barium sulfate precipitate.

Ferric alum. Dissolve 28 g ferric alum crystals in 80 ml of hot water. Cool, filter, and dilute to 100 ml with 6 N nitric acid.

Ferroin. See phenanthroline-ferrous ion indicator.

Methyl orange. Dissolve 0.1 g methyl orange in 100 ml of water.

Methyl red. Dissolve 0.1 g methyl red in 18.7 ml 0.02 N NaOH, and dilute to 250 ml.

Methyl red–bromcresol green mixed indicator. Mix two parts methyl red solution with three parts bromcresol green solution.

Modified methyl orange. Dissolve 0.75 g zylene cyanole FF and 1.5 g methyl orange in 1 liter of water.

Phenanthroline-ferrous ion (Ferroin). Dissolve 0.5 g phenanthroline monohydrate in 100 ml 0.025 M ferrous sulfate solution.

Phenolphthalein. Dissolve 0.1–0.5 g (depending on strength desired) phenolphthalein in 50 ml of alcohol, and add 50 ml of water.

Mathematical Operations

Use of Exponentials

Operations which involve very large or very small numbers may often be simplified by writing the number involved in a "semi-exponential" form, as a power of 10. This permits placing the decimal point at the most convenient position. For example, the number 602,000,000,000,000,000,000,000, which expresses the number of molecules in a mole, is conveniently written as

$$6.02 \times 10^{23}$$

In this form the exponent of 10 indicates the number of places the decimal point must be shifted to write out the number in full. In the number above the decimal point must be shifted 23 places to the *right* in order to write the number in the conventional notation. A negative exponent for 10 indicates the number of places which the decimal point must be shifted to the left.

Examples

$$(a) \ 12{,}345 = 1.2345 \times 10^4 = 12.345 \times 10^3$$
$$= 123.45 \times 10^2, \text{ etc.}$$

$$(b) \ 0.000018 = 18 \times 10^{-6} = 1.8 \times 10^{-5}, \text{ etc.}$$

The following rules govern the common operations which are performed with exponential numbers:

1. To multiply exponentials of the same base, add the exponents algebraically. Multiplication of the non-exponential numbers present is performed in the usual manner.

Examples:

$$(a) \ 10^1 \times 10^2 = 10 \times 100 = 1000 = 10^{1+2} = 10^3$$

$$(b) \quad 6 \times 10^3 \times 2 \times 10^2 \times 3 \times 10^{-7} =$$
$$6 \times 2 \times 3 \times 10^{3+2-7} = 36 \times 10^{-2} = 0.36$$

2. To divide exponentially, subtract the exponents algebraically.

Examples:

$$(a)\ \frac{10^{-2}}{10^3} = 10^{-2-3} = 10^{-5}$$

$$(b)\ \frac{6 \times 10^3 \times 2 \times 10^{-5}}{3 \times 10^{-8}} = \frac{6 \times 2}{3} \times 10^{3-5-(-8)}$$

$$= 4 \times 10^{+3-5+8}$$

$$= 4 \times 10^6$$

3. To raise an exponential to a power, multiply the exponent by the desired power.

Examples:

$$(a)\ (10^2)^2 = (100)^2 = 10000 = 10^4$$

or

$$(b)\ (3 \times 10^{-5})^2 = (3)^2 \times (10^{-5})^2 = 9 \times 10^{-10}$$

4. To extract a root, divide the exponent by the desired root. If the exponent is not divisible by the root to give a whole number, it is convenient to rewrite the number so as to obtain an exponent which is evenly divisible.

Examples:

$$(a)\ \sqrt{10^2} = 10^{2/2} = 10^1 = 10$$

$$(b)\ \sqrt{10^3} = \sqrt{10 \times 10^2} = \sqrt{10} \times \sqrt{10^2}$$

$$= 3.17 \times 10^1 = 31.7$$

$$(c)\ \sqrt[3]{8 \times 10^6} = \sqrt[3]{8} \times \sqrt[3]{10^6}$$

$$= 2 \times 10^2$$

$$(d)\ \sqrt[3]{8 \times 10^5} = \sqrt[3]{800 \times 10^3} = \sqrt[3]{800} \times \sqrt[3]{10^3}$$

$$= 9.28 \times 10^1$$

$$(e)\ \sqrt[3]{8 \times 10^{-4}} = \sqrt[3]{800 \times 10^{-6}} = 9.28 \times 10^{-2}$$

Use of Logarithms

The logarithm (log) of a number is the power to which the base 10 must be raised to give the desired number. Conversely, the antilog of a logarithm is the number whose logarithm has the given value. That is,

$$\log 10 = 1 \qquad \text{antilog } 1 = 10 \text{ or } 10^1 = 10$$

$$\log 100 = 2 \qquad \text{antilog } 2 = 100 \text{ or } 10^2 = 100$$

Every logarithm has two parts, a characteristic and a mantissa. The characteristic is the integral number preceding the decimal point; the mantissa is the fractional portion of the logarithm.

$$\log 200 = 2.30103 \text{ or } 10^{2.30103} = 200$$

In this log the number 2 is the characteristic and the number 0.30103 is the mantissa.

1. Finding a logarithm. Look up the mantissa in a log table, interpolating if necessary. Directions for this operation are given in Appendix IV. Determine the characteristic as follows. Write the number in an exponential form by placing the decimal point after the first digit and multiplying by the proper power of 10. The exponent of 10 is the characteristic of the log.

Examples:

    (a) $\log 2 = 0.30103$ (from log table)

    (b) $\log 20 = \log (2 \times 10^1) = \log 2 + \log 10^1$
                    $= 0.30103 + 1 = 1.30103$

    (c) $\log 0.002 = \log (2 \times 10^{-3}) = \log 2 + \log 10^{-3}$
                    $= 0.30103 + (-3) = \bar{3}.30103 = 7.30103 - 10 \text{ etc.}$

2. Negative mantissas. Log tables give only positive values of the mantissa. If an operation provides a negative logarithm, this must be transformed into a logarithm with a positive mantissa before the antilogarithm can be determined.

Example:

$$\log .002 = \bar{3}.30103$$

This number has a negative characteristic but a positive mantissa. It can be expressed as a negative logarithm by algebraically adding the two parts.

$$\bar{3}.30103 = 3 + 0.30103 = -2.69897$$

Example: Find antilog -2.69897.

Here the mantissa is negative. Rewriting to provide a positive mantissa gives $\bar{3}.30103$, the antilog of which is antilog 0.30103×10^{-3} or 2×10^{-3}.

3. Raising a number to a given power. To raise a number to a power, find the log, multiply the log by the desired power, and find the antilog.

Example: (a) Find $(2 \times 10^4)^2$

        $\log (2 \times 10^4) = 4.30103$

        $2 \times 4.30103 = 8.60206$

        antilog $8.60206 = $ antilog $8 + $ antilog 0.60206
                               $= 10^8 \times 4$

    (b) Find $(0.002)^3$

        $\log 0.002 \qquad = \bar{3}.30103 = -2.69897$

$$3 \times -(2.69897) = -8.09691$$

$$-8.09691 = \bar{9}.90309$$

$$\text{antilog } \bar{9}.90309 = \text{antilog } 0.90309 \times 10^{-9}$$
$$= 8 \times 10^{-9}$$

Optionally, the operation may be performed as follows:

$$\log 0.002 = 0.30103 - 3$$

$$3 \times \log 0.002 = 3 \times (0.30103 - 3) = 0.90309 - 9$$

4. Extraction of a root. To extract a root, divide the log of a number by the desired root and determine the antilog of the resulting number.

Example: Find $\sqrt[3]{0.02}$

$$\sqrt[3]{0.02} = \sqrt[3]{20 \times 10^{-3}} = \sqrt[3]{20} \times \sqrt[3]{10^{-3}}$$
$$= \sqrt[3]{20} \times 10^{-1}$$

$$\log 20 = 1.30103$$

$$\frac{1.30103}{3} = 0.43367$$

$$\text{antilog } 0.43367 = 2.714$$

$$\text{Therefore } \sqrt[3]{0.02} = 2.714 \times 10^{-1} = 0.2714$$

Quadratic Equations

The roots of the equation $ax^2 + bx + c = 0$ are given by the formula:

$$x = \frac{-b \pm \sqrt{b^2 - 4ac}}{2a}$$

Example: (a) Find the roots of $3x^2 + 8x + 2 = 0$.

$$x = \frac{-8 \pm \sqrt{64 - 24}}{6} = \frac{-8 \pm \sqrt{40}}{6} = \frac{-8 \pm 6.3}{6} = \frac{-8 + 6.3}{6} \text{ or } \frac{-8 - 6.3}{6}$$

(b) Solve the following equation:

$$\frac{x^2}{0.1 - x} = 10^{-6}$$

Rearranging:

$$x^2 = 10^{-7} - 10^{-6}x$$

$$x^2 + 10^{-6}x - 10^{-7} = 0$$

Application of the formula gives:

$$x = \frac{-10^{-6} \pm \sqrt{10^{-12} + 4 \times 10^{-7}}}{2}$$

This expression can be simplified by neglecting 10^{-12} (or 0.000000000001) with respect to 4×10^{-7} (or 0.0000004), an approximation which is justifiable, since the error introduced is very small.

$$x = \frac{-10^{-6} \pm \sqrt{40 \times 10^{-8}}}{2} = \frac{-10^{-6} \pm 6.3 \times 10^{-4}}{2}$$

In a chemical problem in which x represents the concentration of some ion, only the positive value has any significance. That is,

$$x = \frac{-10^{-6} + 6.3 \times 10^{-4}}{2} = \frac{-1 \times 10^{-6} + 630 \times 10^{-6}}{2}$$

$$= \frac{629 \times 10^{-6}}{2} = 315 \times 10^{-6} = 3.15 \times 10^{-4}$$

An approximate solution of a quadratic equation may often be obtained by neglecting some small term which is added to or subtracted from a larger number. For example, the equation

$$\frac{x^2}{0.1 - x} = 10^{-6}$$

becomes $\dfrac{x^2}{0.1} = 10^{-6}$ when the x term in the denominator is neglected, or

$$x^2 = 10^{-7} = 10 \times 10^{-8}$$

$$x = 3.16 \times 10^{-4}$$

This answer does not deviate greatly from the answer obtained by exact solution of the equation. For most of the quadratic equations encountered in calculations pertaining to chemical equilibrium, the approximate solution is perfectly valid since the equilibrium constants are not in general known to a precision greater than 10–100 per cent. A term cannot be neglected in all equations, however, since it may cause serious error. For example, in the equation

$$\frac{x^2}{0.1 - x} = 0.1$$

neglecting x leads to a solution

$$x = 0.1, \text{ which is absurd}$$

As a rule, if the solution after neglecting a term x leads to a value for x which is appreciable in comparison with the term from which it has been subtracted or to which it has been added, the approximate solution is not valid.

Logarithms

To find the mantissa of the logarithm of 375.27:
In the table, opposite 375, column "2," find   57426
Difference, column "3" and column "2" = 12.
In the table of proportional parts, headed "12," find
proportional part for 7 (last digit in number)   8.4
  Total, prefixing a decimal point,   .574344
or   .57434
The entire logarithm of 375.27 (rewritten as 3.7527×10^2) is, therefore, 2.57434.

To find the antilogarithm of 7.31478:
First, considering the mantissa: the number whose logarithm in the table is next lower than the given mantissa is 2064, whose logarithm is .31471. The logarithm of the next larger number, 2065, is .31492, giving a difference of .00021. Therefore, the mantissa, .31478, corresponds to a number which is 7/21 or 0.33 digit greater than 2064, or the number is 2064.3. Using the rules given in Appendix III, this gives the number 2.0643×10^7.

N	0	1	2	3	4	5	6	7	8	9
100	00 000	043	087	130	173	217	260	303	346	389
01	432	475	518	561	604	647	689	732	775	817
02	00 860	903	945	988	*030	*072	*115	*157	*199	*242
03	01 284	326	368	410	452	494	536	578	620	662
04	01 703	745	787	828	870	912	953	995	*036	*078
05	02 119	160	202	243	284	325	366	407	449	490
06	531	572	612	653	694	735	776	816	857	898
07	02 938	979	*019	*060	*100	*141	*181	*222	*262	*302
08	03 342	383	423	463	503	543	583	623	663	703
09	03 743	782	822	862	902	941	981	*021	*060	*100
110	04 139	179	218	258	297	336	376	415	454	493
11	532	571	610	650	689	727	766	805	844	883
12	04 922	961	999	*038	*077	*115	*154	*192	*231	*269
13	05 308	346	385	423	461	500	538	576	614	652
14	05 690	729	767	805	843	881	918	956	994	*032
15	06 070	108	145	183	221	258	296	333	371	408
16	446	483	521	558	595	633	670	707	744	781
17	06 819	856	893	930	967	*004	*041	*078	*115	*151
18	07 188	225	262	298	335	372	408	445	482	518
19	555	591	628	664	700	737	773	809	846	882
120	07 918	954	990	*027	*063	*099	*135	*171	*207	*243
21	08 279	314	350	386	422	458	493	529	565	600
22	636	672	707	743	778	814	849	884	920	955
23	08 991	*026	*061	*096	*132	*167	*202	*237	*272	*307
24	09 342	377	412	447	482	517	552	587	621	656
25	09 691	726	760	795	830	864	899	934	968	*003
26	10 037	072	106	140	175	209	243	278	312	346
27	380	415	449	483	517	551	585	619	653	687
28	10 721	755	789	823	857	890	924	958	992	*025
29	11 059	093	126	160	193	227	261	294	327	361
130	394	428	461	494	528	561	594	628	661	694
31	11 727	760	793	826	860	893	926	959	992	*024
32	12 057	090	123	156	189	222	254	287	320	352
33	385	418	450	483	516	548	581	613	646	678
34	12 710	743	775	808	840	872	905	937	969	*001
35	13 033	066	098	130	162	194	226	258	290	322
36	354	386	418	450	481	513	545	577	609	640
37	672	704	735	767	799	830	862	893	925	956
38	13 988	*019	*051	*082	*114	*145	*176	*208	*239	*270
39	14 301	333	364	395	426	457	489	520	551	582
140	613	644	675	706	737	768	799	829	860	891
41	14 922	953	983	*014	*045	*076	*106	*137	*168	*198
42	15 229	259	290	320	351	381	412	442	473	503
43	534	564	594	625	655	685	715	746	776	806
44	15 836	866	897	927	957	987	*017	*047	*077	*107
45	16 137	167	197	227	256	286	316	346	376	406
46	435	465	495	524	554	584	613	643	673	702
47	16 732	761	791	820	850	879	909	938	967	997
48	17 026	056	085	114	143	173	202	231	260	289
49	319	348	377	406	435	464	493	522	551	580
150	17 609	638	667	696	725	754	782	811	840	869
N	0	1	2	3	4	5	6	7	8	9

Prop. Parts

	44	43	42
1	4.4	4.3	4.2
2	8.8	8.6	8.4
3	13.2	12.9	12.6
4	17.6	17.2	16.8
5	22.0	21.5	21.0
6	26.4	25.8	25.2
7	30.8	30.1	29.4
8	35.2	34.4	33.6
9	39.6	38.7	37.8

	41	40	39
1	4.1	4	3.9
2	8.2	8	7.8
3	12.3	12	11.7
4	16.4	16	15.6
5	20.5	20	19.5
6	24.6	24	23.4
7	28.7	28	27.3
8	32.8	32	31.2
9	36.9	36	35.1

	38	37	36
1	3.8	3.7	3.6
2	7.6	7.4	7.2
3	11.4	11.1	10.8
4	15.2	14.8	14.4
5	19.0	18.5	18.0
6	22.8	22.2	21.6
7	26.6	25.9	25.2
8	30.4	29.6	28.8
9	34.2	33.3	32.4

	35	34	33
1	3.5	3.4	3.3
2	7.0	6.8	6.6
3	10.5	10.2	9.9
4	14.0	13.6	13.2
5	17.5	17.0	16.5
6	21.0	20.4	19.8
7	24.5	23.8	23.1
8	28.0	27.2	26.4
9	31.5	30.6	29.7

	32	31	30
1	3.2	3.1	3
2	6.4	6.2	6
3	9.6	9.3	9
4	12.8	12.4	12
5	16.0	15.5	15
6	19.2	18.6	18
7	22.4	21.7	21
8	25.6	24.8	24
9	28.8	27.9	27

* The first two figures of the mantissa are those indicated on the next line of the table.

N	0	1	2	3	4	5	6	7	8	9
150	17 609	638	667	696	725	754	782	811	840	869
51	17 898	926	955	984	*013	*041	*070	*099	*127	*156
52	18 184	213	241	270	298	327	355	384	412	441
53	469	498	526	554	583	611	639	667	696	724
54	18 752	780	808	837	865	893	921	949	977	*005
55	19 033	061	089	117	145	173	201	229	257	285
56	312	340	368	396	424	451	479	507	535	562
57	590	618	645	673	700	728	756	783	811	838
58	19 866	893	921	948	976	*003	*030	*058	*085	*112
59	20 140	167	194	222	249	276	303	330	358	385
160	412	439	466	493	520	548	575	602	629	656
61	683	710	737	763	790	817	844	871	898	925
62	20 952	978	*005	*032	*059	*085	*112	*139	*165	*192
63	21 219	245	272	299	325	352	378	405	431	458
64	484	511	537	564	590	617	643	669	696	722
65	21 748	775	801	827	854	880	906	932	958	985
66	22 011	037	063	089	115	141	167	194	220	246
67	272	298	324	350	376	401	427	453	479	505
68	531	557	583	608	634	660	686	712	737	763
69	22 789	814	840	866	891	917	943	968	994	*019
170	23 045	070	096	121	147	172	198	223	249	274
71	300	325	350	376	401	426	452	477	502	528
72	553	578	603	629	654	679	704	729	754	779
73	23 805	830	855	880	905	930	955	980	*005	*030
74	24 055	080	105	130	155	180	204	229	254	279
75	304	329	353	378	403	428	452	477	502	527
76	551	576	601	625	650	674	699	724	748	773
77	24 797	822	846	871	895	920	944	969	993	*018
78	25 042	066	091	115	139	164	188	212	237	261
79	285	310	334	358	382	406	431	455	479	503
180	527	551	575	600	624	648	672	696	720	744
81	25 768	792	816	840	864	888	912	935	959	983
82	26 007	031	055	079	102	126	150	174	198	221
83	245	269	293	316	340	364	387	411	435	458
84	482	505	529	553	576	600	623	647	670	694
85	717	741	764	788	811	834	858	881	905	928
86	26 951	975	998	*021	*045	*068	*091	*114	*138	*161
87	27 184	207	231	254	277	300	323	346	370	393
88	416	439	462	485	508	531	554	577	600	623
89	646	669	692	715	738	761	784	807	830	852
190	27 875	898	921	944	967	989	*012	*035	*058	*081
91	28 103	126	149	171	194	217	240	262	285	307
92	330	353	375	398	421	443	466	488	511	533
93	556	578	601	623	646	668	691	713	735	758
94	28 780	803	825	847	870	892	914	937	959	981
95	29 003	026	048	070	092	115	137	159	181	203
96	226	248	270	292	314	336	358	380	403	425
97	447	469	491	513	535	557	579	601	623	645
98	667	688	710	732	754	776	798	820	842	863
99	29 885	907	929	951	973	994	*016	*038	*060	*081
200	30 103	125	146	168	190	211	233	255	276	298

Prop. Parts

	29	28
1	2.9	2.8
2	5.8	5.6
3	8.7	8.4
4	11.6	11.2
5	14.5	14.0
6	17.4	16.8
7	20.3	19.6
8	23.2	22.4
9	26.1	25.2

	27	26
1	2.7	2.6
2	5.4	5.2
3	8.1	7.8
4	10.8	10.4
5	13.5	13.0
6	16.2	15.6
7	18.9	18.2
8	21.6	20.8
9	24.3	23.4

	25
1	2.5
2	5.0
3	7.5
4	10.0
5	12.5
6	15.0
7	17.5
8	20.0
9	22.5

	24	23
1	2.4	2.3
2	4.8	4.6
3	7.2	6.9
4	9.6	9.2
5	12.0	11.5
6	14.4	13.8
7	16.8	16.1
8	19.2	18.4
9	21.6	20.7

	22	21
1	2.2	2.1
2	4.4	4.2
3	6.6	6.3
4	8.8	8.4
5	11.0	10.5
6	13.2	12.6
7	15.4	14.7
8	17.6	16.8
9	19.8	18.9

*The first two figures of the mantissa are those indicated on the next line of the table.

N	0	1	2	3	4	5	6	7	8	9
200	30 103	125	146	168	190	211	233	255	276	298
01	320	341	363	384	406	428	449	471	492	514
02	535	557	578	600	621	643	664	685	707	728
03	750	771	792	814	835	856	878	899	920	942
04	30 963	984	*006	*027	*048	*069	*091	*112	*133	*154
05	31 175	197	218	239	260	281	302	323	345	366
06	387	408	429	450	471	492	513	534	555	576
07	597	618	639	660	681	702	723	744	765	785
08	31 806	827	848	869	890	911	931	952	973	994
09	32 015	035	056	077	098	118	139	160	181	201
210	222	243	263	284	305	325	346	366	387	408
11	428	449	469	490	510	531	552	572	593	613
12	634	654	675	695	715	736	756	777	797	818
13	32 838	858	879	899	919	940	960	980	*001	*021
14	33 041	062	082	102	122	143	163	183	203	224
15	244	264	284	304	325	345	365	385	405	425
16	445	465	486	506	526	546	566	586	606	626
17	646	666	686	706	726	746	766	786	806	826
18	33 846	866	885	905	925	945	965	985	*005	*025
19	34 044	064	084	104	124	143	163	183	203	223
220	242	262	282	301	321	341	361	380	400	420
21	439	459	479	498	518	537	557	577	596	616
22	635	655	674	694	713	733	753	772	792	811
23	34 830	850	869	889	908	928	947	967	986	*005
24	35 025	044	064	083	102	122	141	160	180	199
25	218	238	257	276	295	315	334	353	372	392
26	411	430	449	468	488	507	526	545	564	583
27	603	622	641	660	679	698	717	736	755	774
28	793	813	832	851	870	889	908	927	946	965
29	35 984	*003	*021	*040	*059	*078	*097	*116	*135	*154
230	36 173	192	211	229	248	267	286	305	324	342
31	361	380	399	418	436	455	474	493	511	530
32	549	568	586	605	624	642	661	680	698	717
33	736	754	773	791	810	829	847	866	884	903
34	36 922	940	959	977	996	*014	*033	*051	*070	*088
35	37 107	125	144	162	181	199	218	236	254	273
36	291	310	328	346	365	383	401	420	438	457
37	475	493	511	530	548	566	585	603	621	639
38	658	676	694	712	731	749	767	785	803	822
39	37 840	858	876	894	912	931	949	967	985	*003
240	38 021	039	057	075	093	112	130	148	166	184
41	202	220	238	256	274	292	310	328	346	364
42	382	399	417	435	453	471	489	507	525	543
43	561	578	596	614	632	650	668	686	703	721
44	739	757	775	792	810	828	846	863	881	899
45	38 917	934	952	970	987	*005	*023	*041	*058	*076
46	39 094	111	129	146	164	182	199	217	235	252
47	270	287	305	322	340	358	375	393	410	428
48	445	463	480	498	515	533	550	568	585	602
49	620	637	655	672	690	707	724	742	759	777
250	39 794	811	829	846	863	881	898	915	933	950
N	0	1	2	3	4	5	6	7	8	9

Prop. Parts

	22	21
1	2.2	2.1
2	4.4	4.2
3	6.6	6.3
4	8.8	8.4
5	11.0	10.5
6	13.2	12.6
7	15.4	14.7
8	17.6	16.8
9	19.8	18.9

	20
1	2
2	4
3	6
4	8
5	10
6	12
7	14
8	16
9	18

	19
1	1.9
2	3.8
3	5.7
4	7.6
5	9.5
6	11.4
7	13.3
8	15.2
9	17.1

	18
1	1.8
2	3.6
3	5.4
4	7.2
5	9.0
6	10.8
7	12.6
8	14.4
9	16.2

	17
1	1.7
2	3.4
3	5.1
4	6.8
5	8.5
6	10.2
7	11.9
8	13.6
9	15.3

*The first two figures of the mantissa are those indicated on the next line of the table.

Prop. Parts

18	
1	1.8
2	3.6
3	5.4
4	7.2
5	9.0
6	10.8
7	12.6
8	14.4
9	16.2

17	
1	1.7
2	3.4
3	5.1
4	6.8
5	8.5
6	10.2
7	11.9
8	13.6
9	15.3

16	
1	1.6
2	3.2
3	4.8
4	6.4
5	8.0
6	9.6
7	11.2
8	12.8
9	14.4

15	
1	1.5
2	3.0
3	4.5
4	6.0
5	7.5
6	9.0
7	10.5
8	12.0
9	13.5

14	
1	1.4
2	2.8
3	4.2
4	5.6
5	7.0
6	8.4
7	9.8
8	11.2
9	12.6

N	0	1	2	3	4	5	6	7	8	9
250	39 794	811	829	846	863	881	898	915	933	950
51	39 967	985	*002	*019	*037	*054	*071	*088	*106	*123
52	40 140	157	175	192	209	226	243	261	278	295
53	312	329	346	364	381	398	415	432	449	466
54	483	500	518	535	552	569	586	603	620	637
55	654	671	688	705	722	739	756	773	790	807
56	824	841	858	875	892	909	926	943	960	976
57	40 993	*010	*027	*044	*061	*078	*095	*111	*128	*145
58	41 162	179	196	212	229	246	263	280	296	313
59	330	347	363	380	397	414	430	447	464	481
260	497	514	531	547	564	581	597	614	631	647
61	664	681	697	714	731	747	764	780	797	814
62	830	847	863	880	896	913	929	946	963	979
63	41 996	*012	*029	*045	*062	*078	*095	*111	*127	*144
64	42 160	177	193	210	226	243	259	275	292	308
65	325	341	357	374	390	406	423	439	455	472
66	488	504	521	537	553	570	586	602	619	635
67	651	667	684	700	716	732	749	765	781	797
68	813	830	846	862	878	894	911	927	943	959
69	42 975	991	*008	*024	*040	*056	*072	*088	*104	*120
270	43 136	152	169	185	201	217	233	249	265	281
71	297	313	329	345	361	377	393	409	425	441
72	457	473	489	505	521	537	553	569	584	600
73	616	632	648	664	680	696	712	727	743	759
74	775	791	807	823	838	854	870	886	902	917
75	43 933	949	965	981	996	*012	*028	*044	*059	*075
76	44 091	107	122	138	154	170	185	201	217	232
77	248	264	279	295	311	326	342	358	373	389
78	404	420	436	451	467	483	498	514	529	545
79	560	576	592	607	623	638	654	669	685	700
280	716	731	747	762	778	793	809	824	840	855
81	44 871	886	902	917	932	948	963	979	994	*010
82	45 025	040	056	071	086	102	117	133	148	163
83	179	194	209	225	240	255	271	286	301	317
84	332	347	362	378	393	408	423	439	454	469
85	484	500	515	530	545	561	576	591	606	621
86	637	652	667	682	697	712	728	743	758	773
87	788	803	818	834	849	864	879	894	909	924
88	45 939	954	969	984	*000	*015	*030	*045	*060	*075
89	46 090	105	120	135	150	165	180	195	210	225
290	240	255	270	285	300	315	330	345	359	374
91	389	404	419	434	449	464	479	494	509	523
92	538	553	568	583	598	613	627	642	657	672
93	687	702	716	731	746	761	776	790	805	820
94	835	850	864	879	894	909	923	938	953	967
95	46 982	997	*012	*026	*041	*056	*070	*085	*100	*114
96	47 129	144	159	173	188	202	217	232	246	261
97	276	290	305	319	334	349	363	378	392	407
98	422	436	451	465	480	494	509	524	538	553
99	567	582	596	611	625	640	654	669	683	698
300	47 712	727	741	756	770	784	799	813	828	842

Prop. Parts

N	0	1	2	3	4	5	6	7	8	9

*The first two figures of the mantissa are those indicated on the next line of the table.

N	0	1	2	3	4	5	6	7	8	9	Prop. Parts
300	47 712	727	741	756	770	784	799	813	828	842	
01	47 857	871	885	900	914	929	943	958	972	986	
02	48 001	015	029	044	058	073	087	101	116	130	
03	144	159	173	187	202	216	230	244	259	273	
04	287	302	316	330	344	359	373	387	401	416	
05	430	444	458	473	487	501	515	530	544	558	
06	572	586	601	615	629	643	657	671	686	700	
07	714	728	742	756	770	785	799	813	827	841	
08	855	869	883	897	911	926	940	954	968	982	
09	48 996	*010	*024	*038	*052	*066	*080	*094	*108	*122	
310	49 136	150	164	178	192	206	220	234	248	262	
11	276	290	304	318	332	346	360	374	388	402	
12	415	429	443	457	471	485	499	513	527	541	
13	554	568	582	596	610	624	638	651	665	679	
14	693	707	721	734	748	762	776	790	803	817	
15	831	845	859	872	886	900	914	927	941	955	
16	49 969	982	996	*010	*024	*037	*051	*065	*079	*092	
17	50 106	120	133	147	161	174	188	202	215	229	
18	243	256	270	284	297	311	325	338	352	365	
19	379	393	406	420	433	447	461	474	488	501	
320	515	529	542	556	569	583	596	610	623	637	
21	651	664	678	691	705	718	732	745	759	772	
22	786	799	813	826	840	853	866	880	893	907	
23	50 920	934	947	961	974	987	*001	*014	*028	*041	
24	51 055	068	081	095	108	121	135	148	162	175	
25	188	202	215	228	242	255	268	282	295	308	
26	322	335	348	362	375	388	402	415	428	441	
27	455	468	481	495	508	521	534	548	561	574	
28	587	601	614	627	640	654	667	680	693	706	
29	720	733	746	759	772	786	799	812	825	838	
330	851	865	878	891	904	917	930	943	957	970	
31	51 983	996	*009	*022	*035	*048	*061	*075	*088	*101	
32	52 114	127	140	153	166	179	192	205	218	231	
33	244	257	270	284	297	310	323	336	349	362	
34	375	388	401	414	427	440	453	466	479	492	
35	504	517	530	543	556	569	582	595	608	621	
36	634	647	660	673	686	699	711	724	737	750	
37	763	776	789	802	815	827	840	853	866	879	
38	52 892	905	917	930	943	956	969	982	994	*007	
39	53 020	033	046	058	071	084	097	110	122	135	
340	148	161	173	186	199	212	224	237	250	263	
41	275	288	301	314	326	339	352	364	377	390	
42	403	415	428	441	453	466	479	491	504	517	
43	529	542	555	567	580	593	605	618	631	643	
44	656	668	681	694	706	719	732	744	757	769	
45	782	794	807	820	832	845	857	870	882	895	
46	53 908	920	933	945	958	970	983	995	*008	*020	
47	54 033	045	058	070	083	095	108	120	133	145	
48	158	170	183	195	208	220	233	245	258	270	
49	283	295	307	320	332	345	357	370	382	394	
350	54 407	419	432	444	456	469	481	494	506	518	
N	0	1	2	3	4	5	6	7	8	9	Prop. Parts

Prop. Parts

15
1	1.5
2	3.0
3	4.5
4	6.0
5	7.5
6	9.0
7	10.5
8	12.0
9	13.5

14
1	1.4
2	2.8
3	4.2
4	5.6
5	7.0
6	8.4
7	9.8
8	11.2
9	12.6

13
1	1.3
2	2.6
3	3.9
4	5.2
5	6.5
6	7.8
7	9.1
8	10.4
9	11.7

12
1	1.2
2	2.4
3	3.6
4	4.8
5	6.0
6	7.2
7	8.4
8	9.6
9	10.8

* The first two figures of the mantissa are those indicated on the next line of the table.

Proportional Parts

13		12		11		10	
1	1.3	1	1.2	1	1.1	1	1.0
2	2.6	2	2.4	2	2.2	2	2.0
3	3.9	3	3.6	3	3.3	3	3.0
4	5.2	4	4.8	4	4.4	4	4.0
5	6.5	5	6.0	5	5.5	5	5.0
6	7.8	6	7.2	6	6.6	6	6.0
7	9.1	7	8.4	7	7.7	7	7.0
8	10.4	8	9.6	8	8.8	8	8.0
9	11.7	9	10.8	9	9.9	9	9.0

N	0	1	2	3	4	5	6	7	8	9
350	54 407	419	432	444	456	469	481	494	506	518
51	531	543	555	568	580	593	605	617	630	642
52	654	667	679	691	704	716	728	741	753	765
53	777	790	802	814	827	839	851	864	876	888
54	54 900	913	925	937	949	962	974	986	998	*011
55	55 023	035	047	060	072	084	096	108	121	133
56	145	157	169	182	194	206	218	230	242	255
57	267	279	291	303	315	328	340	352	364	376
58	388	400	413	425	437	449	461	473	485	497
59	509	522	534	546	558	570	582	594	606	618
360	630	642	654	666	678	691	703	715	727	739
61	751	763	775	787	799	811	823	835	847	859
62	871	883	895	907	919	931	943	955	967	979
63	55 991	*003	*015	*027	*038	*050	*062	*074	*086	*098
64	56 110	122	134	146	158	170	182	194	205	217
65	229	241	253	265	277	289	301	312	324	336
66	348	360	372	384	396	407	419	431	443	455
67	467	478	490	502	514	526	538	549	561	573
68	585	597	608	620	632	644	656	667	679	691
69	703	714	726	738	750	761	773	785	797	808
370	820	832	844	855	867	879	891	902	914	926
71	56 937	949	961	972	984	996	*008	*019	*031	*043
72	57 054	066	078	089	101	113	124	136	148	159
73	171	183	194	206	217	229	241	252	264	276
74	287	299	310	322	334	345	357	368	380	392
75	403	415	426	438	449	461	473	484	496	507
76	519	530	542	553	565	576	588	600	611	623
77	634	646	657	669	680	692	703	715	726	738
78	749	761	772	784	795	807	818	830	841	852
79	864	875	887	898	910	921	933	944	955	967
380	57 978	990	*001	*013	*024	*035	*047	*058	*070	*081
81	58 092	104	115	127	138	149	161	172	184	195
82	206	218	229	240	252	263	274	286	297	309
83	320	331	343	354	365	377	388	399	410	422
84	433	444	456	467	478	490	501	512	524	535
85	546	557	569	580	591	602	614	625	636	647
86	659	670	681	692	704	715	726	737	749	760
87	771	782	794	805	816	827	838	850	861	872
88	883	894	906	917	928	939	950	961	973	984
89	58 995	*006	*017	*028	*040	*051	*062	*073	*084	*095
390	59 106	118	129	140	151	162	173	184	195	207
91	218	229	240	251	262	273	284	295	306	318
92	329	340	351	362	373	384	395	406	417	428
93	439	450	461	472	483	494	506	517	528	539
94	550	561	572	583	594	605	616	627	638	649
95	660	671	682	693	704	715	726	737	748	759
96	770	780	791	802	813	824	835	846	857	868
97	879	890	901	912	923	934	945	956	966	977
98	59 988	999	*010	*021	*032	*043	*054	*065	*076	*086
99	60 097	108	119	130	141	152	163	173	184	195
400	60 206	217	228	239	249	260	271	282	293	304

Prop. Parts	N	0	1	2	3	4	5	6	7	8	9

*The first two figures of the mantissa are those indicated on the next line of the table.

N	0	1	2	3	4	5	6	7	8	9	Prop. Parts	
400	60 206	217	228	239	249	260	271	282	293	304		
01	314	325	336	347	358	369	379	390	401	412		
02	423	433	444	455	466	477	487	498	509	520		
03	531	541	552	563	574	584	595	606	617	627		
04	638	649	660	670	681	692	703	713	724	735		
05	746	756	767	778	788	799	810	821	831	842		
06	853	863	874	885	895	906	917	927	938	949		**11**
07	60 959	970	981	991	*002	*013	*023	*034	*045	*055	1	1.1
08	61 066	077	087	098	109	119	130	140	151	162	2	2.2
09	172	183	194	204	215	225	236	247	257	268	3	3.3
											4	4.4
410	278	289	300	310	321	331	342	352	363	374	5	5.5
											6	6.6
11	384	395	405	416	426	437	448	458	469	479	7	7.7
12	490	500	511	521	532	542	553	563	574	584	8	8.8
13	595	606	616	627	637	648	658	669	679	690	9	9.9
14	700	711	721	731	742	752	763	773	784	794		
15	805	815	826	836	847	857	868	878	888	899		
16	61 909	920	930	941	951	962	972	982	993	*003		
17	62 014	024	034	045	055	066	076	086	097	107		
18	118	128	138	149	159	170	180	190	201	211		
19	221	232	242	252	263	273	284	294	304	315		
420	325	335	346	356	366	377	387	397	408	418		
21	428	439	449	459	469	480	490	500	511	521		
22	531	542	552	562	572	583	593	603	613	624		**10**
23	634	644	655	665	675	685	696	706	716	726	1	1.0
24	737	747	757	767	778	788	798	808	818	829	2	2.0
25	839	849	859	870	880	890	900	910	921	931	3	3.0
26	62 941	951	961	972	982	992	*002	*012	*022	*033	4	4.0
											5	5.0
27	63 043	053	063	073	083	094	104	114	124	134	6	6.0
28	144	155	165	175	185	195	205	215	225	236	7	7.0
29	246	256	266	276	286	296	306	317	327	337	8	8.0
											9	9.0
430	347	357	367	377	387	397	407	417	428	438		
31	448	458	468	478	488	498	508	518	528	538		
32	548	558	568	579	589	599	609	619	629	639		
33	649	659	669	679	689	699	709	719	729	739		
34	749	759	769	779	789	799	809	819	829	839		
35	849	859	869	879	889	899	909	919	929	939		
36	63 949	959	969	979	988	998	*008	*018	*028	*038		
37	64 048	058	068	078	088	098	108	118	128	137		**9**
38	147	157	167	177	187	197	207	217	227	237	1	0.9
39	246	256	266	276	286	296	306	316	326	335	2	1.8
											3	2.7
440	345	355	365	375	385	395	404	414	424	434	4	3.6
											5	4.5
41	444	454	464	473	483	493	503	513	523	532	6	5.4
42	542	552	562	572	582	591	601	611	621	631	7	6.3
43	640	650	660	670	680	689	699	709	719	729	8	7.2
											9	8.1
44	738	748	758	768	777	787	797	807	816	826		
45	836	846	856	865	875	885	895	904	914	924		
46	64 933	943	953	963	972	982	992	*002	*011	*021		
47	65 031	040	050	060	070	079	089	099	108	118		
48	128	137	147	157	167	176	186	196	205	215		
49	225	234	244	254	263	273	283	292	302	312		
450	65 321	331	341	350	360	369	379	389	398	408		
N	0	1	2	3	4	5	6	7	8	9	Prop. Parts	

*The first two figures of the mantissa are those indicated on the next line of the table.

N	0	1	2	3	4	5	6	7	8	9
450	65 321	331	341	350	360	369	379	389	398	408
51	418	427	437	447	456	466	475	485	495	504
52	514	523	533	543	552	562	571	581	591	600
53	610	619	629	639	648	658	667	677	686	696
54	706	715	725	734	744	753	763	772	782	792
55	801	811	820	830	839	849	858	868	877	887
56	896	906	916	925	935	944	954	963	973	982
57	65 992	*001	*011	*020	*030	*039	*049	*058	*068	*077
58	66 087	096	106	115	124	134	143	153	162	172
59	181	191	200	210	219	229	238	247	257	266
460	276	285	295	304	314	323	332	342	351	361
61	370	380	389	398	408	417	427	436	445	455
62	464	474	483	492	502	511	521	530	539	549
63	558	567	577	586	596	605	614	624	633	642
64	652	661	671	680	689	699	708	717	727	736
65	745	755	764	773	783	792	801	811	820	829
66	839	848	857	867	876	885	894	904	913	922
67	66 932	941	950	960	969	978	987	997	*006	*015
68	67 025	034	043	052	062	071	080	089	099	108
69	117	127	136	145	154	164	173	182	191	201
470	210	219	228	237	247	256	265	274	284	293
71	302	311	321	330	339	348	357	367	376	385
72	394	403	413	422	431	440	449	459	468	477
73	486	495	504	514	523	532	541	550	560	569
74	578	587	596	605	614	624	633	642	651	660
75	669	679	688	697	706	715	724	733	742	752
76	761	770	779	788	797	806	815	825	834	843
77	852	861	870	879	888	897	906	916	925	934
78	67 943	952	961	970	979	988	997	*006	*015	*024
79	68 034	043	052	061	070	079	088	097	106	115
480	124	133	142	151	160	169	178	187	196	205
81	215	224	233	242	251	260	269	278	287	296
82	305	314	323	332	341	350	359	368	377	386
83	395	404	413	422	431	440	449	458	467	476
84	485	494	502	511	520	529	538	547	556	565
85	574	583	592	601	610	619	628	637	646	655
86	664	673	681	690	699	708	717	726	735	744
87	753	762	771	780	789	797	806	815	824	833
88	842	851	860	869	878	886	895	904	913	922
89	68 931	940	949	958	966	975	984	993	*002	*011
490	69 020	028	037	046	055	064	073	082	090	099
91	108	117	126	135	144	152	161	170	179	188
92	197	205	214	223	232	241	249	258	267	276
93	285	294	302	311	320	329	338	346	355	364
94	373	381	390	399	408	417	425	434	443	452
95	461	469	478	487	496	504	513	522	531	539
96	548	557	566	574	583	592	601	609	618	627
97	636	644	653	662	671	679	688	697	705	714
98	723	732	740	749	758	767	775	784	793	801
99	810	819	827	836	845	854	862	871	880	888
500	69 897	906	914	923	932	940	949	958	966	975

Prop. Parts

10
1	1.0
2	2.0
3	3.0
4	4.0
5	5.0
6	6.0
7	7.0
8	8.0
9	9.0

9
1	0.9
2	1.8
3	2.7
4	3.6
5	4.5
6	5.4
7	6.3
8	7.2
9	8.1

8
1	0.8
2	1.6
3	2.4
4	3.2
5	4.0
6	4.8
7	5.6
8	6.4
9	7.2

* The first two figures of the mantissa are those indicated on the next line of the table.

N	0	1	2	3	4	5	6	7	8	9	Prop. Parts	
500	69 897	906	914	923	932	940	949	958	966	975		
01	69 984	992	*001	*010	*018	*027	*036	*044	*053	*062		
02	70 070	079	088	096	105	114	122	131	140	148		
03	157	165	174	183	191	200	209	217	226	234		
04	243	252	260	269	278	286	295	303	312	321		
05	329	338	346	355	364	372	381	389	398	406		
06	415	424	432	441	449	458	467	475	484	492		**9**
07	501	509	518	526	535	544	552	561	569	578	**1**	0.9
08	586	595	603	612	621	629	638	646	655	663	**2**	1.8
09	672	680	689	697	706	714	723	731	740	749	**3**	2.7
											4	3.6
510	757	766	774	783	791	800	808	817	825	834	**5**	4.5
											6	5.4
11	842	851	859	868	876	885	893	902	910	919	**7**	6.3
12	70 927	935	944	952	961	969	978	986	995	*003	**8**	7.2
13	71 012	020	029	037	046	054	063	071	079	088	**9**	8.1
14	096	105	113	122	130	139	147	155	164	172		
15	181	189	198	206	214	223	231	240	248	257		
16	265	273	282	290	299	307	315	324	332	341		
17	349	357	366	374	383	391	399	408	416	425		
18	433	441	450	458	466	475	483	492	500	508		
19	517	525	533	542	550	559	567	575	584	592		
520	600	609	617	625	634	642	650	659	667	675		
21	684	692	700	709	717	725	734	742	750	759		
22	767	775	784	792	800	809	817	825	834	842		**8**
23	850	858	867	875	883	892	900	908	917	925	**1**	0.8
											2	1.6
24	71 933	941	950	958	966	975	983	991	999	*008	**3**	2.4
25	72 016	024	032	041	049	057	066	074	082	090	**4**	3.2
26	099	107	115	123	132	140	148	156	165	173	**5**	4.0
											6	4.8
27	181	189	198	206	214	222	230	239	247	255	**7**	5.6
28	263	272	280	288	296	304	313	321	329	337	**8**	6.4
29	346	354	362	370	378	387	395	403	411	419	**9**	7.2
530	428	436	444	452	460	469	477	485	493	501		
31	509	518	526	534	542	550	558	567	575	583		
32	591	599	607	616	624	632	640	648	656	665		
33	673	681	689	697	705	713	722	730	738	746		
34	754	762	770	779	787	795	803	811	819	827		
35	835	843	852	860	868	876	884	892	900	908		
36	916	925	933	941	949	957	965	973	981	989		
37	72 997	*006	*014	*022	*030	*038	*046	*054	*062	*070		**7**
38	73 078	086	094	102	111	119	127	135	143	151	**1**	0.7
39	159	167	175	183	191	199	207	215	223	231	**2**	1.4
											3	2.1
540	239	247	255	263	272	280	288	296	304	312	**4**	2.8
41	320	328	336	344	352	360	368	376	384	392	**5**	3.5
42	400	408	416	424	432	440	448	456	464	472	**6**	4.2
43	480	488	496	504	512	520	528	536	544	552	**7**	4.9
											8	5.6
44	560	568	576	584	592	600	608	616	624	632	**9**	6.3
45	640	648	656	664	672	679	687	695	703	711		
46	719	727	735	743	751	759	767	775	783	791		
47	799	807	815	823	830	838	846	854	862	870		
48	878	886	894	902	910	918	926	933	941	949		
49	73 957	965	973	981	989	997	*005	*013	*020	*028		
550	74 036	044	052	060	068	076	084	092	099	107		
N	0	1	2	3	4	5	6	7	8	9	Prop. Parts	

* The first two figures of the mantissa are those indicated on the next line of the table.

Prop. Parts	N	0	1	2	3	4	5	6	7	8	9
	550	74 036	044	052	060	068	076	084	092	099	107
	51	115	123	131	139	147	155	162	170	178	186
	52	194	202	210	218	225	233	241	249	257	265
	53	273	280	288	296	304	312	320	327	335	343
	54	351	359	367	374	382	390	398	406	414	421
	55	429	437	445	453	461	468	476	484	492	500
	56	507	515	523	531	539	547	554	562	570	578
	57	586	593	601	609	617	624	632	640	648	656
	58	663	671	679	687	695	702	710	718	726	733
	59	741	749	757	764	772	780	788	796	803	811
	560	819	827	834	842	850	858	865	873	881	889
	61	896	904	912	920	927	935	943	950	958	966
	62	74 974	981	989	997	*005	*012	*020	*028	*035	*043
	63	75 051	059	066	074	082	089	097	105	113	120
	64	128	136	143	151	159	166	174	182	189	197
	65	205	213	220	228	236	243	251	259	266	274
	66	282	289	297	305	312	320	328	335	343	351
	67	358	366	374	381	389	397	404	412	420	427
	68	435	442	450	458	465	473	481	488	496	504
	69	511	519	526	534	542	549	557	565	572	580
	570	587	595	603	610	618	626	633	641	648	656
	71	664	671	679	686	694	702	709	717	724	732
	72	740	747	755	762	770	778	785	793	800	808
	73	815	823	831	838	846	853	861	868	876	884
	74	891	899	906	914	921	929	937	944	952	959
	75	75 967	974	982	989	997	*005	*012	*020	*027	*035
	76	76 042	050	057	065	072	080	087	095	103	110
	77	118	125	133	140	148	155	163	170	178	185
	78	193	200	208	215	223	230	238	245	253	260
	79	268	275	283	290	298	305	313	320	328	335
	580	343	350	358	365	373	380	388	395	403	410
	81	418	425	433	440	448	455	462	470	477	485
	82	492	500	507	515	522	530	537	545	552	559
	83	567	574	582	589	597	604	612	619	626	634
	84	641	649	656	664	671	678	686	693	701	708
	85	716	723	730	738	745	753	760	768	775	782
	86	790	797	805	812	819	827	834	842	849	856
	87	864	871	879	886	893	901	908	916	923	930
	88	76 938	945	953	960	967	975	982	989	997	*004
	89	77 012	019	026	034	041	048	056	063	070	078
	590	085	093	100	107	115	122	129	137	144	151
	91	159	166	173	181	188	195	203	210	217	225
	92	232	240	247	254	262	269	276	283	291	298
	93	305	313	320	327	335	342	349	357	364	371
	94	379	386	393	401	408	415	422	430	437	444
	95	452	459	466	474	481	488	495	503	510	517
	96	525	532	539	546	554	561	568	576	583	590
	97	597	605	612	619	627	634	641	648	656	663
	98	670	677	685	692	699	706	714	721	728	735
	99	743	750	757	764	772	779	786	793	801	808
	600	77 815	822	830	837	844	851	859	866	873	880
Prop. Parts	**N**	**0**	**1**	**2**	**3**	**4**	**5**	**6**	**7**	**8**	**9**

Prop. Parts

	8
1	0.8
2	1.6
3	2.4
4	3.2
5	4.0
6	4.8
7	5.6
8	6.4
9	7.2

	7
1	0.7
2	1.4
3	2.1
4	2.8
5	3.5
6	4.2
7	4.9
8	5.6
9	6.3

* The first two figures of the mantissa are those indicated on the next line of the table.

N	0	1	2	3	4	5	6	7	8	9	Prop. Parts
600	77 815	822	830	837	844	851	859	866	873	880	
01	887	895	902	909	916	924	931	938	945	952	
02	77 960	967	974	981	988	996	*003	*010	*017	*025	
03	78 032	039	046	053	061	068	075	082	089	097	
04	104	111	118	125	132	140	147	154	161	168	
05	176	183	190	197	204	211	219	226	233	240	
06	247	254	262	269	276	283	290	297	305	312	
07	319	326	333	340	347	355	362	369	376	383	
08	390	398	405	412	419	426	433	440	447	455	
09	462	469	476	483	490	497	504	512	519	526	
610	533	540	547	554	561	569	576	583	590	597	
11	604	611	618	625	633	640	647	654	661	668	
12	675	682	689	696	704	711	718	725	732	739	
13	746	753	760	767	774	781	789	796	803	810	
14	817	824	831	838	845	852	859	866	873	880	
15	888	895	902	909	916	923	930	937	944	951	
16	78 958	965	972	979	986	993	*000	*007	*014	*021	
17	79 029	036	043	050	057	064	071	078	085	092	
18	099	106	113	120	127	134	141	148	155	162	
19	169	176	183	190	197	204	211	218	225	232	
620	239	246	253	260	267	274	281	288	295	302	
21	309	316	323	330	337	344	351	358	365	372	
22	379	386	393	400	407	414	421	428	435	442	
23	449	456	463	470	477	484	491	498	505	511	
24	518	525	532	539	546	553	560	567	574	581	
25	588	595	602	609	616	623	630	637	644	650	
26	657	664	671	678	685	692	699	706	713	720	
27	727	734	741	748	754	761	768	775	782	789	
28	796	803	810	817	824	831	837	844	851	858	
29	865	872	879	886	893	900	906	913	920	927	
630	79 934	941	948	955	962	969	975	982	989	996	
31	80 003	010	017	024	030	037	044	051	058	065	
32	072	079	085	092	099	106	113	120	127	134	
33	140	147	154	161	168	175	182	188	195	202	
34	209	216	223	229	236	243	250	257	264	271	
35	277	284	291	298	305	312	318	325	332	339	
36	346	353	359	366	373	380	387	393	400	407	
37	414	421	428	434	441	448	455	462	468	475	
38	482	489	496	502	509	516	523	530	536	543	
39	550	557	564	570	577	584	591	598	604	611	
640	618	625	632	638	645	652	659	665	672	679	
41	686	693	699	706	713	720	726	733	740	747	
42	754	760	767	774	781	787	794	801	808	814	
43	821	828	835	841	848	855	862	868	875	882	
44	889	895	902	909	916	922	929	936	943	949	
45	80 956	963	969	976	983	990	996	*003	*010	*017	
46	81 023	030	037	043	050	057	064	070	077	084	
47	090	097	104	111	117	124	131	137	144	151	
48	158	164	171	178	184	191	198	204	211	218	
49	224	231	238	245	251	258	265	271	278	285	
650	81 291	298	305	311	318	325	331	338	345	351	
N	0	1	2	3	4	5	6	7	8	9	Prop. Parts

Prop. Parts:

	8
1	0.8
2	1.6
3	2.4
4	3.2
5	4.0
6	4.8
7	5.6
8	6.4
9	7.2

	7
1	0.7
2	1.4
3	2.1
4	2.8
5	3.5
6	4.2
7	4.9
8	5.6
9	6.3

	6
1	0.6
2	1.2
3	1.8
4	2.4
5	3.0
6	3.6
7	4.2
8	4.8
9	5.4

* The first two figures of the mantissa are those indicated on the next line of the table.

Prop. Parts	N	0	1	2	3	4	5	6	7	8	9
	750	87 506	512	518	523	529	535	541	547	552	558
	51	564	570	576	581	587	593	599	604	610	616
	52	622	628	633	639	645	651	656	662	668	674
	53	679	685	691	697	703	708	714	720	726	731
	54	737	743	749	754	760	766	772	777	783	789
	55	795	800	806	812	818	823	829	835	841	846
	56	852	858	864	869	875	881	887	892	898	904
	57	910	915	921	927	933	938	944	950	955	961
	58	87 967	973	978	984	990	996	*001	*007	*013	*018
	59	88 024	030	036	041	047	053	058	064	070	076
	760	081	087	093	098	104	110	116	121	127	133
	61	138	144	150	156	161	167	173	178	184	190
	62	195	201	207	213	218	224	230	235	241	247
6	63	252	258	264	270	275	281	287	292	298	304
1 0.6	64	309	315	321	326	332	338	343	349	355	360
2 1.2	65	366	372	377	383	389	395	400	406	412	417
3 1.8	66	423	429	434	440	446	451	457	463	468	474
4 2.4	67	480	485	491	497	502	508	513	519	525	530
5 3.0	68	536	542	547	553	559	564	570	576	581	587
6 3.6	69	593	598	604	610	615	621	627	632	638	643
7 4.2	**770**	649	655	660	666	672	677	683	689	694	700
8 4.8	71	705	711	717	722	728	734	739	745	750	756
9 5.4	72	762	767	773	779	784	790	795	801	807	812
	73	818	824	829	835	840	846	852	857	863	868
	74	874	880	885	891	897	902	908	913	919	925
	75	930	936	941	947	953	958	964	969	975	981
	76	88 986	992	997	*003	*009	*014	*020	*025	*031	*037
	77	89 042	048	053	059	064	070	076	081	087	092
	78	098	104	109	115	120	126	131	137	143	148
	79	154	159	165	170	176	182	187	193	198	204
	780	209	215	221	226	232	237	243	248	254	260
	81	265	271	276	282	287	293	298	304	310	315
5	82	321	326	332	337	343	348	354	360	365	371
1 0.5	83	376	382	387	393	398	404	409	415	421	426
2 1.0	84	432	437	443	448	454	459	465	470	476	481
3 1.5	85	487	492	498	504	509	515	520	526	531	537
4 2.0	86	542	548	553	559	564	570	575	581	586	592
5 2.5	87	597	603	609	614	620	625	631	636	642	647
6 3.0	88	653	658	664	669	675	680	686	691	697	702
7 3.5	89	708	713	719	724	730	735	741	746	752	757
8 4.0	**790**	763	768	774	779	785	790	796	801	807	812
9 4.5	91	818	823	829	834	840	845	851	856	862	867
	92	873	878	883	889	894	900	905	911	916	922
	93	927	933	938	944	949	955	960	966	971	977
	94	89 982	988	993	998	*004	*009	*015	*020	*026	*031
	95	90 037	042	048	053	059	064	069	075	080	086
	96	091	097	102	108	113	119	124	129	135	140
	97	146	151	157	162	168	173	179	184	189	195
	98	200	206	211	217	222	227	233	238	244	249
	99	255	260	266	271	276	282	287	293	298	304
	800	90 309	314	320	325	331	336	342	347	352	358
Prop. Parts	N	0	1	2	3	4	5	6	7	8	9

*The first two figures of the mantissa are those indicated on the next line of the table.

N	0	1	2	3	4	5	6	7	8	9	Prop. Parts
800	90 309	314	320	325	331	336	342	347	352	358	
01	363	369	374	380	385	390	396	401	407	412	
02	417	423	428	434	439	445	450	455	461	466	
03	472	477	482	488	493	499	504	509	515	520	
04	526	531	536	542	547	553	558	563	569	574	
05	580	585	590	596	601	607	612	617	623	628	
06	634	639	644	650	655	660	666	671	677	682	
07	687	693	698	703	709	714	720	725	730	736	
08	741	747	752	757	763	768	773	779	784	789	
09	795	800	806	811	816	822	827	832	838	843	
810	849	854	859	865	870	875	881	886	891	897	
11	902	907	913	918	924	929	934	940	945	950	
12	90 956	961	966	972	977	982	988	993	998	*004	
13	91 009	014	020	025	030	036	041	046	052	057	
14	062	068	073	078	084	089	094	100	105	110	
15	116	121	126	132	137	142	148	153	158	164	
16	169	174	180	185	190	196	201	206	212	217	
17	222	228	233	238	243	249	254	259	265	270	
18	275	281	286	291	297	302	307	312	318	323	
19	328	334	339	344	350	355	360	365	371	376	
820	381	387	392	397	403	408	413	418	424	429	
21	434	440	445	450	455	461	466	471	477	482	
22	487	492	498	503	508	514	519	524	529	535	
23	540	545	551	556	561	566	572	577	582	587	
24	593	598	603	609	614	619	624	630	635	640	
25	645	651	656	661	666	672	677	682	687	693	
26	698	703	709	714	719	724	730	735	740	745	
27	751	756	761	766	772	777	782	787	793	798	
28	803	808	814	819	824	829	834	840	845	850	
29	855	861	866	871	876	882	887	892	897	903	
830	908	913	918	924	929	934	939	944	950	955	
31	91 960	965	971	976	981	986	991	997	*002	*007	
32	92 012	018	023	028	033	038	044	049	054	059	
33	065	070	075	080	085	091	096	101	106	111	
34	117	122	127	132	137	143	148	153	158	163	
35	169	174	179	184	189	195	200	205	210	215	
36	221	226	231	236	241	247	252	257	262	267	
37	273	278	283	288	293	298	304	309	314	319	
38	324	330	335	340	345	350	355	361	366	371	
39	376	381	387	392	397	402	407	412	418	423	
840	428	433	438	443	449	454	459	464	469	474	
41	480	485	490	495	500	505	511	516	521	526	
42	531	536	542	547	552	557	562	567	572	578	
43	583	588	593	598	603	609	614	619	624	629	
44	634	639	645	650	655	660	665	670	675	681	
45	686	691	696	701	706	711	716	722	727	732	
46	737	742	747	752	758	763	768	773	778	783	
47	788	793	799	804	809	814	819	824	829	834	
48	840	845	850	855	860	865	870	875	881	886	
49	891	896	901	906	911	916	921	927	932	937	
850	92 942	947	952	957	962	967	973	978	983	988	
N	0	1	2	3	4	5	6	7	8	9	Prop. Parts

Prop. Parts (6):

	6
1	0.6
2	1.2
3	1.8
4	2.4
5	3.0
6	3.6
7	4.2
8	4.8
9	5.4

Prop. Parts (5):

	5
1	0.5
2	1.0
3	1.5
4	2.0
5	2.5
6	3.0
7	3.5
8	4.0
9	4.5

* The first two figures of the mantissa are those indicated on the next line of the table.

Prop. Parts	N	0	1	2	3	4	5	6	7	8	9
	850	92 942	947	952	957	962	967	973	978	983	988
	51	92 993	998	*003	*008	*013	*018	*024	*029	*034	*039
	52	93 044	049	054	059	064	069	075	080	085	090
	53	095	100	105	110	115	120	125	131	136	141
	54	146	151	156	161	166	171	176	181	186	192
	55	197	202	207	212	217	222	227	232	237	242
	56	247	252	258	263	268	273	278	283	288	293
	57	298	303	308	313	318	323	328	334	339	344
	58	349	354	359	364	369	374	379	384	389	394
	59	399	404	409	414	420	425	430	435	440	445
	860	450	455	460	465	470	475	480	485	490	495
	61	500	505	510	515	520	526	531	536	541	546
	62	551	556	561	566	571	576	581	586	591	596
	63	601	606	611	616	621	626	631	636	641	646
	64	651	656	661	666	671	676	682	687	692	697
	65	702	707	712	717	722	727	732	737	742	747
	66	752	757	762	767	772	777	782	787	792	797
	67	802	807	812	817	822	827	832	837	842	847
	68	852	857	862	867	872	877	882	887	892	897
	69	902	907	912	917	922	927	932	937	942	947
	870	93 952	957	962	967	972	977	982	987	992	997
	71	94 002	007	012	017	022	027	032	037	042	047
	72	052	057	062	067	072	077	082	086	091	096
	73	101	106	111	116	121	126	131	136	141	146
	74	151	156	161	166	171	176	181	186	191	196
	75	201	206	211	216	221	226	231	236	240	245
	76	250	255	260	265	270	275	280	285	290	295
	77	300	305	310	315	320	325	330	335	340	345
	78	349	354	359	364	369	374	379	384	389	394
	79	399	404	409	414	419	424	429	433	438	443
	880	448	453	458	463	468	473	478	483	488	493
	81	498	503	507	512	517	522	527	532	537	542
	82	547	552	557	562	567	571	576	581	586	591
	83	596	601	606	611	616	621	626	630	635	640
	84	645	650	655	660	665	670	675	680	685	689
	85	694	699	704	709	714	719	724	729	734	738
	86	743	748	753	758	763	768	773	778	783	787
	87	792	797	802	807	812	817	822	827	832	836
	88	841	846	851	856	861	866	871	876	880	885
	89	890	895	900	905	910	915	919	924	929	934
	890	939	944	949	954	959	963	968	973	978	983
	91	94 988	993	998	*002	*007	*012	*017	*022	*027	*032
	92	95 036	041	046	051	056	061	066	071	075	080
	93	085	090	095	100	105	109	114	119	124	129
	94	134	139	143	148	153	158	163	168	173	177
	95	182	187	192	197	202	207	211	216	221	226
	96	231	236	240	245	250	255	260	265	270	274
	97	279	284	289	294	299	303	308	313	318	323
	98	328	332	337	342	347	352	357	361	366	371
	99	376	381	386	390	395	400	405	410	415	419
	900	95 424	429	434	439	444	448	453	458	463	468
Prop. Parts	N	0	1	2	3	4	5	6	7	8	9

Prop. Parts

6

1	0.6
2	1.2
3	1.8
4	2.4
5	3.0
6	3.6
7	4.2
8	4.8
9	5.4

5

1	0.5
2	1.0
3	1.5
4	2.0
5	2.5
6	3.0
7	3.5
8	4.0
9	4.5

4

1	0.4
2	0.8
3	1.2
4	1.6
5	2.0
6	2.4
7	2.8
8	3.2
9	3.6

* The first two figures of the mantissa are those indicated on the next line of the table.

N	0	1	2	3	4	5	6	7	8	9	Prop. Parts
900	95 424	429	434	439	444	448	453	458	463	468	
01	472	477	482	487	492	497	501	506	511	516	
02	521	525	530	535	540	545	550	554	559	564	
03	569	574	578	583	588	593	598	602	607	612	
04	617	622	626	631	636	641	646	650	655	660	
05	665	670	674	679	684	689	694	698	703	708	
06	713	718	722	727	732	737	742	746	751	756	
07	761	766	770	775	780	785	789	794	799	804	
08	809	813	818	823	828	832	837	842	847	852	
09	856	861	866	871	875	880	885	890	895	899	
910	904	909	914	918	923	928	933	938	942	947	
11	952	957	961	966	971	976	980	985	990	995	
12	95 999	*004	*009	*014	*019	*023	*028	*033	*038	*042	**5**
13	96 047	052	057	061	066	071	076	080	085	090	1 0.5
14	095	099	104	109	114	118	123	128	133	137	2 1.0 / 3 1.5
15	142	147	152	156	161	166	171	175	180	185	4 2.0
16	190	194	199	204	209	213	218	223	227	232	5 2.5 / 6 3.0
17	237	242	246	251	256	261	265	270	275	280	7 3.5
18	284	289	294	298	303	308	313	317	322	327	8 4.0
19	332	336	341	346	350	355	360	365	369	374	9 4.5
920	379	384	388	393	398	402	407	412	417	421	
21	426	431	435	440	445	450	454	459	464	468	
22	473	478	483	487	492	497	501	506	511	515	
23	520	525	530	534	539	544	548	553	558	562	
24	567	572	577	581	586	591	595	600	605	609	
25	614	619	624	628	633	638	642	647	652	656	
26	661	666	670	675	680	685	689	694	699	703	
27	708	713	717	722	727	731	736	741	745	750	
28	755	759	764	769	774	778	783	788	792	797	
29	802	806	811	816	820	825	830	834	839	844	
930	848	853	858	862	867	872	876	881	886	890	
31	895	900	904	909	914	918	923	928	932	937	**4**
32	942	946	951	956	960	965	970	974	979	984	1 0.4
33	96 988	993	997	*002	*007	*011	*016	*021	*025	*030	2 0.8 / 3 1.2
34	97 035	039	044	049	053	058	063	067	072	077	4 1.6
35	081	086	090	095	100	104	109	114	118	123	5 2.0
36	128	132	137	142	146	151	155	160	165	169	6 2.4
37	174	179	183	188	192	197	202	206	211	216	7 2.8 / 8 3.2
38	220	225	230	234	239	243	248	253	257	262	9 3.6
39	267	271	276	280	285	290	294	299	304	308	
940	313	317	322	327	331	336	340	345	350	354	
41	359	364	368	373	377	382	387	391	396	400	
42	405	410	414	419	424	428	433	437	442	447	
43	451	456	460	465	470	474	479	483	488	493	
44	497	502	506	511	516	520	525	529	534	539	
45	543	548	552	557	562	566	571	575	580	585	
46	589	594	598	603	607	612	617	621	626	630	
47	635	640	644	649	653	658	663	667	672	676	
48	681	685	690	695	699	704	708	713	717	722	
49	727	731	736	740	745	749	754	759	763	768	
950	97 772	777	782	786	791	795	800	804	809	813	
N	0	1	2	3	4	5	6	7	8	9	Prop. Parts

* The first two figures of the mantissa are those indicated on the next line of the table.

Prop. Parts	N	0	1	2	3	4	5	6	7	8	9
	950	97 772	777	782	786	791	795	800	804	809	813
	51	818	823	827	832	836	841	845	850	855	859
	52	864	868	873	877	882	886	891	896	900	905
	53	909	914	918	923	928	932	937	941	946	950
	54	97 955	959	964	968	973	978	982	987	991	996
	55	98 000	005	009	014	019	023	028	032	037	041
	56	046	050	055	059	064	068	073	078	082	087
	57	091	096	100	105	109	114	118	123	127	132
	58	137	141	146	150	155	159	164	168	173	177
	59	182	186	191	195	200	204	209	214	218	223
	960	227	232	236	241	245	250	254	259	263	268
	61	272	277	281	286	290	295	299	304	308	313
5	62	318	322	327	331	336	340	345	349	354	358
1 0.5	63	363	367	372	376	381	385	390	394	399	403
2 1.0	64	408	412	417	421	426	430	435	439	444	448
3 1.5	65	453	457	462	466	471	475	480	484	489	493
4 2.0	66	498	502	507	511	516	520	525	529	534	538
5 2.5	67	543	547	552	556	561	565	570	574	579	583
6 3.0	68	588	592	597	601	605	610	614	619	623	628
7 3.5	69	632	637	641	646	650	655	659	664	668	673
8 4.0	**970**	677	682	686	691	695	700	704	709	713	717
9 4.5	71	722	726	731	735	740	744	749	753	758	762
	72	767	771	776	780	784	789	793	798	802	807
	73	811	816	820	825	829	834	838	843	847	851
	74	856	860	865	869	874	878	883	887	892	896
	75	900	905	909	914	918	923	927	932	936	941
	76	945	949	954	958	963	967	972	976	981	985
	77	98 989	994	998	*003	*007	*012	*016	*021	*025	*029
	78	99 034	038	043	047	052	056	061	065	069	074
	79	078	083	087	092	096	100	105	109	114	118
	980	123	127	131	136	140	145	149	154	158	162
	81	167	171	176	180	185	189	193	198	202	207
4	82	211	216	220	224	229	233	238	242	247	251
1 0.4	83	255	260	264	269	273	277	282	286	291	295
2 0.8	84	300	304	308	313	317	322	326	330	335	339
3 1.2	85	344	348	352	357	361	366	370	374	379	383
4 1.6	86	388	392	396	401	405	410	414	419	423	427
5 2.0	87	432	436	441	445	449	454	458	463	467	471
6 2.4	88	476	480	484	489	493	498	502	506	511	515
7 2.8	89	520	524	528	533	537	542	546	550	555	559
8 3.2	**990**	564	568	572	577	581	585	590	594	599	603
9 3.6	91	607	612	616	621	625	629	634	638	642	647
	92	651	656	660	664	669	673	677	682	686	691
	93	695	699	704	708	712	717	721	726	730	734
	94	739	743	747	752	756	760	765	769	774	778
	95	782	787	791	795	800	804	808	813	817	822
	96	826	830	835	839	843	848	852	856	861	865
	97	870	874	878	883	887	891	896	900	904	909
	98	913	917	922	926	930	935	939	944	948	952
	99	99 957	961	965	970	974	978	983	987	991	996
	1000	00 000	004	009	013	017	022	026	030	035	039
Prop. Parts	N	0	1	2	3	4	5	6	7	8	9

*The first two figures of the mantissa are those indicated on the next line of the table.